4th International Conference on Science, Technology, and Interdisciplinary Research (IC-STAR 2018)

IOP Conference Series: Materials Science and Engineering
Volume 532

Tanjung Pandan, Indonesia
29 - 30 August 2018

Editors:

Ardian Ulvan
Dewi Agustina Iryani
Melvi Ulvan
Hery Dian Septama
Irza Sukmana
Ahmad Kafrawi Nasution

ISBN: 978-1-5108-8918-7
ISSN: 1757-8981

Printed from e-media with permission by:

Curran Associates, Inc.
57 Morehouse Lane
Red Hook, NY 12571

Some format issues inherent in the e-media version may also appear in this print version.

Printed by Curran Associates, Inc. (2019)

For permission requests, please contact the Institute of Physics at the address below.

Institute of Physics
Dirac House, Temple Back
Bristol BS1 6BE UK

Phone: 44 1 17 929 7481
Fax: 44 1 17 920 0979

techtracking@iop.org

Additional copies of this publication are available from:

Curran Associates, Inc.
57 Morehouse Lane
Red Hook, NY 12571 USA
Phone: 845-758-0400
Fax: 845-758-2633
Email: curran@proceedings.com
Web: www.proceedings.com

Table of contents

Volume 532

The 4th International Conference on Science, Technology and Interdisciplinary Research (IC-STAR 2018)

29–30 August 2018, Bangka-Belitung Province, Indonesia

Accepted papers received: 16 April 2019
Published online: 13 June 2019

Preface

The 4th International Conference on Science, Technology and Interdisciplinary Research (IC-STAR 2018)

Peer review statement

Papers

Processing, Engineering and Application

Kinetics of Fly Ash Geopolymerization using Semi Quantitative Fourier-Transform Infrared Spectroscopy (FTIR); Corr Data

Himawan Tri Bayu Murti Petrus, Arie Satya Adelizar, Achmad Widiyatmoko, Muhammad Olvianas, Wisnu Suprapta, Indra Perdana, Agus Prasetya and Widi Astuti.....1

Performance Evaluation of a Liquid-Liquid Cylindrical Cyclone (LLCC) for Oil-Water Separation

I A Dharma, A Widyaparaga and F I Faza.....10

Warehouse Layout Designing of Slab Using Dedicated Storage and Particle Swarm Optimization

Y Muharni, Kulsum and M Khoirunnisa.....16

Calcination process on chanos chanos forsk (CCF) of milkfish bones to get Hydroxyapatite (HAp) as Composites Application

Agus Pramono, Fatah Sulaiman, Suryana and Anistasia Milandia.....27

Graphical Analysis Springs Dampers System on the A Freight Cars

Indrawan Arifianto, Rahmat Alamtaha, Ary Syahriar and Dwi Astharini.....35

Two-Stage Algorithm for the Open Vehicle Routing Problem

A Imran.....43

Effect of adhesive thickness on the strength of steel-composite joints aged in wet environment

S Sugiman, S Salman, P D Setyawan and S Sulardjaka.....49

Effect of fiber length on the mechanical properties and water absorption of bamboo fiber/polystyrene-modified unsaturated polyester composites

S Sugiman, P D Setyawan and B Anshari.....58

Single stage dimethyl ether plant model based on gasification of palm empty fruit bunch

R Heryadi, A S Uyun, E Yandri, S M Nur and K Abdullah.....69

Noise reduction in near infrared-based glucose concentration measurement sensor

K Prawiroredjo, E S Julian and G Tjahjadi.....78

Design and Mechanism Analysis of Paddlewheel Aerator with Movable Blades

Samsul Bahri, Radite Praeko Agus Setiawan, Wawan Hermawan and Muhammad Zairin Junior.....85

The effect of time and number of balls on shaker milling process in hydroxyapatite powder synthesis

A Fadli, S R Yenti, R Rasyidin and M Sari.....94

Extraction of ElectroEncephaloGraph (EEG) Signal Using the Subband Coefficient of Wavelet Transform on Cursor Moves

H Hindarto and A Muntasa.....101

Friction welding of AZ31-SS316L for partially-degradable orthopaedic pins

A K Nasution, P Nawangsari, A Junaidi and H Hermawan.....109

Multidisciplinary

Colour Behaviour and Stability of Anthocyanin in the Blue Flower of *Stachytarpheta jamaicensis* and *Hydrangea macrophylla*

AM Marpaung, I Zhang and H Sutanto.....118

A mathematical model of the intravenous glucose tolerance test illustrating an *n*-order decay rate of plasma insulin in healthy and type 2 diabetic subjects

A Kartono, Nurullaeli, H Syafutra, S T Wahyudi and T Sumaryada.....124

Functional cake characteristics of modified arrowroot starch (MAS) with the gelatinization-retrograde method

D Damat, A Tain, H Handjani, U Chasanah and D D Siskawardani.....133

Bioactive Compounds of Lemongrass (*Cymbopogon citratus*) essential oil from different parts of the plant and distillation methods as natural antioxidant in broiler meat

E S Hartatie, I Prihartini, W Widodo and A Wahyudi.....139

Integration Distance Similarity with Keyword Algorithm for Improving Cohesion between Sentences in Text Summarization

Rizki Darmawan and Adi Wijaya.....145

A New Geological Map of Formation Distribution on Southern Part of South China Sea:; Natuna Island, Indonesia

H Kausarian, S Lei, G T Lai, Y Cui and A Suryadi.....152

Molecular Reidentification of Endophytic Fungus LBKURCC40 Isolated from Dahlia's Tuber (*Dahlia Variabilis*) Using the Sequence of ITS rDNA

Nabella Suraya, Finna Piska, Nova Wahyu Pratiwi, Aulia Ardhi and Saryono.....160

Developing organic feed stuff as the main component of broiler chickens' feed with canvas model in East Java: *A Case Study of Analysis on Organic Rice and Corn in Malang and Tulungagung*

A Sutanto, W Widodo and I D Rahayu.....167

Agronomic character of several local genotypes of sorghum (*Sorghum bicolor* l.) at East Java

Sulistyawati, Dyah Roeswitawati, Jabal Tarik Ibrahim and Maftuchah.....172

Chemistry and Structure Characterization of Bamboo Pulp with Formacell Pulping

S Hidayati, E Suroso, W Satyajaya and D A Iryani.....177

Electrical Resistivity Imaging (ERI) and Induced Polarization (IP) Survey to Solve Water Drought Problem at Alor Gajah, Melaka, Malaysia

A Suryadi, Batara and S N Amir.....185

Synthesization and characterization of chlorhexidine/calcium phosphate/poly(lactic-*co*-glycolic acid) bone cement

N M Daud, Z Tapsir, A Idris, A K Nasution and S Saidin.....192

Analysis and Charactrezation

Optimization of Bioactive Metabolite Production From *Fusarium oxysporum* LBKURCC41

S Marlinda, H Y Teruna, N W Pratiwi, A Ardhi and Saryono.....198

Isolation and Identification of Bacteria and Actinomycetes Isolated from Wilting Banana Plants (*Musa* Sp.)

Saryono, P. Finna, P. Usman, P. N. Wahyu and A. Aulia.....207

IOP Conf. Series: Materials Science and Engineering **532** (2019) 011001 doi:10.1088/1757-899X/532/1/011001

IOP CONFERENCE SERIES: MATERIAL SCIENCES AND ENGINEERING

PAPER • OPEN ACCESS

The 4th International Conference on Science, Technology, and Interdisciplinary Research (IC-STAR)

IOP Conf. Series: Materials Science and Engineering **532** (2019) 011001 doi:10.1088/1757-899X/532/1/011001

PREFACE

The 4th International Conference on Science, Technology and Interdisciplinary Research (IC-STAR 2018) was organized by University of Lampung, in Tanjung Pandan, Belitung, Indonesia, on August, 29th - 30th 2017. Following the successful of previous events, IC-STAR 2018 shared and highlighted the original and novel ideas, significant advances, the technological state-of-the-art, current status, and future challenges on all aspects related to the research and development in the field of science, engineering and technology, and their inter-discipline cross-linked.

IC-STAR 2018 accomplished as a forum for information dissemination, gathering, and networking for academia, researchers, engineers, and professionals within the areas. The conference also provided advanced orientation and understanding for related industries and academia to looking across industrial partnerships, business strategic, and policy and regulations, including the environmental awareness and the mitigation of natural disasters. One of the most important outcomes of IC-STAR 2018 was the understanding to emphasize Krakatau and its potential disaster as a common issue and interest.

We commit on increasing the quality of the papers through a fruitful peer-review. This year we used the service of EDAS for the conference management system, by which the acceptance rate was 35%. The accepted papers were presented in the conference. The peer-review processes have been done based on the originality, impact, novelty, and the close relation to the area of the conference. The presented papers were selected selected from IC-STAR 2018 and its conference partner.

We are pleased and thankful to all distinguish keynote speakers, authors, and reviewers for their contribution that have made this conference possible. We do hope that you will enjoy this publication and find it as a useful guide and reference.

IOP Conf. Series: Materials Science and Engineering **532** (2019) 011001 doi:10.1088/1757-899X/532/1/011001

IC-STAR 2018 Committee

Advisory Board

Prof. Dr. Ir. Hasriadi Mat Akin – *Rector of University of Lampung*
Prof. Akio Tomiyama – *Kobe University, Japan*
Prof. Hadi Nur – *Universiti Teknologi Malaysia, Malaysia*
Prof. Pavel Ripka – *Czech Technical University in Prague*
Prof. Toshiro Ohashi – *Hokkaido University, Japan*

Steering Committee

Prof. Bujang Rahman – *Vice Rector, University of Lampung, Indonesia*
Prof. Dr. Suharno – *Dean of Faculty of Engineering, University of Lampung, Indonesia*
Dr. Warsono – *Director of Research Institute and Community Service of University of Lampung*
Dr. Irza Sukmana – *University of Lampung, Indonesia*
Dr. Ahmad Zaenuddin – *University of Lampung, Indonesia*
Prof. Ing. Boris Simak – *Czech Technical University in Prague, Czech Republic*
Prof. Norma Alias – *Universiti Teknologi Malaysia, Malaysia*
Mr. M. Ikhsan – *PT. Telkom, Indonesia*
Dr. Jiri Krejci – *Vodafone, Czech Republic*
Mr. Terklin Sinulingga – *General Electric, United Kingdom*

Organizing Committee

Dewi Agustina Iryani – *Conference Chair*
A. Yudi Eka Risano – *Secretary, Public Relation, and Sponsorship*
Melvi – *Treasurer*
Wahyu Eko Sulistyono – *Technical and Scientific Administration*
Novri Tanti – *Logistics and Appurtenance, Social and Excursion Program*
Hery Dian Septama – *Publication and Circulation*

Technical Program Committee (TPC)

Ardian Ulvan – *University of Lampung, Indonesia – Chair*
Robert Bestak – *Czech Technical University in Prague, Czech Republic– co-Chair*
Denni Kurniawan – *Curtin University Sarawak, Malaysia – co-Chair*

TPC Member

Joni Agustian – *University of Lampung, Indonesia*
Anna Antonyova – *University of Presov, Slovakia*
Naseem Ahmed – *Indian Institute of Technology Roorkee, India*
Fitri Arnia – *Syiah Kuala University, Indonesia*

IOP Conf. Series: Materials Science and Engineering **532** (2019) 011001 doi:10.1088/1757-899X/532/1/011001

Ferry Burhan – *10 November Institute of Technology, Indonesia*
Homero Toral-Cruz – *University of Quintana Roo, Quintana Roo, Mexico*
Joy Rizky Pangestu Djuansjah – *Universiti Teknologi Malaysia, Malaysia*
Taha Abd. El-Rahman – *University of Sherbrooke, QC, Canada*
Edno Gentilho – *Instituto Federal de Educação Ciência e Tecnologia do Paraná, Brasil*
Ainul Ghurri – *Udayana University, Indonesia*
Agus Haryanto – *University of Lampung, Indonesia*
Khomaini Hasan – *Biosciences and Biotechnology Research Centre, Indonesia*
Udin Hasanudin – *University of Lampung, Indonesia*
Basem Abu Izneid – *Effate University, Saudi Arabia*
Ismail Tasib Karas – *Karabuk University, Turkey*
Naoya Kasai – *Yokohama National University*
Mamoru Kikumoto – *Yokohama National University*
Irdika Mansur – *Bogor Institute of Agriculture, Indonesia*
Antonin Mazalek – *University of Defence, Brno, Czech Republic*
Christia Meidiana – *Brawijaya University, Indonesia*
Lee Meyer – *University of Kentucky, USA*
Khairul Munadi – *Syiah Kuala University, Indonesia*
Noor Danish Mundari – *LPU, India*
Kaneko Nabuhiro -*Yokohama National University, Japan*
Poppy Puspitasari – *Universitas Negeri Malang, Indonesia*
Chithra Reghuvaran – *National Institute of Technology, Rourkela, Odisha, India*
Uceu Fuad Hassan R.S. – *GKSS, Germany*
Ahmad Saudi Samosir – *University of Lampung, Indonesia*
Shirley Savetlana – *University of Lampung, Indonesia*
Marialisa Scata – *Universitas di Catania, Italy*
Takahasi Susume – *Kyoei University, Japan*
Irfan Syamsuddin – *Makassar State Polytechnic, Indonesia*
Gunasekar Thangarasu – *University Teknologi Petronas, Malaysia*
Josaphat Tetuko – *Chiba University, Japan*
Agung Trisetyarso – *Binus University, Indonesia*
Mokhamad Fakhrul Ulum – *Bogor Institute of Agriculture, Indonesia*
Mustafa Usman – *University of Lampung, Indonesia*
Zuzana Vranova – *University of Defense, Brno, Czech Republic*
Dedy H.B. Wicaksono – *Universiti Teknologi Malaysia, Malaysia*

IOP Conf. Series: Materials Science and Engineering **532** (2019) 011001 doi:10.1088/1757-899X/532/1/011001

IOP Conf. Series: Materials Science and Engineering **532** (2019) 011002 doi:10.1088/1757-899X/532/1/011002

Peer review statement

All papers published in this volume of *IOP Conference Series: Materials Science and Engineering* have been peer reviewed through processes administered by the proceedings Editors. Reviews were conducted by expert referees to the professional and scientific standards expected of a proceedings journal published by IOP Publishing.

IOP Conf. Series: Materials Science and Engineering **532** (2019) 012001 doi:10.1088/1757-899X/532/1/012001

Kinetics of Fly Ash Geopolymerization using Semi Quantitative Fourier-Transform Infrared Spectroscopy (FTIR); Corr Data

Himawan Tri Bayu Murti Petrus[1,4*], Arie Satya Adelizar[1,2], Achmad Widiyatmoko[1], Muhammad Olvianas[1], Wisnu Suprapta[1], Indra perdana[1,4], Agus Prasetya[1,4], and Widi Astuti[3]

[1]Department of Chemical Engineering (Sustainable Mineral Processing Research Group), Universitas Gadjah Mada, Jl. Grafika 2 Yogyakarta 55281, Indonesia.
[2]Graduate School of Chemical Engineering, Universitas Gadjah Mada, Jl. Grafika 2 Yogyakarta 55281, Indonesia.
[3]Research Unit for Mineral Technology, Indonesian Institute of Sciences (LIPI), Jl. Ir. Sutami Km. 15, Tanjung Bintang, Lampung Selatan, Indonesia
[4]Unconventional Geo-resources Research Center, Faculty of Engineering, Universitas Gadjah Mada, Jl. Grafika No. 2, Yogyakarta 55281, Indonesia

[*]Corresponding Author: bayupetrus@ugm.ac.id

Abstract. Fly ash utilization is very urgent to be conducted alas contaminating the environment. One of the promising utilizations is geopolymer formation for material construction. In this study, geopolymer formation was studied using fly ash from PLTU Tanjung Jati Jepara. This study included kinetics prediction using solid state isoconversion kinetics that has been developed based on nucleation, geometric contraction, and diffusion models. Semi quantitative FTIR analysis (corr) has been applied to predict the kinetics of the fly ash geopolymerization in curing temperature of 30o and 90°C. In order to be able to verify the kinetics model time series data were taken in interval of 1,3,5,14,21, and 28 days. It is the diffusion model that represents the fly ash geopolymerization with R^2 of 0.9437.

1. Introduction

The needs of energy cannot be avoided nowadays. The increase of energy needs will be significant with the increase of population. Fig. 1 shows the increase of electricity consumption worlwidely in many sectors.

IOP Conf. Series: Materials Science and Engineering **532** (2019) 012001 doi:10.1088/1757-899X/532/1/012001

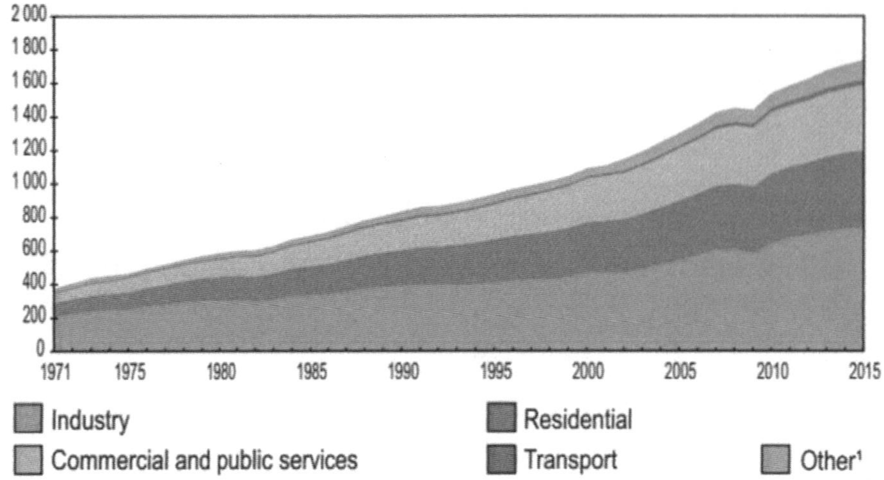

Fig. 1. Total electricity consumption in many sectors worlwidely, [1].

In the year of 2015, it was recorded that the world's electricity needs reached the value of 24.255 TWh supplied with coal based power generation as it can be seen in Fig. 2.
Indonesia is facing similar condition at which of about 50% of the energy supply will be generated from the coal buring power generation in order to fulfill the 35 GW in 2025. Along with this coal based energy supply increases there will be of about 8.31 million tonnes of fly ash in 2019 with predicted increase of about 5% per year [2].

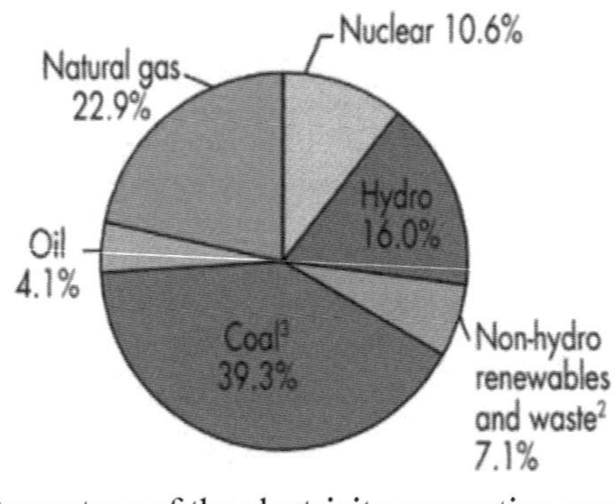

Fig. 2. Percentage of the electricity generation sources, [1].

Due to its potential danger to the environment, some research on fly ash utilization have been developed, i.e., cenospheres recovery using both wet and dry separation methods [3]-[5]; adsorbent [6]-[10]; and material construction in the form of pozzolanic [11]; concrete [12]-[14]; and geopolymer [15]-[16]. Among other utilizations, geoplymer possesess some advantages due to its possibility to overpower the regular concrete, i.e., the hardening time is faster, the tensile strength as well as compresive strength are higher, and most of all geopolymer has good resistancy to corrosion. Although many studies on geopolymer have been conducted, the kinetics study has not been well developed. One of the methods to observe the growth of geopolimerization is by using FTIR [17]. Thus, in this study we would like to predict the kinetics of fly ash geopolymerization using semi quantitative approach from FTIR Corr data.

2. Materials and Method

A. *Materials.*

Fly ash from the coal burning electricity plant generation in PLTU Tanjung Jati Jepara has been obtained and used as the raw material for the geopolymer through all the experiments. Other materials, such ash NaOH and sodium silicate, were supplied by PT. Brataco Chemical with analytical grade concentration.

B. *Method*

1. *Raw material preparation*

Fly ash was dried in an oven with temperature of 50°C for 5 hours. It was then milled and sieved to result product below 300 mesh. The over 300 mesh product will be remilled and sieved.

2. *Chemical composition analysis*

Chemical composition of fly ash is crusial to distinguish due to certain proportion of SiO_2 and Al_2O_3 should be fulfilled for geopolimerization. In order to provide chemical composition of the fly ash used in this experiment, energy dispersive X-ray spectroscopy (EDX) was conducted. The chemical composition of the fly ash is shown in Fig. 3.

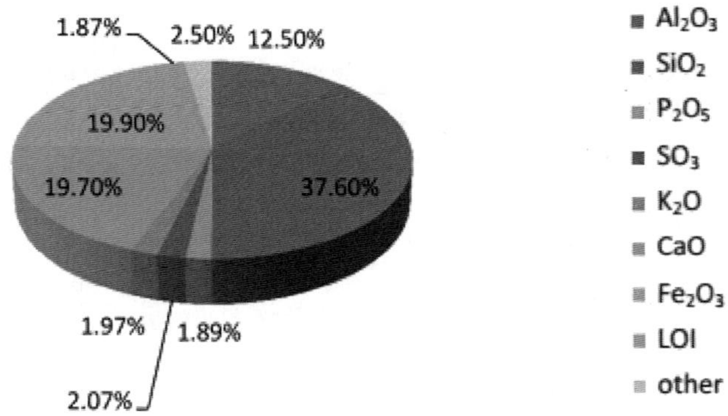

Fig. 3. Chemical composition of fly ash

From Fig. 3, it can be seen that the composition of fly ash from PLTU Tanjung Jati Jepara is dominated with silica. The ratio of SiO_2/Al_2O_3 is 3.01 in the range value that is optimum for geopylimerization in the range of 3,00 – 3,80 [18]-[20].

3. *Geopolimerization*

The -300 mesh fly ash was thoroughly mixed with NaOH and natrium silicate (alkaline activator) forming paste. The concentration of NaOH used in this geopolymerization is 12 M. Ensuring the homogenized geopolymer paste by agitation, the paste was then poured into cubical cast with the size of 5x5x5 cm^3 and let it solidify for 2 days. Each cubical cast would comsumed 50 mL of alkaline activator.

Further treatment was conducted by curing the samples for 8 hours at certain temperature (30° and 90°C). The samples were prepared and taken for further analysis using compresive strngth and Fourier-transform infrared spectroscopy (FTIR) in certain time interval of 1,3,5,14,21, and 28 days. Three geopolymer samples were prepared for each data point to ensure the reliability.

4. *Analytical methods*

Geopolymer samples were analyzed using Universal Testing Machine Avery-Denison to measure the compresive strength of each data point. After compresive strength analysis, the sample was then milled to have powdery sample and further analyse using FTIR to observed the growth of geopolymer bonding. The fingerprint of geopolymerization reaction is similar to that of zeolite which is determined by the Si-O-Al formation (aluminosilicate). The wavelength for the Si-O-T (T: Si or Al) bonding is detected in the range of 900-1300 cm-1. However there are 2 other peaks showing the growth of the geopolimer as shown in Table 1.

IOP Conf. Series: Materials Science and Engineering **532** (2019) 012001 doi:10.1088/1757-899X/532/1/012001

Table 1. FTIR spectra for geopolymerization of fly ash [21]

Wavelength	**Functional groups**
900-1300 cm^{-1}	Si-O-T (T: Si, Al Tetrahedral) *assymetric stretching*
1650 cm^{-1}	H-OH *bending*
3400-3650 cm^{-1}	-OH *symmetric and asymmetric stretching*

Based on the peak of the aluminosilicate wavelength the correlation value (Corr) will be calculated relative to the zeolit peaks in accordance with equation 1 and 2 [22].

$$r = \frac{\Sigma(a_i - \bar{a})(b_i - \bar{b})}{\sqrt{\Sigma(a_i - \bar{a})^2(b_i - \bar{b})^2}} \quad (1)$$

$$Corr = \frac{999\,(r + 1)}{2} \quad (2)$$

5. *Kinetics approach*

Kinetics of fly ash geopolymerization will be verified using iso-conversion approach on solid state reaction as formulated in equation 3.

$$\frac{d\alpha}{dt} = A \exp\left(\frac{-E}{RT}\right) f(\alpha) \quad (3)$$

α is defined as followed (equation 4),

$$\alpha = \frac{x_0 - x_t}{x_0 - x_\infty} \quad (4)$$

χ can represent any quantitative value showing the physical change phenomenon. In this study the χ value will be represented by the semi-quantitative FTIR data (Corr) [23]. There are several derivations of isoconversion approach based on the background phenomenon as tabulated in Table 2.

Table 2. Derivative model of solid state reaction [23]

Kinetics	**Differential** $f(\alpha) = 1/k\ d\alpha/dt$	**Integral** $g(\alpha) = kt$
Nucleation Model		
Avrami-Erofeyev (A2)	$2(1-\alpha)[-\ln(1-\alpha)]^{1/2}$	$[-\ln(1-\alpha)]^{1/2}$
Avrami-Erofeyev (A3)	$3(1-\alpha)[-\ln(1-\alpha)]^{2/3}$	$[-\ln(1-\alpha)]^{1/3}$
Avrami-Erofeyev (A4)	$4(1-\alpha)[-\ln(1-\alpha)]^{3/4}$	$[-\ln(1-\alpha)]^{1/4}$
Prout-Tompkins (B1)	$\alpha(1-\alpha)$	$\ln[\alpha/(1-\alpha)] + c^{\alpha}$
Geometry Contraction Model		
Contracting area (R2)	$2(1-\alpha)^{1/2}$	$1 - (1-\alpha)^{1/2}$
Contracting volume (R3)	$3(1-\alpha)^{2/3}$	$1 - (1-\alpha)^{1/3}$
Diffusion Model		
1-D diffusion (D1)	$1/(2\alpha)$	α^2
2-D diffusion (D2)	$-[1/\ln(1-\alpha)]$	$((1-\alpha)\ln(1-\alpha)) + \alpha$
3-D diffusion-Jander (D3)	$[3(1-\alpha)^{2/3}]/[2(1-(1-\alpha)^{1/3})]$	$(1-(1-\alpha)^{1/3})^2$
Ginstling-Brounshtein (D4)	$3/[2((1-\alpha)^{-1/3}-1)]$	$1 - (2/3)\alpha - (1-\alpha)^{2/3}$
Order Reaction Model		

IOP Conf. Series: Materials Science and Engineering **532** (2019) 012001 doi:10.1088/1757-899X/532/1/012001

Zero-order (F0/R1)	1	α
First-order (F1)	$(1-\alpha)$	$-\ln(1-\alpha)$
Second-order (F2)	$(1-\alpha)^2$	$[1/(1-\alpha)] - 1$
Third-order (F3)	$(1-\alpha)^3$	$(1/2)[(1-\alpha)^{-2} - 1]$

The corr semi quantitaive FTIR data will be calculated in accordance with the kinetics models that have been developed. The kinetics of fly ash geopolymerization will be determined by the most fitted model for solid state reaction as tabulated in Table 2.

3. Results and Discussion

Raw material characterization is an important stage that has to be accomplished in order to predict the potential utilization of the fly ash. As it has been discussed previously that fly ash from PLTU Tanjung Jati Jepara has SiO_2/Al_2O_3 ratio in the order of 3.01 which is potential to utilize as geopolymer raw material. Further, analysis using FTIR it shows that the existence of Si-O-Al in the fly ash matrix is relatively low in comparison to the standard zeolite used in this experiment, as seen in Fig. 4.a.

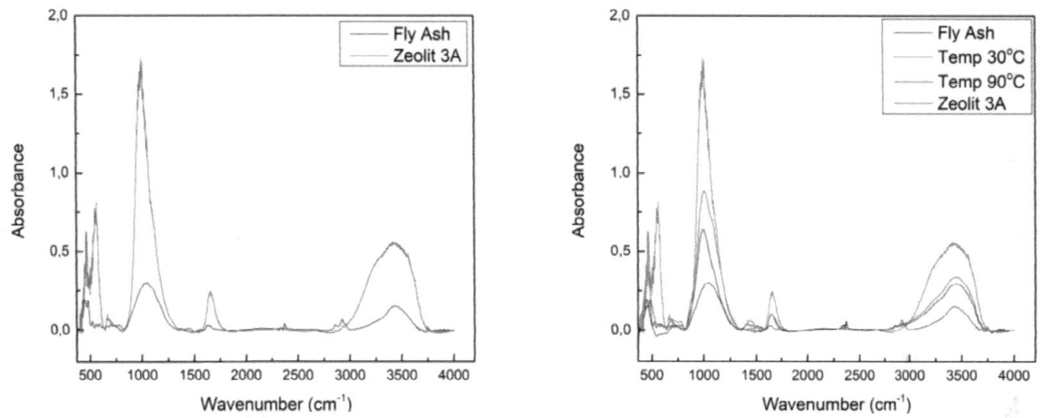

(a) (b)

Fig. 4. FTIR characterization of fly ash and geopolymer samples (curing temperature of 30° and 90°C) in comparison with zeolite.

After treatment with curing temperature of 30° and 90°C it can be seen that the absorbance of the peaks showing the existence of auminosilicate (Si-O-Al) increases along with the higher curing temperature (Fig. 4.b.). The increase of this absorbance representing the higher conversion of aluminosilicate formation is in line with the increase of compressive strength of the geopolymer samples (Fig. 5)

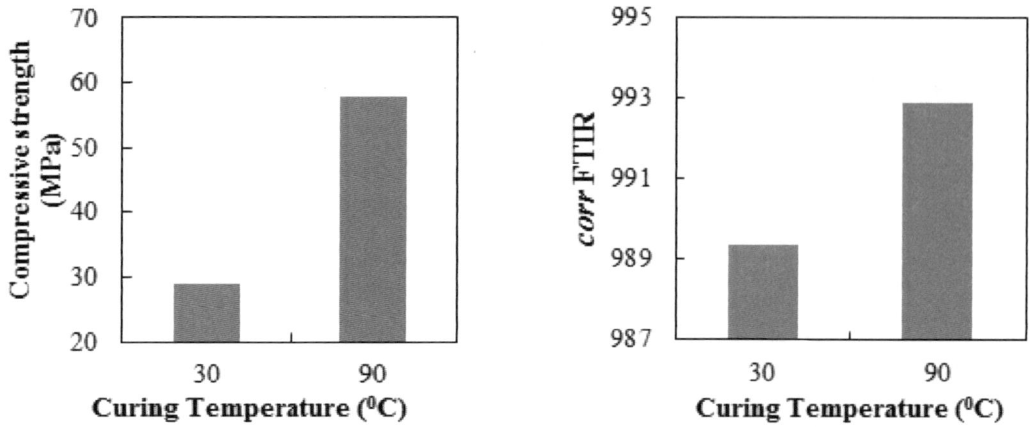

(a) (b)

Fig. 5. The comparison between compresive strength and Corr FTIR on fly ash geopolymer.

The compressive strength of fly ash geopolymer sample represents the degree of geopolymerisation at where formation of aluminosilicate takes place (Olvianas et al., 2017(b)). It can be seen from Fig. 5.a., the compressive strength increases from 29 MPa to 58 MPa along with the increase of curing temperature from 30° to 90°C. The increase of compressive strength is in accordance with the corr from the FTIR from 989.5 to 992.8. This phenomenon shows that the physical characteristic of geopolymer is affected by the degree of polymerization or formation of aluminosilicate matrix. Having this supportive macro and micro-phenomenon on fly ash geopolymer, we would like to understand the mechanism of fly ash geopolymerization by fitting the existing solid state reaction as tabulated in TABLE II with time series data that we have conducted on fly ash geopolymerization in two curing temperatures (30° and 90°C) (Fig. 6).

The highest corr for any FTIR data is 1000 showing that the correlation with the associated peaks is at the maximum value similar to that of 100% of conversion. As it can be seen in Fig. 6, the corr value increases with time in both curing temperature of 30° and 90°C. These data series of corr was then calculated to fit the kinetic model of fly ash geopolymerization. In solid state reaction, there are several kinetic models that have been developed. Eliminating the reaction controlling model due to its fast reaction step [24], the avrami, contracting volume, and diffusion model were analyzed to fit the corr data shown in Fig. 6.

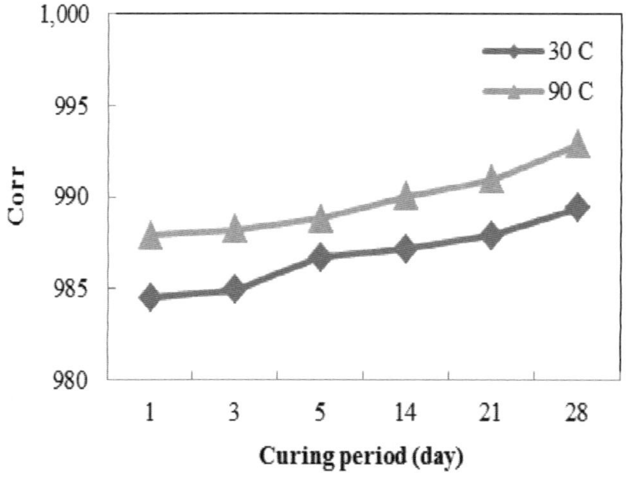

IOP Conf. Series: Materials Science and Engineering **532** (2019) 012001 doi:10.1088/1757-899X/532/1/012001

Fig. 6. Corr for each data point taken as a function of time sequence (both curing temperature 30° and 90°C)

From the kinetic model fitting as shown in Fig. 7, it is found that the diffusion model fits the fly ash geopolymerization model with average R^2 of 0.9437. In comparison to that of the avrami model with R^2 of 0.9409 and the contracting volume with R^2 of 0.9427. The finding is actually aligned with the mechanism of the geopolymerization proposed by Davidovits (1999).in which the steps of geopolymerization includes dissolution, reaction, gel formation, and solidification. Indeed the proposed mechanism merely includes diffusion in the process.

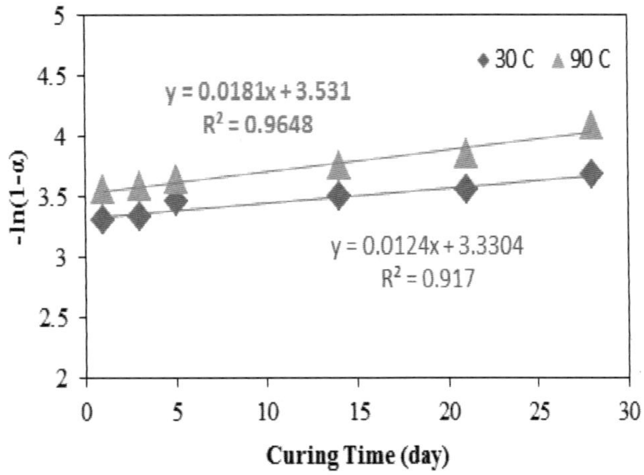

(a)

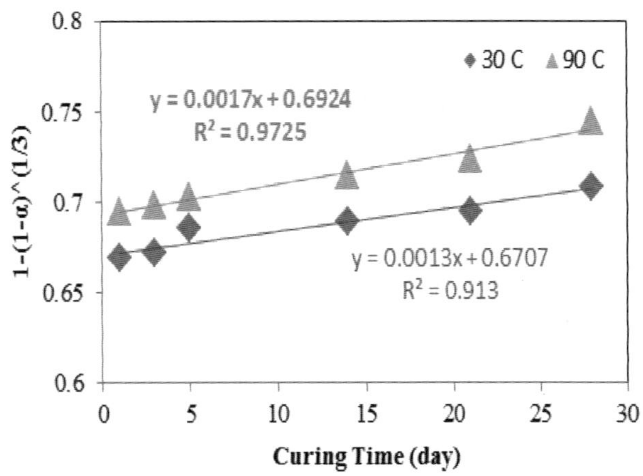

(b)

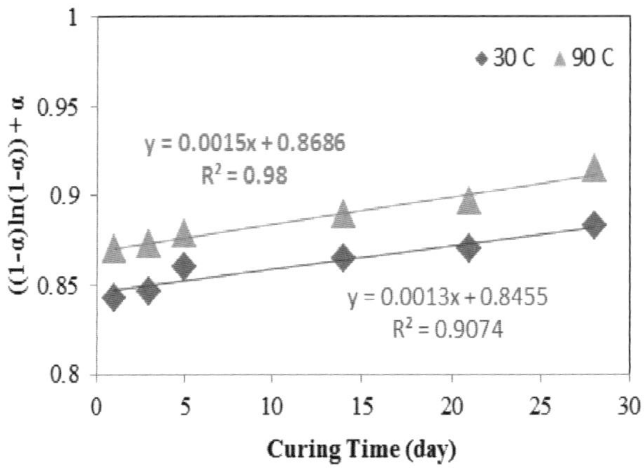

(c)

Fig. 7. Kinetic model of fly ash geopolymer (a) Avrami model (b) volume contraction model and (c) diffusion model.

4. Conclusion

Geopolymer has successfully formed using fly ash from PLTU Tanjung Jati Jepara with the highest compresive strength of 58 MPa resulted in curing temperature of 90°C for 28 days. The increase of compresive strength as a function of temperature and series of time sequence is in accordance with the semi quantitative FTIR analysis represented by corr value. The kinetics of fly ash geopolymerization has fitted with the diffusion model resembled with the proposed geopolymer mechanism by Davindovits (1999) in which diffusion control in the process including dissolution, gel formation, and solidification. The diffusion model fits the phenomenon with R^2 of 0.9437.

5. Acknowledgement

Highly appreciation to the Minister of Higher Education of Indonesia for the financial support to conduct this project under PTUPT scheme with contract number 1860/UN1/DITLIT/DIT-LIT/LT/2018.

6. References

[1] International Energy Agency (IEA). (2017, September). Key World Energy Statistics in 2017. Dipetik July 1, 2018, dari https://www.iea.org/publications/freepublications /publication/KeyWorld2017.pdf

[2] Direktorat Jenderal Ketenagalistrikan. (2016, February 16). Kementerian Energi dan Sumber Daya Mineral, Direktorat Jenderal Ketenagalistrikan. Dipetik April 12, 2017, dari Pemerintah Dorong Pemanfaatan Limbah PLTU: http://www.djk.esdm.go.id/index.php/detail-berita?ide=4060

[3] Hirajima, T., Oosako, Y., Nonaka, M., Petrus, H., Sasaki, K., and Ando, T., Recovery of hollow and spherical particles from coal fly ash by wet separation process, Journal of MMIJ 124 (12), 2008, pp. 878-884

[4] Hirajima, T., Petrus, H., Oosako, Y., Nonaka, M., Sasaki, K., and Ando, T., Recovery of cenospheres from coal fly ash using a dry separation process: Separation estimation and potential application, International Journal of Mineral Processing 95 (1-4), 2010, pp. 18-24.

[5] Petrus, H., Hirajima, T., Oosako, Y., Nonaka, M., Sasaki, K., and Ando, T., Performance of dry-separation processes in the recovery of cenospheres from fly ash and their implementation in a recovery unit, International Journal of Mineral Processing 98 (1-2), 2011, pp. 15-23.

[6] Davini, P., 1996, Investigation of the SO_2 adsorption properties of $Ca(OH)_2$-fly ash systems, Fuel, vol. 75, 1996, pp. 713–716.

[7] Lu, G.,Q., and Do, D., D., 1991, Adsorption properties of fly ash particles for NOx removal from flue gases, Fuel Process Technol., vol. 27, 1991, pp. 95–107.
[8] Dasmahapatra, G., P., Pal, T., K., Bhadra, A., K., and Bhattacharya, B., 1996, Studies on separation characteristics of hexavalent chromium from aqueous solution by fly ash, Sep.Sci.Technol., vol. 31, 1996, pp. 2001–2009.
[9] Alinnor, I., J., 2007, Adsorption of heavy metal ions from aqueous solution by fly ash, Fuel, vol. 86, 2007, pp. 853–857.
[10] Kapoor, A., and Viraraghavan, T., 1992, Adsorption of mercury from wastewater by fly ash, Adsorpt.Sci.Technol., vol. 9, 1992, pp. 130–147
[11] Singh, M., and Garg, M., 1999, Cementitious binder from fly ash and other industrial wastes, Cem.Concr.Res., vol. 29, 1999, pp. 309–314.
[12] Siddique, R., 2004, Performance characteristics of high-volume Class F fly ash concrete, Cem.Concr.Res., vol. 34, 2004, pp. 487–493.
[13] Oner, A., Akyuz, S., and Yildiz, R., 2005, An experimental study on strength development of concrete containing fly ash and optimum usage of fly ash in concrete, Cem.Concr.Res., vol. 35, 2005, pp.1165–1171.
[14] Bouzoubaa, N., Zhang, M., H., and Malhotra, V., M., 2001, Mechanical properties and durability of concrete made with high-volume fly ash blended cements using a coarse fly ash, Cem.Concr.Res., vol. 31, 2001, pp.1393–1402.
[15] Wilson, M., Cabrera, J., G., and Zou, Y., 1994, The process and mechanism of alkali-silica reaction using fused silica as the reactive aggregate, Adv.Cem.Res., vol. 6, 1994, pp.117–25.
[16] Davidovits, J., Chemistry of Geopolymeric Systems, Geopolymer 99 International Conference, 1999.
[17] Olvianas, M., Najmina, M., Prihardana, B.S.L., Sutapa, F.A., Nurhayati, A., and Petrus, H.T.B.M., Study on the Geopolymerization of Geothermal Silica and Kaolinite, Advanced materials Research, 1112, 2017, pp. 528-532.
[18] De Silva, P., Sagoe-Crenstil, K., dan Sirivivatnanon, V., Kinetics of Geopolymerization: Role of Al_2O_3 and SiO_2. Cement and Concrete Research, vol. 37, 2007, pp. 512-518.
[19] Petrus, H.T.B.M., Hulu, J., Malinda, E., and Prakosa, R.A., 2016, Effect of Bentonite Addition on Geopolymer Concrete from Geothermal Silica., Material Science Forum, 2016, p. 841.
[20] Olvianas, M., Widiyatmoko, A., and Petrus, H.T.B.M., IR spectral similarity studies of geothermal silica-bentonite based geopolymer, AIP Conference Proceedings, 2017, 1887 (1).
[21] Krol, M., Minkiewicz, J., dan Mozgawa, W., IR Spectroscopy Studies of Zeolites in Geopolymeric Materials derived from Kaolinite, Journal of Molecular Structure, 2016, pp. 200-206.
[22] Varmuza, K., Demuth, W., dan Karlovits, M., Structural and Spectral Similarity, CHEM 02 - Biannual Conference on Chemistry. Cairo: Chemistry Department Cairo University, 2002.
[23] Khawam, A., dan Flanagan, D., Solid-State Kinetic Models: Basics and Mathematical Fundamentals, J. Phys. Chem. B, 110(35), 2006, pp. 17315-17328.
[24] Xu, H., dan Van Deventer, J., The Geopolymerisation of Alumino-Silicate Minerals, International Journal of Mineral Processing, 2000, pp. 247-266.

Performance Evaluation of a Liquid-Liquid Cylindrical Cyclone (LLCC) for Oil-Water Separation

I A Dharma[1], A Widyaparaga[2, 3, 4], F I Faza[2]

[1] Department of Mechanical Engineering, Universitas Islam Indonesia, Jalan Kaliurang km. 14.5, 55584, Yogyakarta, Indonesia
[2] Department of Mechanical and Industrial Engineering, Universitas Gadjah Mada, Jalan Grafika No. 2, 55281, Yogyakarta, Indonesia

[3] Centre for Energy Studies, Universitas Gadjah Mada, Sekip Blok K1.A, 55281, Yogyakarta, Indonesia

[4] E-mail: adhika@ugm.ac.id

Abstract. In this experimental study, the type of Hydrocyclone Separators that used is Liquid-Liquid Cylindrical Cyclone (LLCC). The objective of this experimental study is to understand the influences of operating and design parameters such as inlet velocity, split- ratio, and vortex finder diameter towards the result of separation process (LLCC's performance). The separation performance of LLCC is determined by measuring the watercut that goes through the lower output called as underflow. The examined inlet mixture velocity in this experimental study are 1 m/s and 1.2 m/s with split-ratio variation of 5%, 15%, 30%, 45%, 60%, 75%, and 90%. The diameters of vortex finder that used in this experimental study are 18 mm, 22 mm, and 27 mm. All variables are tested under a constant inlet oil volume fraction 25%. From the experiment with those various variables, LLCC can achieve 99% of watercut with different Optimum Split Ratio (OSR) for each inlet mixture velocity and vortex finder diameter.

1. Introduction

Indonesian territory is a strategic area for trading activities and sea transportation. It has an impact on the higher number of ship passing through the Indonesian seas and increasing the volume of waste oil. Higher waste pollution will damage the environment. [1]

This study develops a new method that is cheaper and more efficient to prevent the waste pollution by separating the oil content in the water so that LLCC (Liquid Liquid Cylindrical Cyclone) separators have been designed. However, the LLCC is not to replace the conventional separators. LLCC or generally called Hydrocyclone Separators is often used as a last step to clean up a small quantity of oil remaining in the water to the minimum level after the primary separation so water can legally to be discharged into the sea. [2] Some studies have been carried out to study the performance and behaviour of LLCC with different design and different operating parameters [3]-[6], to investigate the flow characteristics [7], and to develop the control systems [8]-[9].

Unfortunately, each particular LLCC design has its own behavior. It make LLCC difficult to operate. Thus, a standard is required between non-dimensional design parameters and the operating parameters to obtain best performance. Those proposed standard is called LLCC Behaviour Map. To the best of the author's knowledge, LLCC Behaviour Map has not been systematically studied yet. However, despite its helpfulness, its not easy to build this map. The big experimental database of LLCC systems for oil and water separation is needed to supporting LLCC Behaviour Map. Therefore, this study is conducted as a first step to develop the established LLCC Behaviour Map. The purpose of this study is to obtain some experimental data related to the performance of LLCC to supporting the big experimental database of LLCC.

2. Methods

The experimental facility was established in Fluid Mechanics and Heat and Mass Transfer Laboratory, Department of Mechanical and Industrial Engineering, Faculty of Engineering, Universitas Gadjah Mada. Fig. 1 shows a schematic of the flow loop system. Kerosene and water were selected as the initial working fluids. Water and kerosene have a density of 997 kg / m^3 and 820 kg / m^3 and dynamic viscosity of 0.00086 kg / ms and 0.00164 kg / ms respectively at 27° C. A loading dye (red) was added to the kerosene to improve flow visualization.

As already shown in Figure 1, water and kerosene were stored in different tanks, wherein each tank has a maximum capacity of 250 liters. Each tank was connected to one pump that are equipped with by-pass line. Water and kerosene were pumped (P-1 and P-2) from the tank to the mixture tube. The flow rate of water and kerosene were set by using gate valve (V-2 and V-4). The water flow rate was measured by using a rotameter (I-1) that mounted on water line. While the kerosene flow rate was measured by using a turbine digital flowmeter (I-4) that mounted on kerosene line. Check valve (V-5 and V-6) were mounted for each line of water and kerosene to prevent backflow. After passing through the mixture tube, water and kerosene will be mixed and entering into LLCC through tangential inlet. The oil-rich flow and the water-rich flow will be out through the overflow and the underflow respectively. The valves (V-9 and V-11) in both the overflow and underflow allow to control the flow rates of fluid leaving from the LLCC. After being separated, both the overflow stream and the underflow stream flow into the separator to the further separation under the gravitational force. Finally, the water and kerosene will be returned to the tank by using circulation pump respectively (P-3 and P-4). At the overflow and inlet there are sample lines that is used to take samples. The flow in overflow is dominantly by oil, therefore the samples taken are to calculate the oil volume fraction. This sample is put on cylinder glass and left resting for 10-15 minutes (until water and oil separated by gravitational force). A camera with frame rate video recording 1000 – 3000 fps, and maximum resolutions 1280 x 800 pixels is placed in the front of LLCC to take the visualization (photographs) data. 50 Watt LED lamps were used as a light source.

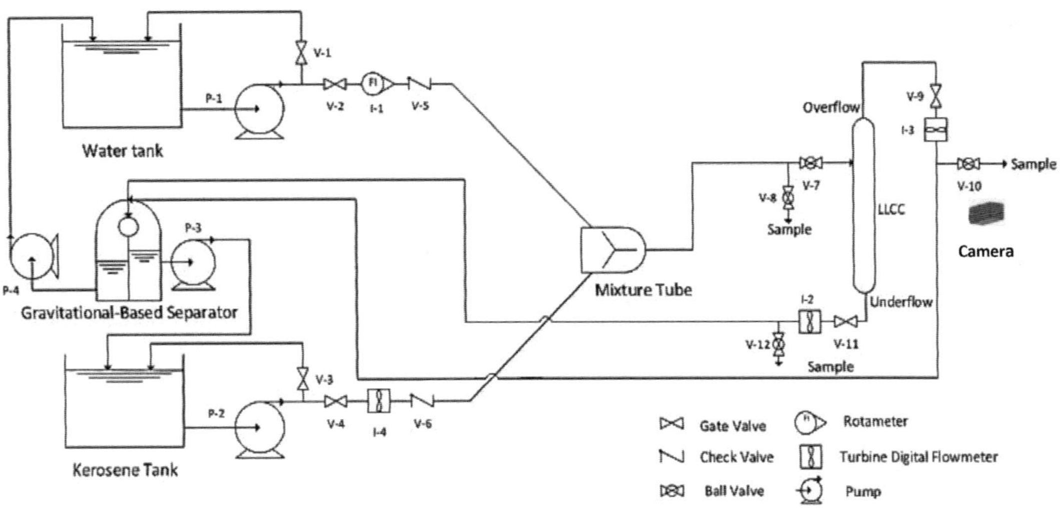

Figure 1. Schematic Diagram of the LLCC Experimental Facility.

The dimension of LLCC is given in Figure 2. LLCC was made from transparent acrylic pipes to enable the visual observation of separating process. The vertical pipe of LLCC is a 1300 mm height (*H*) and 50 mm *ID* with 16 mm *ID* horizontal tangential inlet (*d*). The inlet is attached to the LLCC 300 mm below its top (*h*). The lower section below the inlet is 1000 mm long. The underflow line is an 18 mm *ID* pipe.

The dimension of LLCC is given in Figure 2. LLCC was made from transparent acrylic pipes to enable the visual observation of separating process. The vertical pipe of LLCC is a 1300 mm height (*H*) and 50 mm *ID* with 16 mm *ID* horizontal tangential inlet (*d*). The inlet is attached to the LLCC 300 mm below its top (*h*). The lower section below the inlet is 1000 mm long. The underflow line is an 18 mm *ID* pipe.

This study proposed non-dimensional parameters for the LLCC design. The non-dimensional design parameters of LLCC is shown in Table 1.

Table 1. Non-dimensional Design Parameters of the LLCC Comparing to the Other Studies.

Non-dimensional parameters	This Study	Oropeza Vazquez, et al. [4]	H.-f. Liu, et al. [3]
d/D	0,32	0,53 with istalled nozzle 25% of the in- let full cross sectional area)	0,44 with istalled nozzle in the inlet (20% of the in- let full cross sectional area)
h/H	0,23	0,47	0,11
D	50 mm	50,8 mm (2 inch)	50 mm
H	1300 mm	2413 mm (95 inch)	900 mm

Comparing to the other studies, the LLCC design that used in this study has different non-dimensional parameters. LLCC behavior map is constructed from the relationship between non-dimensional parameters design, operating parameters, and separation performance of LLCC. Hence, this study has an added value for the development of this map. In this case, the effect of other parameters such as oil viscosity (type of oil that used), pipe friction, temperature, can be ignored.

The operating parameters in this study were inlet mixture velocity, inlet oil volume fraction, and split-ratio. The examined inlet mixture velocity were 1 m/s dan 1,2 m/s and throughout the experiment, the inlet oil volume fraction was kept constant at 25%. Split-ratio is varied at 5%, 15%, 30%, 45%, 60%, 75% and 90%. These operating parameters were tested on three variations of vortex finder diameter (D_o), that are 18 mm, 22mm, and 27 mm with Vortex Finder Depth (*L*) is 0 mm. Split-Ratio (*SR*) is defined as the ratio between the flow rate in the overflow (Q_o) and the flow rate in the inlet (Q_i). Split-ratio is calculated by using equation as follow:

$$SR = \frac{Qo}{Qi} \tag{1}$$

The separation performance of LLCC was calculated by measuring watercut (water content) in underflow.

Kerosene and water were selected as the initial working fluids. Kerosene was selected as working fluids on grounds of easy availability, non-hazardous, non-explosive, fast separation, appropriate optical characteristics, non-degrading properties, and has low emulsification. A dye (red) was added to the kerosene to improve flow visualization of the separation phenomena.

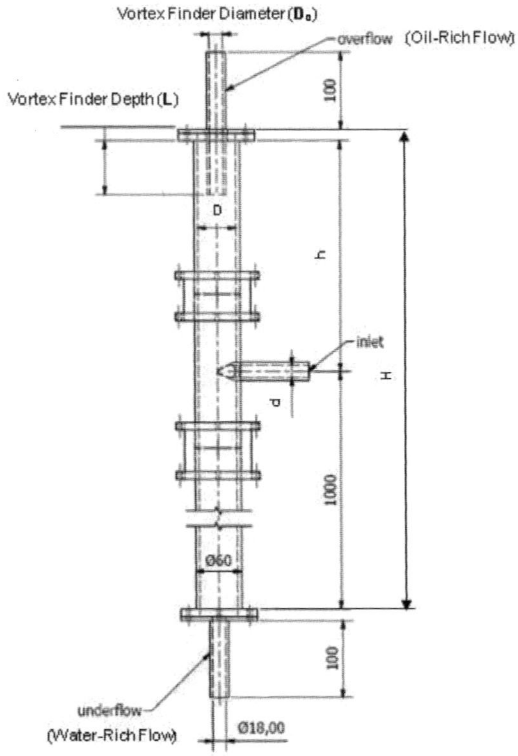

Figure 2. The Dimension of LLCC.

3. Results and Discussion

The experimental results on the effect of split-ratio (*SR*) and inlet velocity (V_m) to watercut percentage on the underflow will be discussed. The experimental data of this study is represented generally in Figure 3.

Those figures below showed the effect of the split-ratio and inlet mixture velocity (V_m) to the underflow watercut of the separation process performed by LLCC on each variation of the vortex finder diameter. The change of watercut is greatly influenced by the split-ratio. The increasing split-ratio will raise the watercut until its maximum point. The higher split-ratio that is operated will created the aperture of gate valve more opened on overflow than underflow. This causes a high flow (Q_o) that passes through overflow. Thus, the most of fluids entering LLCC will flowing through overflow, especially oil that have lower density, and make the watercut value in high split-ratio can reached 99%, as well otherwise.

To obtained the maximum watercut value (up to 99%), LLCC should be operated before the oil content move to the underflow or before oil drag occurs. The LLCC's operating limit before the occurrence of oil drag is called optimum split-ratio (OSR). Therefore, the LLCC must operate in higher OSR to obtained the best separation performance. Figure 4 showed the separation behaviour inside LLCC. The photograph indicated the occurring of oil drag.

On the other hand, Inlet mixture velocity (V_m) affected the OSR. From the graphs above, it can be seen that higher inlet mixture velocity changed the value of OSR. For higher inlet mixture velocity, the maximum value of underflow watercut occurred at lower OSR. This is happened because high flow velocity causes the oil to push toward the underflow resulting in lowering of the quality of the watercut.

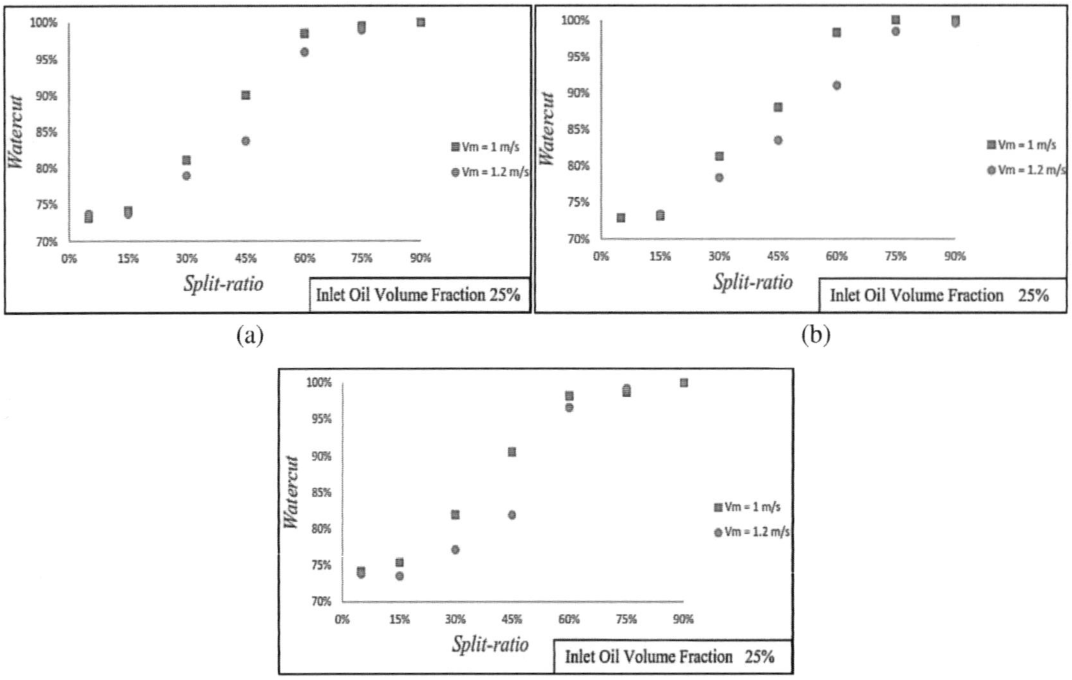

Figure 3. The Effect of Inlet Mixture Velocity and Split-Ratio to the Underflow Watercut. (a) D_o=18 mm; (b) D_o=22 mm; (c) D_o=27 mm

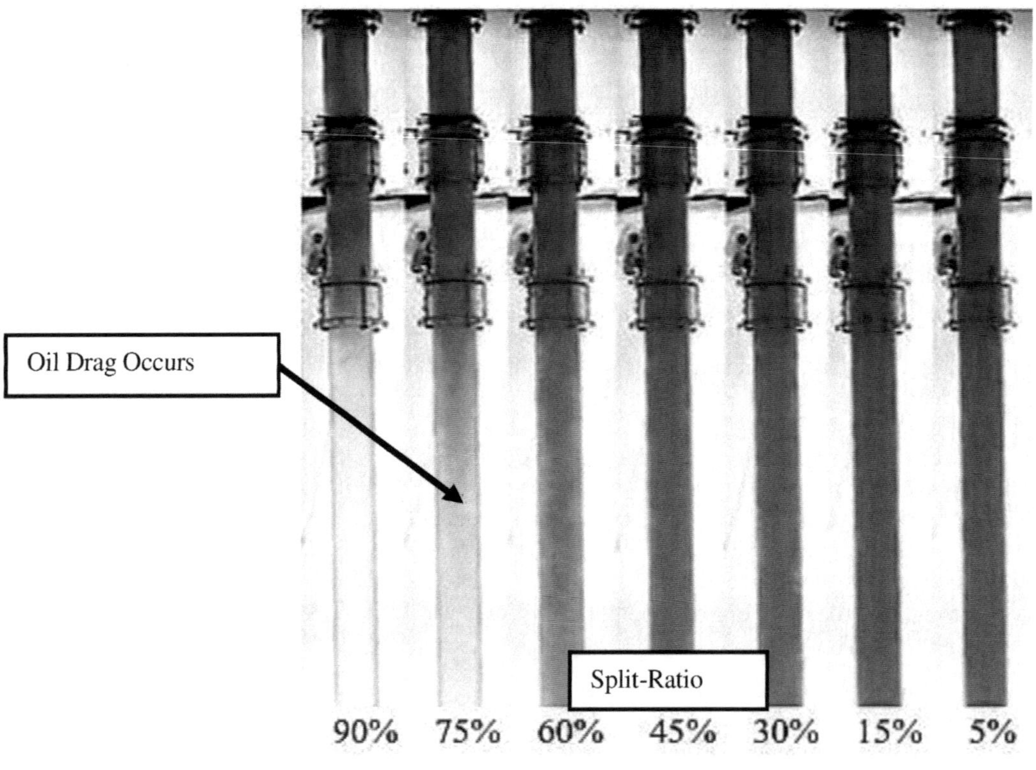

Figure 4. The photographs of separation behaviour inside LLCC under different split-ratio (D_o = 22 mm, Vm = 1 m/s)

4. Conclusion

From the result and discussions, the main conclusion can be drawn that the watercut value obtained from the LLCC separation process reached 99%. The LLCC should be operated in the higher of OSR to obtained the best performance of LLCC. The increasing of inlet mixture velocity would reduce the value of OSR. Separation phenomena in visualization data (photographs) explained how the oil drag occurs.

References

[1] Rifki Rezvani, "Analisa Penerapan Dissolved Air Flotation Sebagai Metode Alternatif Penanganan Limbah Kapal Pada Rancangan Port Reception Facility Di Pelabuhan Tanjung Perak Surabaya," Undergraduate Thesis, Shipping Systems Engineering, Institut Teknologi Sepuluh Nopember, Surabaya, 2006.

[2] A. C. Stones, "Oil/Water separation in a novel cyclone, "PhD. Thesis, School of Engineering, Cranfield University, Bedford, 2007.

[3] H.-f. Liu, J.-y. Xu, J. Zhang, H.-q. Sun, J. Zhang and Y.-x. Wu, "Oil/Water Separation in A Liquid-Liquid Cylindrical Cyclone," *Journal of Hydrodynamics,* vol. 24, no. 1, pp. 116-123, 2012.

[4] C. Oropeza-Vazquez, E. A. L. Gomez, S. Wang, R. Mohan, O. Shoham and G. Kouba, "Oil-Water Separation in a Novel Liquid-Liquid Cylindrical Cyclone (LLCC) Compact Separator-Experiment and Modelling," *Journal of Fluids Engineering,* pp. 553-563, 2004.

[5] I. A. Dharma, F. Arffan, A. R. Prambudi, A. Widyaparaga, I. Pranoto, and Khasani, "Development of liquid-liquid cylindrical cy[1-9]clone (LLCC) separator for oil-water separation," 2016, p. 040013.

[6] O. M. Escobar, "Performance Evaluation of a Modified Liquid-Liquid Cylindrical Cyclone (LLCC) Separator," Master Thesis, Discipline of Petroleum Engineering, University of Tulsa, Tulsa, 2005.

[7] N. Zhou, Y. Gao, W. An, and M. Yang, "Investigation of velocity field and oil distribution in an oil–water hydrocyclone using a particle dynamics analyzer," *Chemical Engineering Journal*, vol. 157, no. 1, pp. 73–79, Feb. 2010.

[8] R. S. Mathiravedu, S. Wang, R. S. Mohan, O. Shoham and J. D. Marrelli, "Performance and Control of Liquid-Liquid Cylindrical Cyclone Separators," *Journal of Energy Resources Technology,* pp. 1-9, 2010.

[9] S. Wang, "Control System Analysis of Gas-Liquid Cylindrical Cyclone Separators," Master Thesis, Discipline of Petroleum Engineering, University of Tulsa, Tulsa, 1997.

Warehouse Layout Designing of Slab Using Dedicated Storage and Particle Swarm Optimization

Y Muharni[1], Kulsum[2], M Khoirunnisa[3]

Abstract— Warehouse is a supporting facility and has the important role in the production system. Goods warehouse arrangement is an activity based on related/rationale reason. There are several policies in random storage, fixed or dedicated storage, and class-based storage. We conducted this study in XYZ Company which the main business is producing steel. In this research, dedicated storage and particle swarm optimization will be used in the layout design of slab raw material warehouse to obtain the best layout to minimize material handling cost. Based on the result of the research, particle swarm optimization method gives the best layout result with the least material handling cost compared to the existing layout and layout using dedicated storage method.

1. Introduction

A warehouse is one of supporting and an important part of a production system. Good conditions and arrangements in the warehouse are expected to avoid corporate losses and minimize the costs incurred and speed up operations and services at the warehouse. Storage process can be performed by different storage policies. The most used and preferred policies can be given as randomized storage policy. the dedicated storage policy, and class-based storage policy. The randomized storage policy is performed by the allocation of the storage location based on the available space at the time of the storage job. Storage decision is left to the operator in another word. A dedicated storage policy determines a particular predetermined location for each product to be stored. A class-based storage policy is a common and shared policy between randomized and dedicated policies. It divides goods into classes based on some criteria and each class is assigned a block of storage locations. This policy can be called as ABC zoning [1]. Hot Strip Mill (HSM) is one of XYZ which produces the hot rolled coil with the raw material of slab steel. In the period 2016-2017. XYZ has to order raw material of slab steel as much as 2,777,089 tons, in 2018. PT. XYZ plans to increase its production capacity so as to support its new warehouse to store raw material of slab steel bar.

PT. XYZ has been doing soil dredging which will be used as outer warehouse 04 to store 20 types of grades and length groups of raw materials of different slabs with 83,070.12 m2.

[1]Y. Muharni is with the Universitas Sultan Ageng Tirtayasa in Cilegan. Banten Province. Indonesia (0254-399552. yusmuharni@gmail.com)

[2]Kulsum is with the Department of Industrial Engineering in the Universitas Sultan Ageng Tirtayasa in Cilegan. Banten Province. Indonesia (0254-399552. yusmuharni@gmail.com) (kulsumkumio@yahoo.com)

[3]M. Khairunnisak is with Production System Laboratory at Industrial Engineering Department. Universitas Sultan Ageng Tirtayasa (mayakhairunnisak7@gmail.com)

IOP Conf. Series: Materials Science and Engineering **532** (2019) 012003 doi:10.1088/1757-899X/532/1/012003

Currently, outer warehouse 04 has no specific rules applied to arrange the placement and preparation of raw material steel slabs (slab). As there is rules, the raw material storage is arbitrary. in order to have an efficient warehouse layout which saves storage space, has short material handling distance and low material handling costs, the method of dedicated storage and particle swarm optimization (PSO) are applied to obtain the best layout.
The dedicated storage method of compiling the product is based on the comparison of the activity of each product to the required space required by the product and then obtained the order of products from the largest to the decimated. Whereas. A particle swarm optimization (PSO) algorithm as a metaheuristic algorithm was developed for determining the optimal layout.

2. Literature Review

2.1. Warehouse
The function of a production warehouse is to store raw materials, work-in-process and finished products, associated with a manufacturing and/or assembly process. Raw materials and finished products may be stored for long periods. This occurs for example when the procurement batch of incoming parts is much larger than the production batch, or when the production batch exceeds the customer order quantity of finished products [2]. The warehouse design is a complex task because of the interaction and relationships between each of the activities in warehouses [3]. The warehouse plays role in supporting a company's supply chain success. The mission of warehouse is to effectively ship product in any configuration to the next step in the supply chain without damaging or altering the product's basic form. Numerous steps must be accomplished and hence there are key warehousing opportunities to address. Doing that will optimize the methods used to achieve the mission. If the warehouse cannot process orders quickly, effectively, and accurately. Then a company's supply chain optimization efforts will suffer. All warehousing opportunities, including improving order picking operations, utilizing cross-docking, increasing productivity, utilizing space and increasing value-added services [4].

2.2. Storage Policy
Storage process can be performed by different storage policies. The most used and preferred policies can be given as randomized storage policy, a dedicated storage policy, and class-based storage policy.

Randomized Storage Policy: The randomized storage policy is performed by the allocation of the storage location based on the available space at the time of the storage job. Storage decision is left to the operator in another word. With a pure randomized storage system each unit of particular product is equally likely to be retrieved when retrieval operation is performed; likewise, each empty storage slot is equally likely to be selected for a storage operation is performed.

Dedicated Storage Policy: With dedicated storage policy a particular set of storage slots or location is assigned to the specific product; hence. a number of slots equal to the maximum inventory level for the product must be provided. The Warehouse layout problem considered involves the assignment of products to storage locations in the warehouse. One of the advantages of dedicated storage is the data-handling efficiency due to the fixed addressing of storage items [5]. In order to minimize the total expected distance travelled the following approach is taken. [4]

$$\sum_{j=1}^{n} \sum_{k=1}^{q} \frac{T_j}{S_j} \qquad (1)$$

The following notation is used:

q = number of storage locations

n = number of products

Sj = number of storage locations required for product j

Tj = number of trips in/out of storage for product j. that is,the throughput of product j

An approach is presented for determining the optimum dedicated storage; rectilinear travel is assumed. The rectilinear travel can be formulated as follows:

$$dij = |xi\text{-}xj| + |yi\text{-}yj| \quad (2)$$

The following notation is used:

*d*ij : distance Traveled

*x*i : coordinate x for point 1

*x*j : coordinate y for I/O

*y*i : coordinate x for point 1

*y*j : coordinate y for I/O

Class-Based Storage Policy: A class-based storage policy is a common and shared policy between randomized and dedicated policies. It divides goods into classes based on some criteria and each class is assigned a block of storage locations. This policy can be called as ABC zoning [1]. Each class is then assigned to a dedicated area of the warehouse. Storage within an area is random. The advantage of this policy is that fast moving products can be stored close to the depot while the flexibility and high storage space utilization of random storage are applicable [6].

2.3. Particle Swarm Optimization

Particle Swarm Optimization (PSO) is a stochastic optimization technique and also a population-based search algorithm first proposed by Eberhart and Kennedy (1995). It is a population-based search method that is inspired by the social behavior of organisms such as bird flocking or fish schooling. A brief and complete survey on PSO mechanism. technique. and application is provided by Kennedy and Eberhart (2001) [7].

Each particle has got a velocity. a memory. and informants. The positions and the velocity of the first population of particles are generated randomly. To decide to move. the particle needs four pieces of information: her actual velocity. her best performance and the best performance of her neighbors. It is based on three parameters w. c1. and c2 which allow them respectively either to follow its own path. retrace his steps. or follow the best neighbor [8].

In PSO. each particle included by social structure keeps in mind its best position and uses this as a factor affecting its speed. A particle gains speed toward its individual best position considering with how far away from that point. It also shows the same behavior for the global best position. In other words. while it is scanning the surface. it is affected by the global best position and adjusts its own speed. In the situation of that. it is far from the global best position. there will be a higher change in its speed and direction [1].

Each individual (particle)'s speed changes according to the formula:

$$V_{ij} = wt\text{-}1\ vt\text{-}1_{ij} + c_1 r_1\,(p_t\text{-}1_{ij} - x_t\text{-}1_{ij}) + c_2 r_2\,(g_t\text{-}1_j - xt\text{-}1_{ij}) \quad (3)$$

The following notation is used:
vtij : velocity of particle i. area j on t-1th iteration
wt-1 : inertia weight pada iterasi t-1

xtij : position value of particle i. area j on tth iteration
xt-1ij : position value of particle i. area j on t-1th iteration
pt-1ij : personal best of particle i. area j on t-1th iteration
gt-1j : global best of particle i. area j on t-1th iteration
c1 . c2 : cognitive and social parameter
r1 . r2 : random uniform [0.1]

Inertia value of the equation changes on each iteration. This change is based on the logic of decreasing from the value determined to a minimum value according to inertia function. The objective is to converge the created speed by diminishing on the further iterations; hence more similar results can be obtained.

Inertia function is obtained as follows:

$$w_t = w_t\text{-}1x\ \alpha \tag{4}$$

The following notation is used:
wt : inertia weight tth iteration
wt-1 : inertia weight t-1th iteration
α : decrement factor

Positions of the particles change by speeds as shown in Eq. (5)

$$x_{tij} = x_t\text{-}1_{ij} + v_{tij} \tag{5}$$

3. Research method

In this section, the stage in problem-solving is described in the following flowchart.

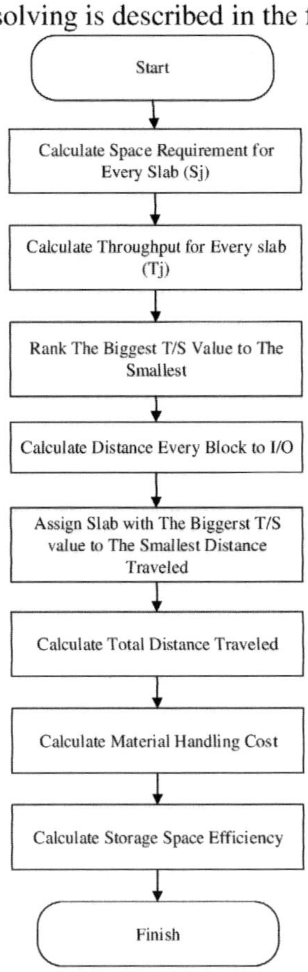

Figure 1. Flow Chart of Dedicated Storage Method

In proposal 2 layout of outer warehouse 04 is utilizing particle swarm optimization method. The problem-solving path using the particle swarm optimization method is as shown in Figure 2. The PSO method for designing the outer warehouse layout 04 is implemented using MATLAB software. The parameters used are the population (swarm size) of 50 particles. Maximum generation is 1000 Cognitive parameters (c1) and social parameters (c2) are 2 and inertia weight (w_t) of 0.9.

4. Results

The following is the results obtained on proposals 1 and 2 of the outer warehouse layout 04 using dedicated storage and particle swarm optimization methods

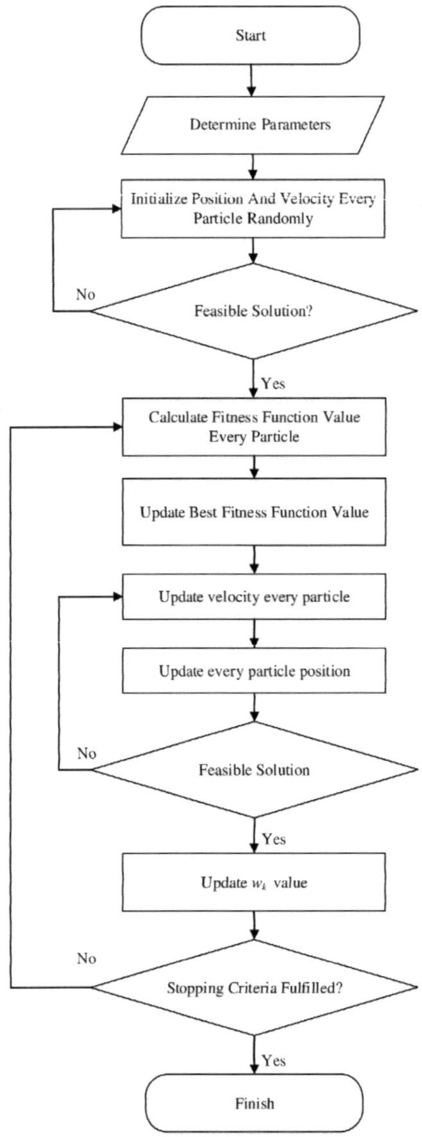

Figure 2. Flow Chart of Particle Swarm Optimization Method

4.1 Dedicated Storage

Layout placement using dedicated storage method is based on the comparison between throughput with space requirement every type of slab.

Table 1. Ranking of throughput over space requirement

No.	Slab	Space Requirement	Throughput	T/S
1	F1.1	4	24	6
2	G.1	2	4	2
3	J.1	4	7	1.75
4	D1.3	88	151	1.715909
5	C.1	71	117	1.647887
6	A.3	46	74	1.608696
7	F1.3	120	156	1.3
8	J.3	20	26	1.3
9	A.1	73	91	1.246575
10	K.1	18	22	1.222222
11	M1.3	78	93	1.192308
12	H2.3	64	76	1.1875
13	D.3	57	60	1.052632
14	K.3	91	88	0.967033
15	C.3	429	396	0.923077
16	G1.3	46	42	0.913043
17	H.3	72	65	0.902778
18	M1.1	51	46	0.901961
19	D1.1	48	41	0.854167
20	G.3	711	539	0.758087

The outer warehouse layout 04 is designed based on the result of the throughput ratio and the space requirement from the largest to the smallest and closest to the outer door of the warehouse 04. The following is the layout of proposal 1 outsourced warehouse 04 using the dedicated storage method.
The design of the second proposal of warehouse layout is using the particle swarm optimization method which requires several steps. such as preparing the total distance input matrix for each slab. determines parameters for particle swarm optimization algorithm. The results obtained from MATLAB as shown in Figure 3.

```
ans =

  Columns 1 through 8

        1000           5           2          19          11          12           4          14

  Columns 9 through 16

           1          15           7          10          20           6           3          16

  Columns 17 through 22

          13          17           9           8          18       24198

  My PSO program took 6.082945 seconds to run.
```

Figure 3. Output Particle Swarm Optimization

Based on the Figure above. we obtained the sequence of location based on the best objective fitness value should be placed close to the outer door of the output 04 or the entrance of the HSM warehouse location are: 5 – 2 – 19 – 11 – 12 – 4 – 14 – 1 – 15 – 7 – 10 – 20 – 6 – 3 – 16 – 13 – 17 – 9 – 8 – 18 with the types : K.3 – F1.1 – F1.3 – K.1 – H.3 – J.3 – C.1 – D1.1 – M1.3 – G.1 – D1.3 – J.1 – A.1 – H2.3 – C.3 – G1.3 – G.3 – M1.1 – A.3 – D.3. The objective fitness value for each of generation is shown in Figure 4.

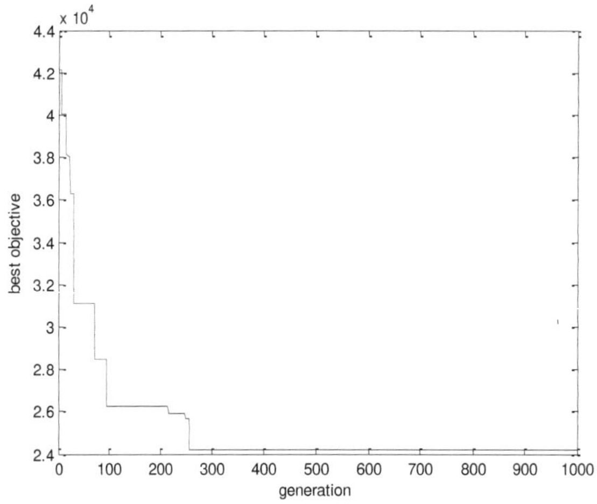

Figure 4. Objective Fitness Every Generation of PSO

4.2 Total Distance Travelled

The calculation of distance travelled using the rectilinear method. That distance is measured along the path by using a perpendicular (orthogonal) line with each other. Here is the distance of each area trip to the input point on the existing layout:

Table 2. Total distance each area to input point in existing layout

From	X	Y	To	X	Y	Distance (m)
Input WH.04	0	286.70	A.1	625.89	26.54	886.048
	0	286.70	A.3	576.89	25.40	838.188
	0	286.70	C.1	531.39	26.45	791.638
	0	286.70	C.3	368.56	27.24	628.015
	0	286.70	D.3	224.55	27.50	483.746
	0	286.70	D1.1	93.00	91.60	288.098
	0	286.70	D1.3	359.74	53.10	593.334
	0	286.70	F1.1	370.54	52.20	605.031
	0	286.70	F1.3	328.04	58.50	556.236
	0	286.70	G.1	285.54	50.10	522.130
	0	286.70	G.3	124.34	54.46	356.580
	0	286.70	G1.3	513.89	54.45	746.140
	0	286.70	H.3	60.50	91.60	255.598

	0	286.70	H2.3	660.55	30.17	917.072
	0	286.70	J.1	278.54	52.20	513.030
	0	286.70	J.3	68.38	50.25	304.822
	0	286.70	K.1	669.89	54.45	902.138
	0	286.70	K.3	624.38	53.40	857.675
	0	286.70	M1.1	565.89	54.45	798.138
						0

Table 3. Total distance each area to output point in existing layout

From	X	Y	To	X	Y	Distance (m)
A.1	625.89	26.54		659.39	0.00	60.041
A.3	576.89	25.40		659.39	0.00	107.901
C.1	531.39	26.45		659.39	0.00	154.451
C.3	368.56	27.24		659.39	0.00	318.074
D.3	224.55	27.50		659.39	0.00	462.343
D1.1	93.00	91.60		659.39	0.00	657.991
D1.3	359.74	53.10		659.39	0.00	352.755
F1.1	370.54	52.20		659.39	0.00	341.058
F1.3	328.04	58.50		659.39	0.00	389.853
G.1	285.54	50.10	Output	659.39	0.00	423.959
G.3	124.34	54.46	WH 04	659.39	0.00	589.509
G1.3	513.89	54.45		659.39	0.00	199.949
H.3	60.50	91.60		659.39	0.00	690.491
H2.3	660.55	30.17		659.39	0.00	31.329
J.1	278.54	52.20		659.39	0.00	433.059
J.3	68.38	50.25		659.39	0.00	641.267
K.1	669.89	54.45		659.39	0.00	64.949
K.3	624.38	53.40		659.39	0.00	88.414
M1.1	565.89	54.45		659.39	0.00	147.951
M1.3	56.50	102.76		659.39	0.00	705.654
			TOTAL			6861

Table 4. Total distance each area to output point in based on PSO algorithm

From	X	Y	To	X	Y	Distance (m)
A.1	533.39	4.20		659.39	0.00	130.201
A.3	570.89	25.40		659.39	0.00	113.900
C.1	683.77	20.52		659.39	0.00	44.895
C.3	315.54	27.50	Output	659.39	0.00	371.356
D.3	559.89	53.40	WH.04	659.39	0.00	152.901
D1.1	263.39	4.20		659.39	0.00	400.201
D1.3	622.89	26.45		659.39	0.00	62.950
F1.1	651.89	4.20		659.39	0.00	11.701

F1.3	505.89	26.45	659.39	0.00	179.951
G.1	644.89	2.10	659.39	0.00	16.601
G.3	171.33	49.03	659.39	0.00	537.092
G1.3	341.39	4.20	659.39	0.00	322.201
H.3	400.04	58.50	659.39	0.00	317.856
H2.3	611.89	53.40	659.39	0.00	100.901
J.1	637.89	4.20	659.39	0.00	25.701
J.3	614.89	4.20	659.39	0.00	48.701
K.1	527.39	54.45	659.39	0.00	186.450
K.3	434.30	12.91	659.39	0.00	237.994
M1.1	185.54	26.45	659.39	0.00	500.306
M1.3	670.14	53.40	659.39	0.00	64.149
		TOTAL			3826.006

4.3 Material Handling Cost

The calculation for the total cost of material handling for an existing condition.

Table 5. Material Handling Cost of Existing Layout

From	To	Distance (m)	Freq.	MH	Cost	Total Cost (IDR)
A.1		60.041	41			2.893.365
A.3		107.901	31			3.931.473
C.1		154.451	55			9.984.401
C.3		318.074	213			79.630.173
D.3		462.343	29			15.759.118
D1.1		657.991	3			2.320.124
D1.3		352.755	61			25.291.393
F1.1		341.058	4			1.603.460
F1.3		389.853	76			34.824.475
G.1	Output WH.0 4	423.959	1	Trailer Truck	IDR. 1.175	498.303
G.3		589.509	247			171.142.125
G1.3		199.949	21			4.935.248
H.3		690.491	32			25.970.358
H2.3		31.329	19			699.621
J.1		433.059	2			1.017.998
J.3		641.267	10			7.537.171
K.1		64.949	6			458.032
K.3		88.414	40			4.156.729
M1.1		147.951	23			3.999.597
M1.3		705.654	23			9.076.091
TOTAL						IDR. 415.729.255

4.4 Space Efficiency

The calculation of storage space efficiency is

$$Percent\ of\ Efficiency == \frac{Total\ Block\ Area}{area\ of\ space}\ x\ 100\%$$

Based on the above calculation obtained the percentage of storage space efficiency on the existing outer warehouse layout 04 is 46.704%. The following is a calculation of the area of each slab area in the proposed layout 1 outside warehouse 04.

Table 6. Space Efficiency of First Alternative Layout

Area	Area				Total Area
	I	II	III	IV	
A.1	1041.6	0	0		1041.6
A.3	655.2	0	0		655.2
C.1	737.111907	273	0		1010.111907
C.3	4641	0	0		4641
D.3	764.396325	0	0		764.396325
D1.1	655.2	0	0		655.2
D1.3	1228.511907	0	0		1228.511907
F1.1	58.8	0	0		58.8
F1.3	1719.9	0	0		1719.9
G.1	29.4	0	0		29.4
G.3	4368.74991	3003	2457		9828.74991
G1.3	655.2	0	0		655.2
H.3	1092	0	0		1092
H2.3	873.6	0	0		873.6
J.1	58.8	0	0		58.8
J.3	323.4	0	0		323.4
K.1	245.7	0	0		245.7
K.3	491.4	764.4	0		1255.8
M1.1	737.1	0	0		737.1
M1.3	1083.6	0	0		1083.6
Total					27958.07

5. Analysis and discussion

The total distance obtained in layout proposal 1 using dedicated storage method is 18992.036 m with material handling cost IDR. 340.062.757, the percentage of storage space efficiency obtained in a layout by using dedicated storage method is 46.685%. Placement of all types of slab using particle swarm optimization method based on the best solution search process in the form of moving material handling distance with parameters that have been determined. The sequence generated in the best solution (global best) will serve as a sequence on the placement of the slab in the layout. The first sequence of the slab will be placed at the location which has the smallest distance that is close to the outer warehouse output door 04. The total distance obtained in the layout using the particle swarm optimization method is 18955.780 m with material handling cost of IDR. 334.217.964 and the percentage of storage space efficiency of 48.189%.In the existing layout. The distance obtained is

IOP Conf. Series: Materials Science and Engineering **532** (2019) 012003 doi:10.1088/1757-899X/532/1/012003

18945.092 m which is the smallest distance. However, material handling cost is IDR 415.729.255. Thus, the best layout in this research is the proposed 2 layout using particle swarm optimization method because it produces the lowest material handling cost.

6. Conclusion

The best layout in this research is the layout of proposal 2 using particle swarm optimization method because it produces the lowest material handling cost that is IDR 334,217, 964.

References

[1] Semih Onut. Umut R. Tuzkaya. Bilgehan Dogac."A particle swarm optimization algorithm for the multiple-level warehouse layout design problem" ELSEVIER. 54. pp. 783–799. 2008.
[2] Rouwenhorst. B., Reuter. B., Stockrahm. V., Van Houtum. G. J., Mantel. R. J., & Zijm. W. H. M. Warehouse design and control: Framework and literature review. European Journal of Operational Research. 122. 515–533. 2000.
[3] Sooksaksun. N and Kachitvichyanukul. V. Particle Swarm Optimization for Warehouse Design Problem. Asia Pacific Regional Meeting of International Foundation for Production Research. Pathumthani : Asian Institute of Technology. 2010.
[4] Tompkins. J. A. et al. Facilities Design. Third Edition. Singapore : John Wiley & Sons (ASIA) Pte Ltd.2003
[5] M.K. Lee and E. A. Elsayed." Optimization of warehouse storage capacity under a dedicated storage policy" International Journal of Production Research. Vol. 43. No. 9. pp.1785–1805. 1 May 2005.
[6] Felix T.S. Chan. H.K. Chan. "Improving the productivity of order picking of a manual-pick and multi-level rack distribution warehouse through the implementation of class-based storage" ELSEVIER. 38. pp. 2686–2700. 2011
[7] Kennedy. J.. Eberhart. R. Swarm intelligence. San Francisco: Morgen Kaufmann Publishers.2001
[8] N. Yalaoui. L. Amodeo.H. Mahdi.F. Yalaoui. " Hybrid method to solve a facility layout problem: Genetic Algorithm – Particle Swarm Optimization" Proceedings of the 13th IFAC Symposium on Information Control Problems in Manufacturing Moscow. Russia. June 3-5. 2009

IOP Conf. Series: Materials Science and Engineering **532** (2019) 012004 doi:10.1088/1757-899X/532/1/012004

Calcination process on chanos chanos forsk (CCF) of milkfish bones to get Hydroxyapatite (HAp) as Composites Application

Agus Pramono[1], Fatah Sulaiman[2], Suryana[1] and Anistasia Milandia[1]
[1]Department of Metallurgy Engineering, University of Sultan Ageng Tirtayasa
[2]Department of Chemical Engineering, University of Sultan Ageng Tirtayasa
Jl. Jenderal Sudirman Km 3, Kotabumi, Cilegon, Banten 42435, Indonesia

E-mail: agus.pramono@untirta.ac.id

Abstract. The material hydroxyapatite (HAp) from fish bone, $Ca_5(PO_4)_3(OH)_2$ used as a material of composite. HAp is an inorganic material that shows good biocompatibility for bone-connection material. It has excellent ability to form direct bonds with hard tissue in bone, when used as bone filler or as a ceramic coating on surgical implants. In this work, investigation focused on the suitability of fish bone material as a hydroxyapatite matrix with two additive elements ie magnesium (Mg), titanium (Ti) or copper (Cu) or another kind of coupling agent. Aluminium (Al) powder as a reinforcement in the composite materials. Investigation of morphological and structural properties of HAp samples derived from bone fish-chanos-chanos forsk (CCF) traced by microscope optical (M.O) to show the interconnected relative morphology grains to form complex compact matrix in hybrid composite materials. Analysis of sample elements using XRD to reveal the decomposition occurring at each calcination temperature. As well as a Finite Element Machine, type ANSYS R.18.2 trace simulation to ascertain how much actual temperature is occurring in the sample, it is necessary to trace the heat transformation that occurs within the HAp based composites of material. The test variable is the use of intermediate compositions, between the use of wetting agents between Mg-Cu, Mg-Ti or another combination of coupling agent. By high score software of XRD tracing to show how the hexagonal/monolith structures on HAp powder diffuses well against the wetting and reinforcing elements, so the process of the calcination will be applied for hybrid composite.

1. Introduction

Hydroxyapatite (HAp) is a crystalline molecule essentially composed of phosphorus and calcium, $Ca_5(PO_4)_3(OH)_2$ molecular formula. The 65% molecule consists of the mineral fraction present in the human bone. This material is also present in human tooth structure especially in dentine and enamel, this material is important in health world application. Bone of material has hydroxyapatite content, one of which is fish of bone, specially chanos chanos forsk (CCF). The Nutritional content of fish bone of CCF, based on laboratory test results found to contain calcium (Ca) as much as 4.7756% and phosphorus (P) as much as 1.3125%. Ca content in most milkfish found on the bone. Fish bone is one of the wastes that has not been utilized properly even though the CCF of milkfish bone contains of high calcium wastes are produced every day to reach 15-25 kg or about 5-12 tons per year. In Bone fish of CCF contains Ca 4%, P 3%, and 32% protein. Thus, the need for new product processing for waste making can be used optimally. One effort to utilize the waste by processing bone fish waste, but in the body of milkfish is suspected contain heavy metal copper (Cu).

This occurs from the shifting of Cu elements in the water content such as in the pond, the feasibility of inland ponds for milkfish culture can be determined the content of Cu in the body of milkfish ranging from <0.01 to 3.28 mg/kg, while the Cu content in pond water is <0.01 mg/l was undetectable [1-3]. Milkfish bones of CCF can be used as composite materials, milkfish bones such as: phosphorus, calcium and copper are elements that metallurgically can be combined with magnesium (Mg) as a wetting element that will provide an interface between the mixed material, made into a composite material with a material content of more than one type of matrix, this is referred to as hybrid-composite.

Currently, hybrid-composite material becomes the trend of future material technology because it can be applied to various types of components with high strength and mild density, so that the use of bone fish bone will lead to the eco-material that demands the development of material novelty without leaving the impact of pollution in the process of manufacture [3-4]. This research is an early synthesis of how the calcination process is done on the bone of CCF into hydroxyapatite, due to several advantages. The use of temperature following the ideal calcining temperature of bone of CCF between 600-700^{0}C. After becoming hydroxyapatite, it will be combined by metal material of using two type matrix or reinforcement for hybrid composites application.

2. Procedure Experimental

2.1 Heat Treatment on Milkfish bones of CCF

The initial stage was the washing of milkfish bone using water and soap to remove the fat and dirt that stick to the bones of milkfish, after a clean bone boiled until boiling for± 60 minutes. The next step is the heating process (calcination) to get hydroxyapatite, the heating was applied at a temperature of 500-700^{0}C. The reading scale inside furnace was 600^{0}C but outside the temperature of 700^{0}C by thermocouple reading scale, so it is necessary to search the actual temperature, using finite element method (FEM) by the new latest type 18.2 of ANSYS software. The heating process for obtaining hydroxyapatite shown in Fig. 1.

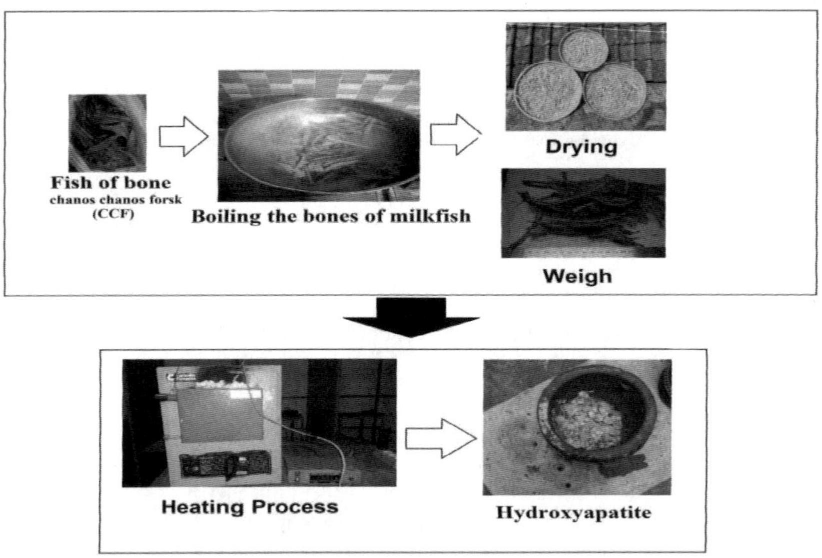

Figure 1. The calcination process on the milkfish bone of CCF become hydroxyapatite.

2.2 Characterization for obtaining hydroxyapatite of milkfish of bones (CCF)

The first test is the actual temperature accuracy tracking, this test used FEM software, to trace the actual temperature on the samples of milkfish bone (CCF). Temperatures need cross-check to get the actual temperature of the sample as a standard method of calculation of bone fish calcination, so that with the actual temperature standard can be used standard operates temperature. Micro-structural testing was performed to obtain an analysis of morphological changes in bone fish of CCF structure as well as bone defects such as porosity, segregation, or dirt on the bones of milkfish. This needs to be done to obtain an analysis related to the possibility of bone fish as a hydroxyapatite bone combined with a metal element as a composite base material. XRD testing is applied to obtain hydroxyapatite elements and possibly other luggage elements. using high score converters to determine the type of element or percentage of hydroxyapatite compound obtained. Among the tests will be applied a comprehensive analysis so that for the next stage can be obtained accurate hypothesis related to the application.

3. Hydroxyapatite as a Result Product

The heating process of bone fish of CCF calcination produces HAp. Formation of HAp depends on the suitability of temperature processes, on each heating from 1 to 5 hours resulted in different HAp in terms of physical and colour, at temperature 600^0C duration 2 hours produced HAp which still have black colour, suspected as carbon burning at CCF, evidenced in Fig. 2 in terms of colour and fig. 4 on microstructural observation results. Too longer the heating-up time, black colour turns brown until 4 hours. Within 5 hours produced bright of white colour without black or brown, this proves hydroxyapatite metamorphosed perfectly, but over than 5 hours, the overall bone of the fish was burning, this proves a transformation over than 5 hours, turning into total carbon. Changes of bone fish transformation presented in Fig. 2.

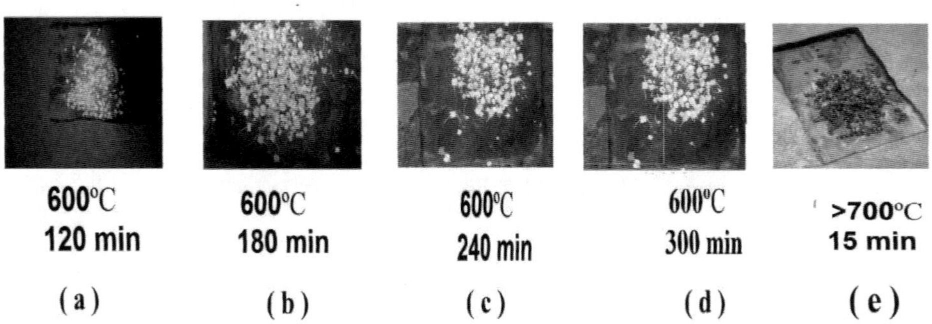

Figure 2. The hydroxyapatite colour change based on the duration of heating.

4. Result and Discussion

4.1 Simulation of samples Hydroxyapatite by FEM Ansys 18.2

Finite element machine (FEM) was applied to trace the temperature inside the furnace, CCF fish bone was inserted into the dies, and the temperature absorbed inside the CCF from the furnace to the inside of dies is presented in figure 3. In Figure 3 a. Temperature of 600^0C 1 hour indicates imperfect calcination, there was still black colour between the pieces of HAp in some parts, the temperature absorbed based on FEM still too low temperatures: 518^0C while at the base of the dies, the place where the bone was heated absorbed ±421-453^0C. Duration of 3 hours of heating is shown in Fig 3 b. Optimum temperature in the dies reaches 581^0C, at the bottom of the dies, where the CCF is placed, temperature are ±558-568^0C, if seen from fig 2 b. there is still black and brown colour indicating the hydroxyapatite is not perfect. Heating 4 hours fig 3 c. optimum temperature in the dies reaches 595^0C, on the bottom in dies where the CCF is placed, the temperature rises ±590-592^{0}C, seen fig. 2 c shown the black colour is not exist but brown colour still exist with a small amount. The carbon from the burning has changed to hydroxyapatite, brown colour is a transformation that indicates near perfect hydroxyapatite.

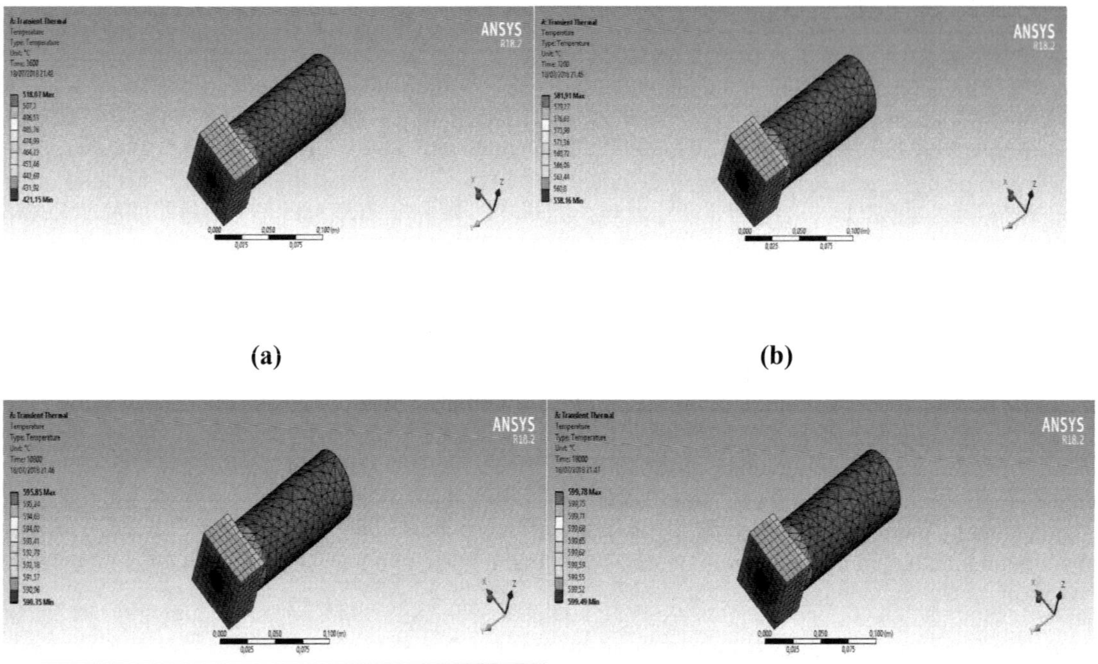

(a) (b) (c) (d)

Figure 3. Heating of CCF into a 600^{0}C: (a) temperature for 2 hours (b). temperature for 3 hours c). temperature for 4 hours d). temperature for 5 hours.

Heating 5 hours in fig 3 d. the optimum temperature of the dies reaches 599^{0}C, at the bottom on the dies, where the CCF is placed, the temperature is 599^{0}C as well, this condition indicates that the heating is evenly distributed in all parts, seen from 2 d, no black and brown color. Complete combustion can affect the perfect transformation of hydroxyapatite.

4.2 Microstructures Analysis

The micro structure of the bone before heated was observed, in bones there was a grey colour that indicates the bone is still mushy, figure 4 a. CCF requires heating to transform into hydroxyapatite. Heating was try carried out at 600^{0}C for 1 hour. Figure 4 b. Heating for 1 hour there were some porosity in some connection bones, black whitish is the transformation from bone into CCF, up to 2 hours of warming 600^{0}C in the picture 4 c. Increased to the duration of 3 hours in figure 4.d some of the parts exposed to excessive heat into black carbon and transformation of white HAp is not perfect. On duration 4 hours 600^{0}C image 4 e. carbon is reduced and dispersed into freckles between the surface of the HAp structure. The 5-hour warming time of 600^{0}C is the ideal time, shown in Fig. 4 f. there is no carbon and the color turns brown and the presence of HAp is able to cover the porosity and blow holes into HAp structure grains.

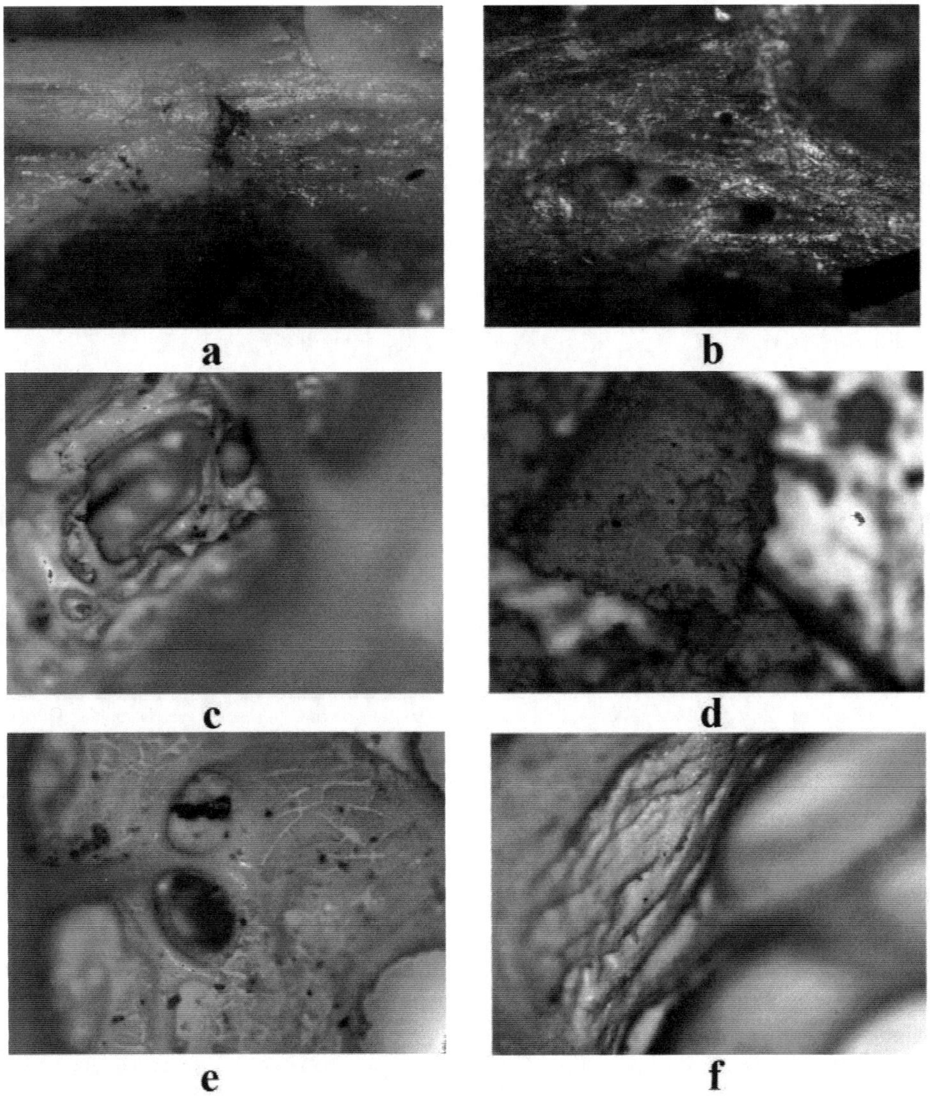

Figure 4. The Microstructure of CCF (a). CCF Non-heating (b). 600^{0}C 1-hour (c). 600^{0}C 2 hours (d). 600^{0}C 3 hours (e). 600^{0}C 4 hours (f). 600^{0}C 5 hours

4.3 Characterizations of XRD

XRD testing results prove that CCF contain of HAp, very dominant, accompanied by Ca and P. The HAp compound is $Ca_5(PO_4)3.OH_2$ [3] while Ca and P, are elements that have been bound in HAp. HAp can be CO_3 substituted properly with Tricalcium Phosphate (TCP) if the calcination temperature increases from 700-1000°C. The consolidated results show that the extracted HAp powder shows a crystallinity between 50 to 70% corresponding to the high palladium HAp phase crystallinity [5]. As in some other metal content such as Copper (Cu) or Lead (Pb) is the result of the absorption of the water environment where the milkfish live, such as the concentration of Pb in milkfish because in water there is a concentration of Pb [6]. Cu and Pb in fish bone tissue will interfere with integration with other metal elements in the composition, because it will make the segregation defect, so that the element is avoided in composite integration. The results of XRD will be recommended for hybrid composite with Al as a reinforcement, magnesium and titanium as a wetting agent and HAp as matrix.

The process to be used is Self-Propagating high temperature synthesis (SHS), with the help of pressure as an additional procedure [7]. Calcination of milkfish bones for 1 hour produced HAp as much as 39.6% HAp of monolith structure and 60.4% of HAp hexagonal structure, shown fig 5. a. In calcination, 3 hours in fig 5 b. resulted HAp of monolith structured as much as 39.3%, but remaining of 60.7% was dominated by carbonate. While 5 hour calcination of milkfish bones produces 49.4% HAp of monolithic structure and 50.6% HAp hexagonal structure presented in fig 5 c.

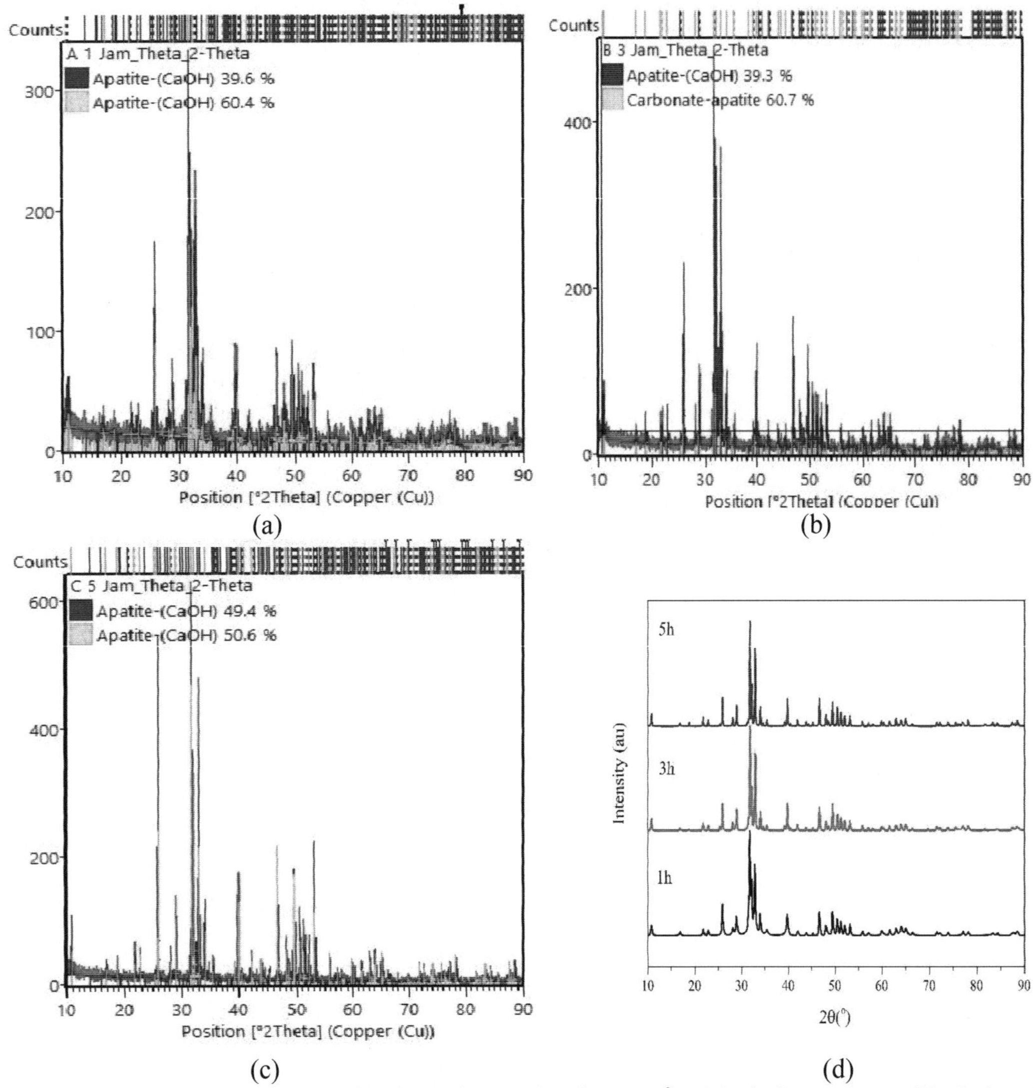

Figure 5. Results of XRD calcination of CCF heating 600^{0}C (a) during 1 hour (b) during 3 hour (c) during 5 hours (d) Result of XRD analysis using high scores to obtain HAp of monolith structure.

The structure of HAP monolith structured is easier to react with metal elements as composite materials, compared to the hexagonal HAp of structure. Whereas carbonates are more likely to form fragility in composite materials. Calcination of 1 hour and 3 hours have not produced an optimal monolith structure compared to 5 hours calcination. The calcination process of milkfish bone during 5 hours produced a monolithic HAp structure that easily reacts with metal compounds as composite base material. Based on microstructural point of view, structural HAp contains calcium phosphate compounds, with characteristics similar to needle-shaped ceramics with angstrom size. this structure is very suitable to be used as matrix on concentrate of composite material [8]. Figure 6 shown data of XRD by high score software processing, milkfish CCF of bone structure presents the percentage of different elements, depending on the environment and the initial synthesis process. The black color from peak, indicated the bone of CCF was not produce the dominant HAp, the lack of heating time (1-hour calcination) makes the heat insufficient for the process of monolith structure collision, at the red peak, 3 hours warming of the HAp partially becomes carbonate. lack of heating time HAp is transformed into carbon which tends to be detrimental as an element. 5 hours is enough time to transform into a monolith HAp. This monolith structure can be used as the basis for composite formation.

5. Conclusions

Hydroxyapatite (HAp) is a crystalline molecule essentially composed of phosphorus and calcium, $Ca_5(PO_4)3OH_2$ molecular formula. The 65% molecule consists of the mineral fraction present in the human bone. The heating process (calcination) to get hydroxyapatite, the heating was applied at a temperature of 500-700^{0}C. The reading scale inside furnace was 600^{0}C but outside the temperature of 700^{0}C by thermocouple reading scale. The heating process of bone fish calcination produces hydroxyapatite (HAp), formation of HAp depends on the suitability of temperature processes, on each heating from 1 to 5 hours resulted in different HAp in terms of physical and colour. Five-hour heating time of 600^{0}C was ideal time, because there was no carbon and the colour turn brown and the presence of HAp was able to cover the porosity and blow holes into HAp structure grains. HAp product of CCF will be recommended for hybrid composite of ceramic matrix composites (CMC) with Al as a reinforcement, Mg/Cu as a wetting agent and HAp as matrix. The process to be used is Self-Propagating high temperature synthesis (SHS), With the help of pressure as an additional procedure, namely Exothermic Pressing (EP).

References

[1] R. Petit. 1999. The use of hydroxyapatite in orthopaedic surgery: A ten-year review. European Journal of Orthopaedic Surgery & Traumatology Volume 9, Issue **2**, pp. 71–74.

[2] Roedy Budirahardjo. 2010. Sisik Ikan Sebagai bahan yang berpotensi Mempercepat proses penyembuhan Jaringan Lunak Rongga Mulut, Regenerasi Dentin Tulang Alveolar. Bagian Pedodonsia FKG Universitas Jember. Stomatognatic (J.K.G Unej) Vol. **7** No. 2: pp. 136-140.

[3] Nana Kariada, Tri Martuti. Kandungan Logam Berat Cu Dalam Ikan Bandeng. 2012. Studi Kasus Di Tambak Wilayah Tapak Semarang. Prosiding Seminar Nasional Pengelolaan Sumberdaya Alam dan Lingkungan Semarang, 11 September 2012.

[4] Agus Pramono, The composites of technology as a trend of the future. Kompasiana. https://www.kompasiana.com/pramono_agus/554f82c9aa333112e1f8b511e.5 March 2014.

[5] Jess Gambe, Reynaldo Vequizo , Arnold Alguno, Simon Jude Burgos, Rolando Candidato, Bianca Rae Sambo and Majvell Kay Odarve. 2013. Nano-Sized and CO_3-Substituted Hydroxyapatite (HAp) from Thermally-Processed Milkfish (Chanos Chanos) Bones with Crystallinity between 50-70%. The Third International Congress on Interdisciplinary Research and Development, 30 - 31 May, Thailand.

[6] Tarzan Purnomo and Muchyiddin. 2007. Analysis of Lead Content (Pb) on Milkfish (Chanos chanos Forsk). Neptunus, Vol. 14, No. **1**, pp. 68-77.

IOP Conf. Series: Materials Science and Engineering **532** (2019) 012004 doi:10.1088/1757-899X/532/1/012004

[7] Agus Pramono, Lembit Kommel, Lauri Kollo and Renno Veinthal. 2016. The Aluminum Based Composite Produced by Self Propagating High Temperature Synthesis. Materials Science (MEDŽIAGOTYRA). Vol. 22, No. **1**. pp. 40-43.

[8] Grimm, M.J. 2009. Orthopaedic Biomaterials. In Standard Handbook for Biomedical Engineering and Design, 2nd Ed. (Edited by Kutz, M.), McGraw-Hill, New York.

IOP Conf. Series: Materials Science and Engineering **532** (2019) 012005 doi:10.1088/1757-899X/532/1/012005

Graphical Analysis Springs Dampers System on the A Freight Cars

Indrawan Arifianto[1] , Rahmat Alamtaha[1], Ary Syahriar[1], Dwi Astharini[1]
[1]Department of Electrical Engineering, University of Al Azhar Indonesia, Jakarta, Indonesia

E-mail: indrawanarifianto1997@gmail.com

Abstract. In mechanical system where spring and damper are used often create unwanted vibration, due to external forces. Unwanted vibration may cause further noise and other damages such as engine break, etc. To reach system stability, we model of coupling series of freight car and analyze them using root loci and plot waves. From the result we can analyze what is the maximum stability by altering spring and damper constant. It is found when the spring and damper constant increase, the stability will be fast.

1. Introduction

Any repetitive motion in the same time interval is called periodic motion or harmonic motion. If a particle in a periodic motion moves back and forth through the same trajectory the motion is called the oscillation motion. The oscillation motion is one form of a system model with a spring-damper. This response is caused by an external force.

Mechanical systems that work with a damper-spring system often cause unwanted vibrations. Unwanted vibrations often cause problems such as noise, equipment damage and uncomfortable condition for passengers (in a vehicle).

Therefore the model of a system needs to be built to determine the characteristics of a particular system. By knowing the characteristics of a system then we can provide solutions to control this content to fit the desired criteria. This paper discusses mechanical coupling consisting of springs and dampers which are used to hold the freight train to keep the distance between carriage 1 with another carriage when one of the carriages stops or moves.

The effect and magnitude of spring damper system will be discussed. The analysis is performed any changing the values of spring and damper to reach faster stability. All simulation were done using Simulink MATLAB structure.

2. Basic Theory

Simple harmonic oscillation is an oscillatory motion of an object that is affected by a linear restoring force and is not frictionless so that it does not experience power dissipation. Simple harmonic oscillations can also be interpreted as a vibrating system in which the restoring force is directly proportional to its negative deviation. Oscillations like this usually occur in springs that provide force or load.

If the initial length of the spring does not change, then it can be assumed that the variable x is a function of the spring configuration that is derived, the initial relation decreases to [2]:

$$f_s = -kx \quad (1)$$

where is the restoring force, is the compression or extension distance (the change in length from the free length), and is the spring constant, or stiffness, which is defined to be always positive. If we connect it with acceleration, it will be obtained:

$$ma = kx = m\frac{d^2x}{dt^2} \quad (2)$$

So,

$$a = \frac{d^2x}{dt^2} = -\left(\frac{k}{m}\right) \quad (3)$$

A negative sign is a restoring force or defined as a force acting in the direction of returning the mass of an object to its equilibrium position. The spring element provides an oscillating force in response to displacement. Whereas, the damping element is an element that rejects the relative velocity above it which makes the oscillation stop. Damping modeling as a linear function of relative velocity and no effect is at all the effect of damping strength, the linear damping force model as a function of the relative disadvantage of v is [4]

$$f_d = cv = c\frac{dx}{dt} \quad (4)$$

Where is the damping coefficient. The units of are force / velocity. Damping causes the mechanical energy of the system to decrease within a certain time interval. There are 4 characteristics of the oscillation system based on the natural response:
1. Excessive damped odor (Overdamped response).
2. Low damped odor (Underdamped response).
3. Critically damped response (Critically damped response)
4. Oscillation is not muffled (Un damped Response)

The method used to solve the feedback system such as the spring damper system is bode plot method, the Nyquist method, and the root locus method [3]. In this paper the system used is analyzing bode plot method and root locus method. The bode plot consists of two graphs, the plot of the logarithm of the magnitude of the sinusoidal transfer function and the plot of the phase angle, both plotted against frequency on a logarithmic scale. The main advantage of using a bode plot is that the multiplication of magnitude produced can be added [7].

The basic factors that very frequently occur in an arbitrary transfer function G $(j\omega)$ H $(j\omega$)are [7] :
1. Gain K.
2. Derivative and Integral factors $(j\omega)^{\pm 1}$
3. First-order factors $(1 + j\omega T)^{\pm 1}$
4. Quadratic factors $(1 + \frac{2\zeta j\omega}{\omega_n} + (\frac{2\zeta j\omega}{\omega_n})^2)^{\pm 1}$

The root locus method can predict effects at the location of closed loop poles by varying the gain value or adding open loop poles and open loop zeros [7]. The root-locus method proved to be very useful in designing linear control systems, because open loop and zero poles can be modified to meet system performance response specifications [3]. The closed-loop transfer function is

$$\frac{C(s)}{R(s)} = \frac{G(s)}{1+G(s)H(s)} \quad (5)$$

Here we assume that G (s) H (s) is the polynomial ratio in s. R (s) is the input of the system and C (s) is the output of the system.

IOP Conf. Series: Materials Science and Engineering **532** (2019) 012005 doi:10.1088/1757-899X/532/1/012005

3. Research Methods

Script structure which is the result of review and study of theory or concept written as follows:

3.1. Identification of errors

Identification of errors concerning the damping of the system with springs and dampers obtained from the results of the joint discussion. Damping system using springs and dampers is rarely used whereas spring and damper systems are the most dominant mechanical systems. For example: engine failure due to excessive vibration and in the long run.

3.2. Implementation of the design

Implementation of the design of the damp with spring and damper is done by system identification and modeled into mathematical equations. Then converted by Laplace transform into transfer function so that it can be modeled with MATLAB program precisely Simulink. New process of system analysis done. After that varied values of existing parameters. From that process, a comparison of the results was obtained.

4. Modeling System

Here is a real picture of the coupling system of a number of freight cars.

Figure 1. Coupling system of series of freight cars. [8]

The above figure is a mechanical coupling commonly used to hold the freight cars in the following picture. This system can be modeled as two masses, a spring, a damper, and thrust and pull forces on each mass. And here is a clutch system model of a series of freight cars.

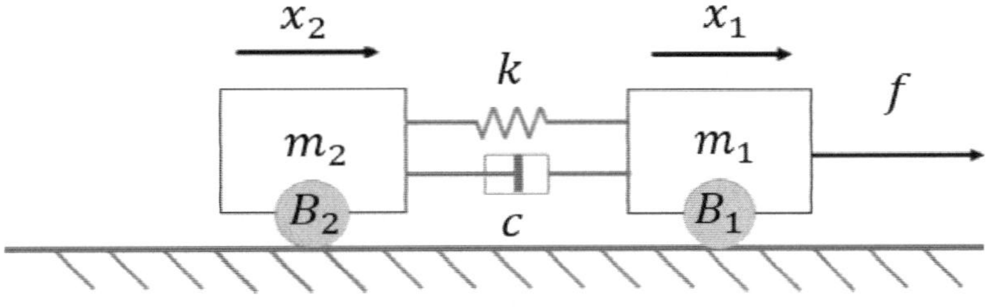

Figure 2. Model spring-damper coupling series of freight cars.

where:

f = Input force of freight cars (N)
m_1 = Mass of freight cars (kg)
m_2 = Mass of the car being coupled (kg)
B_1 = Friction of freight car (N s/m)
B_2 = Friction of the car being coupled (Ns/m)
x_1 = Freight car distance (output) (m)
x_2 = Car being coupled distance (output) (m)
c = Constant of damper (Ns/m)
k = Constant of spring (N/m)

Completion of a system with many orders is more complicated than a system that is modeled with two orders, but the completion of a system with many orders can be solved using the concept of two orders [1]. From the spring-damper system model is made Mathematical equations as follows:

The equation of m_1 :

$$f(t) = m_1 \frac{d^2x_1(t)}{dt^2} + B_1 \frac{dx_1(t)}{dt} + c\left(\frac{dx_1(t)}{dt} - \frac{dx_2(t)}{dt}\right) + k(x_1(t) - x_2(t)) \tag{6}$$

The equation of m_2:

$$c\left(\frac{dx_1(t)}{dt} - \frac{dx_2(t)}{dt}\right) + k(x_1(t) - x_2(t)) = m_2 \frac{d^2x_2(t)}{dt^2} + B_2 \frac{dx_2(t)}{dt} \tag{7}$$

Apply the Laplace transform to this equation, using zero initial condition, to obtain transfer function. Substituting equation 1 and equation 2 to obtain $\frac{X_1(s)}{F(s)}$ and $\frac{X_2(s)}{F(s)}$ and if $V_1 = \frac{X_1(s)}{s}$ and $V_2 = \frac{X_2(s)}{s}$, so the equation as follows:

$$\frac{V_1(s)}{F(s)} = \frac{(m_1s^3+(B_2+c)s^2+ks)}{(m_1s^3+(c+B_1)s^2+k)^2-(cs+k)^2} \tag{8}$$

$$\frac{V_2(s)}{F(s)} = \frac{(cs^2+ks)}{(m_2s^3+(c+B_1)s^2+k)^2-(cs+k)^2} \tag{9}$$

5. Result and Discussion

The mass spring damper system is used to design coupling system of series of freight train. In table 1 used several parameters to be simulated using MATLAB so it can be known the influence of springs and dampers on of the freight train. And the result of the graph also can be used to analyze about stability, damping ratio, and others.

Table 1. Variations Input Parameters Of Spring And Damper Constants

No	Parameter	m_1=m_2 (kg)	k	B_1=B_2=c (N.s/m)
1	Parameter 1 (P1)	1x10^4	20x10^3	30x10^3
2	Parameter 2 (P2)	1x10^4	30x10^3	40x10^3
3	Parameter 3 (P3)	1x10^4	40x10^3	50x10^3

The following is a simulation of the equation result so that the speed graph is obtained when driving, when there is braking and the distance between mass 1 with mass 2 with parameters already given.

IOP Conf. Series: Materials Science and Engineering **532** (2019) 012005 doi:10.1088/1757-899X/532/1/012005

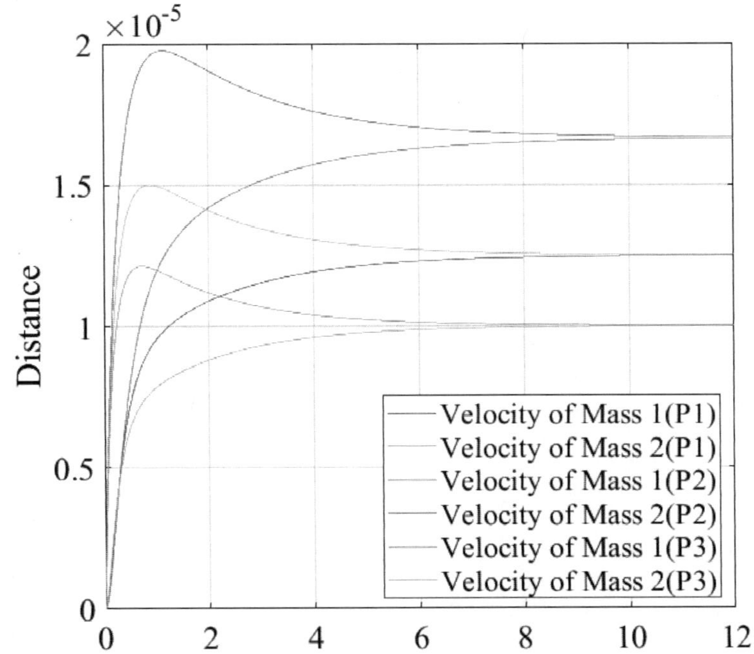

Figure 3. Graph of speed simulation with three different parameters.

From Figure 3 we see the output of mass velocity 1 and mass 2. Although in the initial conditions the speeds of m_1 and m_2 are different, but in steady state the two masses have the same speed. From Figure 3 it can also be seen that by using large dumper and spring constant values, the time needed to reach steady state is faster.

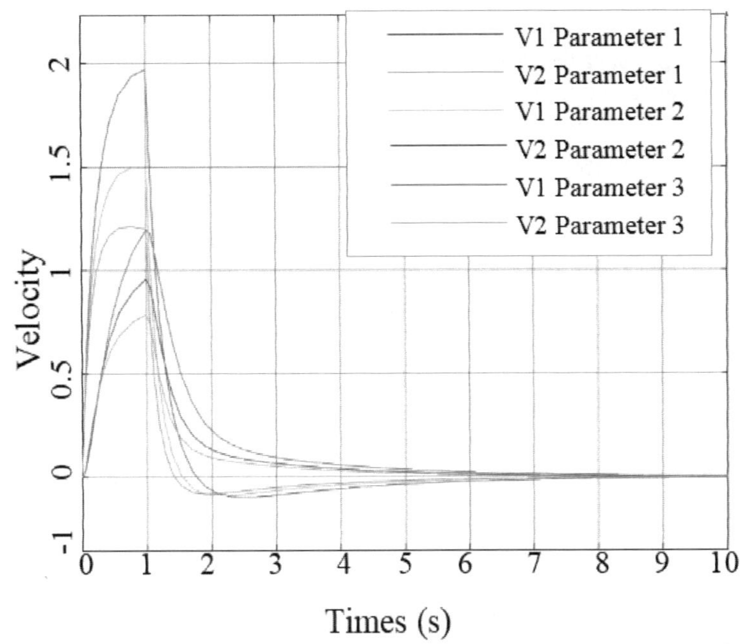

Figure 4. Graph of speed comparison simulation when Braking Occurs with three different parameters.

Figure 4 shown the output of mass velocity 1 and mass 2 when braking occurs. It is apparent that the mass velocity of 1 is decreased faster than mass velocity 2 but in steady state the two masses have the same velocity. From Figure 4 it can also be seen that by using large values, the time required to get to steady state is faster.

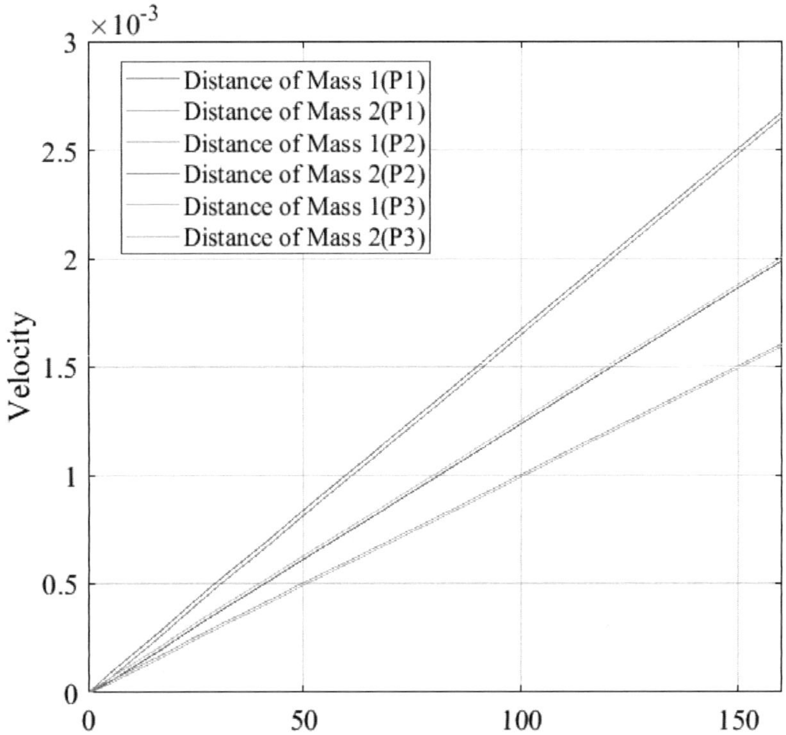

Figure 5. Graph of the distance between mass 1 and mass 2.

From Figure 5 that the distance difference between the two masses tends to be stable or in other words there is still a distance between mass 1 and mass 2.

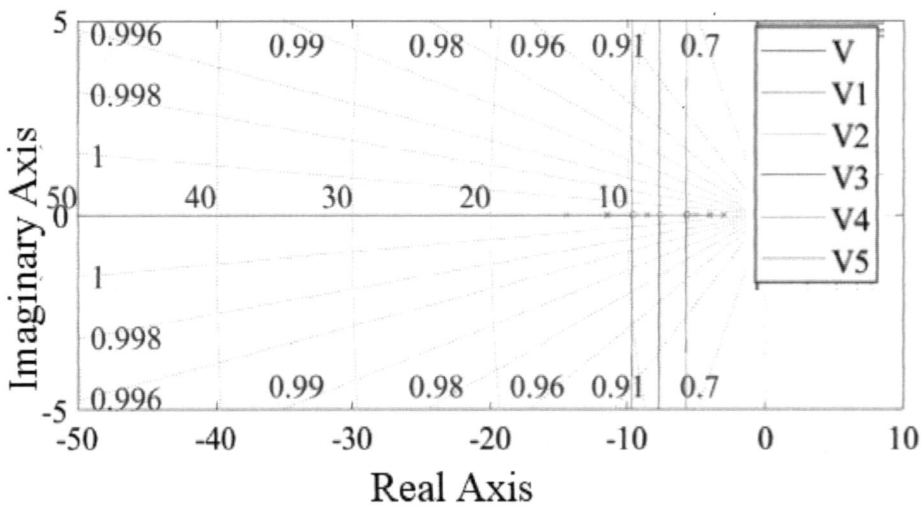

Figure 6. Root Locus with variation damping coefficient

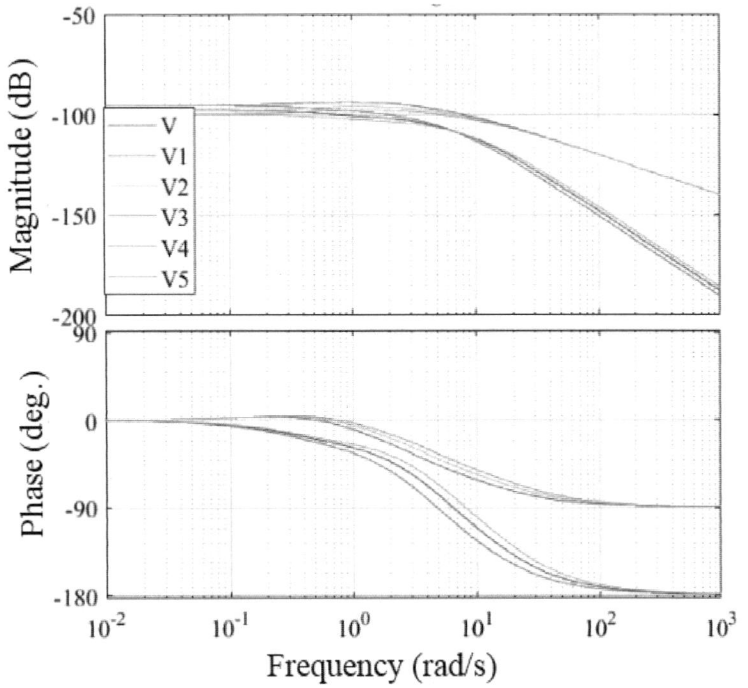

Figure 7. Bode Plot with variation damping coefficient

From Figure 6 and Figure 7, by looking at the location of the roots of the characteristic equation and the bode plot. It is seen that in mass 1 by looking at the position of the roots having a complex root value, the system will tend to be underdamped (overshoot) and at mass 2 with the corresponding spring and damper values of the parameters get the over damped system response. The results can also be seen that the response of the system tends to be more stable because it has a pole and zero value on the left. As well as the frequency response analysis using a bode plot. The greater the damping coefficient and the spring then the system will tend to be more quickly stable.

6. Conclusion

From the results of the graph analysis can be concluded that the damped oscillations are influenced by the spring constants and the steady constants. From the graph also, we can know that with the constant increase of constant spring and damper the higher attenuation is proved by the decreasing graph value in graph 3 we also know the effect of damper counter to steady state time faster.

Acknowledgments

Manuscript received August 15th, 2018. This work was supported in part by Research and Public Service Institute of University of Al Azhar Indonesia for publication funding and Secretary of Faculty Science and Engineering University of Al Azhar Indonesia

References

[1] Benaroya, Haym Han, Nagurka, Mark L., "Mechanical vibration: analysis, uncertainties, and control," CRC Press 2018
[2] Jorge Angeles, "Dynamic Response of Linear Mechanical Systems_ Modeling, Analysis and Simulation-Springer," US, 2012

[3] To, Cho W. S, "Introduction to Dynamics and Control in Mechanical Engineering Systems," Wiley-ASME, 2016
[4] William Palm III "System Dynamics," McGraw-Hill Education. 2013
[5] C. Mei, "On Teaching the Simplification of Block Diagrams," *Int. J. Engng Ed. Vol. 18, No. 6, pp. 697-703, 2002*
[6] P. Flores, J. Ambro sio, J.C.P. Claro, H.M. Lankarani, C.S. Koshy, "A study on dynamics of mechanical systems including joints," *Sience Direct*, Mechanism and Machine Theory 41, pp 247–261, 2006.
[7] Katsuhiko Ogata, "Modern Control Engineering," ,Prentice Hal, 2010
[8] Mobil Truk Besar Gandeng Luar Negri | Mobil Truk. (n.d.). Retrieved November 27, 2018, from http://trukaz.com/326/mobil-truk-besar-gandeng-penting-diketahui.html/mobil-truk-besar-gandeng-luar-negri

IOP Conf. Series: Materials Science and Engineering **532** (2019) 012006 doi:10.1088/1757-899X/532/1/012006

Two-Stage Algorithm for the Open Vehicle Routing Problem

A Imran[1]
[1]Industrial Engineering Department, Institut Teknologi Nasional, Jl.PHH Mustopa 23, Bandung 40124, Indonesia

E-mail: arifimr@yahoo.com

Abstract. Open Vehicle Routing Problem (OVRP) is a problem that is often found in transportation/distribution problems. On this issue it is known that some consumers are served by several homogeneous vehicles that depart from the depot where after visiting the last consumer the vehicle does not go back to the depot as in the OVRP the vehicle is not possessed by the company. The distribution cost is calculated from the depot until delivery to the last consumer. OVRP will be solved using the 2-stage method. The first stage is the initial construction phase of the solution while the second stage is the quality improvement phase of the solution. The Nearest Neighbor and 2-Opt algorithm are applied to generate the initial solution. The second stage uses Multi-Level Algorithm. The algorithm is equipped with four local searches. The data set from the literature is used to test the algorithm.

1. Introduction

This paper is an initial investigation (work in progress) to solve the Open Vehicle Routing Problem (OVRP). The OVRP is a problem that is often found in transportation/ distribution problems. The OVRP is a variant of the well-known Capacitated Vehicle Routing Problem (CVRP). In the CVRP, the objective is to find the shortest distance in serving the customers. Here, we need to consider that consumers are served by several homogeneous vehicles that start and return to the depot. The difference with the OVRP is in the OVRP the vehicle does not go back to the depot after visiting the last consumer as the they are not owned by the company. The distribution cost is calculated from the depot until the delivery to the last consumer is done. The aim of the OVRP is to minimize the number of vehicles used and produce the shortest distance in serving the consumer.

There are several methods/algorithms that have been developed to tackle the OVRP. The OVRP was first addressed by Sariklis and Powell [20] when they are solving the distribution management issues. They proposed the Cluster First, Route Second (CFRS) based algorithm. Brandao [3] and Fu et al. [6] proposed Tabu Search algorithms. Tarantilis et al. [23] applied a Threshold Accepting based algorithm called the List Based Threshold Accepting (LBTA) to tackle the OVRP data set and the actual problem (distributing goods from a warehouse to customers throughout the greater Athens Metropolitan area). Pisinger and Ropke [15] used the Adaptive Large Neighborhood Search (ALNS). They also applied their algorithm to solve several variants of the VRP and it produced excellent results. Li et al. [12] developed a Record to Record (RTR) Travel based algorithm to address the OVRP. They also created and solved large data set for the OVRP. Letchford et al. [10] proposed an exact method to solve the OVRP. They used the Branch and Cut algorithm. A variable neighborhood search based approach was used by Fleszar et al. [5]. An Integer Linear Programming (ILP) improvement procedure was proposed by Salari et al. [17]. MirHassani and Abolghasemi [13] proposed a Particle Swarm Optimization based algorithm to tackle the OVRP. Pichpibul and

Kawtummachai [14] developed Clark and Wright algorithm based method. They obtained good results. Sedighpur et al. [21] applied a hybrid ant colony optimization based algorithm. The results produced by Pichpibul and Kawtummachai [14] are discussed in Sorensen et al. [22]. Sorensen et al. [22] found that the results are questionable.

The applications of the OVRP in the real problem can be seen in the distribution problem of lubricants in Greece (Repoussis et al. [16]), the problem for determining school bus routes in Hong Kong (Li and Fu [11]), the train service problem through the Channel Tunnel (Fu and Wright [7]) and in determining the delivery routes for FedEx aircraft (Bodin et al. [2])

The paper is written as follows, the introduction is given in section 1. Section 2 explain of the proposed method. In section 3 the results obtained by the proposed method are presented and discussed. In the final section, the conclusions and suggestions for improving the results are given.

2. The proposed algorithm

In this study, the OVRP will be solved using the 2-stage method. The first stage is the initial construction phase of the solution, while the second stage is the quality improvement phase of the solution. In the first stage, the nearest neighbor algorithm followed by applying 2-Opt intra-route local search are used to generate the initial solution. In the second stage, the solution generated by the first stage is used as an input Multi-Level heuristic based algorithm of Salhi and Sari [19] are implemented.

Four local searches such as 2 types 1-0 insertions and 2 types swap (1-1 insertion) are applied in the framework of Multi-Level Algorithm to enhance the solution quality. The first type is a local search for improving the route by considering one route (intra-route) while the second type is an improvement procedure that consider two routes (inter-route) when conducting 1-0 and 1-1 move.

In the case for solving the vehicle routing problem and its variants, the application of the Multi-Level algorithm can be found in Imran et al. [9] for Heterogeneous Fleet Vehicle Routing Problem (HFVRP), Salhi et al. [18] for the Multi-Depot Heterogeneous Vehicle Routing Problem (MDHFVRP), Salhi and Sari [19] for MDHFVRP and Multi-Depot Vehicle Routing Problem (MDVRP) and in Imran et al. [8] for the Heterogeneous Fixed Fleet Vehicle Routing Problem (HFFVRP).

1-0 insertion consider removing a customer to a new feasible position while 1-1 insertion (swap) swap position of a customer with another customer. A move is taken when it can produce better solution. The illustration of 1-0 and 1-1 moves for inter-route case is given by Figure 1, Figure 2 and Figure 3. In Figure 1 we have two routes R1 and R2. R1 have 5 customers (c1, c2, c3, c4, c5) and R2 also have 5 customers (c6, c7, c8 c9, c10).

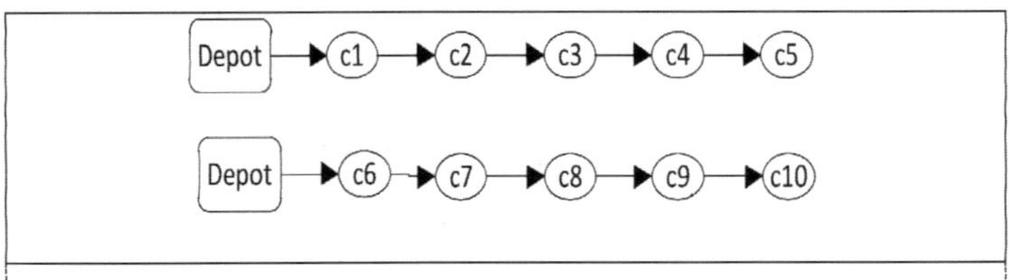

Figure 1. Route R1 and R2

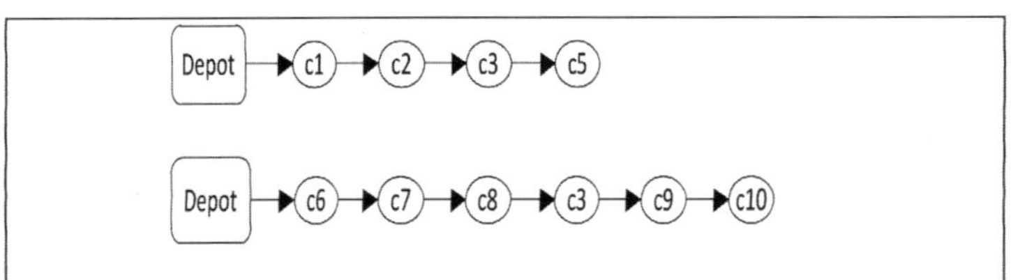

Figure 2. 1-0 insertion inter route.

In Fig 2 customer c3 of R1 is inserted between customer c8 and c9 of R2.

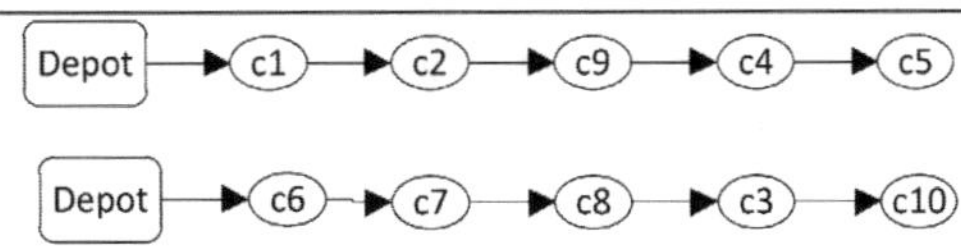

Figure 3. Swap 1-0 inter route.

In Figure 3, customer c3 of R1 is swapped with c9 of R2.

The local searches are used in Multi-Level algorithm of Salhi and Sari [19]. The search begins by generating the initial solution by using the Nearest Neighbor Algorithm. The algorithm picks the closest customer to the depot as the next customer. It continues by selecting the nearest customer to the last customer picked in the route. The process is done by considering vehicle capacity. The algorithm stops after all customers are allocated. The routes obtained is improved by applying 2-opt algorithm. The Multi-Level algorithm starts by using the first local search L_1 to obtain the best solution s''. If it is found that s'' is smaller than s, then $s = s''$ and the search returns to L_1, otherwise the second local search (L_2) is used to find s''. If s'' is smaller than s, then $s = s''$ and the search goes back to L_1, otherwise the third local search search (L_3) is applied to find s''. The search is repeated until L_4 cannot obtain a better solution. The best solution is recorded. The algorithm stops when all neighborhoods are used and there is no better solution is found.

The steps of the proposed algorithm are given in Figure 4.

Define a set of local searches L_j, for $j = 1, \ldots, j_{\max}$
Generate an initial solution s,
set $s_{best} = s$.
$j = 1$
(a) Apply local search #1 (L_1).
If the local optimum s'' is smaller than the incumbent s
Set $s \leftarrow s''$ and $j \leftarrow j+1$
(b) Apply local search #2 (L_2).
If the local optimum s'' is smaller than the incumbent s
Set $s \leftarrow s''$ and $j \leftarrow 1$;
Else
Set $j \leftarrow j+1$
If ($j<j_{max}$)
go to (a)
Else
Stop

Figure 4. The Proposed algorithm.

Where, j_{max} is the maximum number of local search used in search process.

The number of vehicles used (nv) is determined by calculating the minimum number of vehicles by using Equation 1.

$$nv = \sum_{i=1}^{n} {Qi}/{C} \tag{1}$$

The minimum number of vehicles (nv) is the ratio between the sum of the demand (Q_i) and the vehicle capacity (C). After that, the route is formed with respect to the limits of vehicle capacity and other restrictions.

3. Results

The proposed algorithm was implemented in C++ and executed by using the computer with Intel core i3 processor and 1 GB of RAM. Five data from a data set of Augerat et al. [1] are solved to testing the algorithm. The solutions obtained and the computation time used are then compared with the results obtained by Letchford et al. [10] and Pichpibul and Kawtummachai [14] to see the quality of the proposed algorithm.

The results of the test are presented in Table 1 and Table 2. Table 1 contains solutions obtained by the proposed algorithms and solutions obtained by Letchford et al. [10] and Pichpibul and Kawtummachai [14]. Table 2 contains the computational time used by the two stage algorithm and by the similar paper with Table 1. From Table 1 it can be seen that the proposed method produced one solution similar to the best known solution (#1) and solutions #2 #3 #4 and #5 are not as good as the best known solutions or the solutions obtained by Letchford et al. [10] and Pichpibul and Kawtummachai [14].

In terms of the computation time, the proposed algorithm can be said has used a reasonable computation time when compared with the computation time of other algorithm. However, we need to solve large data set in order to observe the computation time.

Table 1. Solutions (Augerat et al. data set [1])

No	#Customer	Best Solution	Letchford et al. [10]	Pitchibul and Kawtummachai [14]	Proposed Algorithm
1	16	235.06	235.06	235.06	235.06
2	19	168.57	168.57	168.57	169.92
3	20	170.28	170.28	170.28	174.20
4	21	163.88	163.88	163.88	164.36
5	22	167.19	167.19	167.19	185.05

Table 2. Computation time (in second)

No	#Customer	Letchford et al. [10]	Pitchibul and Kawtummachai [14]	Proposed Algorithm
1	16	0.09	1.17	1.35
2	19	0.2	1.57	0.79
3	20	0.2	1.71	1.47
4	21	0.02	1.73	0.94
5	22	0.03	1.89	0.83

4. Conclusion

In this study, a two stage algorithm is developed to address the OVRP. The proposed algorithm is tested to solve the data set from the literature. However, the quality of the solutions of the proposed algorithm needs to be improved as they are not as good as the solutions obtained by the two methods that are used for comparison. For further research, the use of more neighborhood and other local search types such as 2-Opt local search (intra and inter-route), Or-Opt local search (intra-and inter-route) and perturbation local search can be applied in order to solve larger data set and to improve the solution quality. Adding a procedure that provides the possibility to obtain new solutions as inputs if the proposed algorithm is repeated from the beginning of the algorithm can also be introduced.

5. References

[1] Augerat P, Belenguer J, Benavent E, Corbern A, Naddef D and Rinaldi G 1995 Computational results with a branch and cut code for the capacitated vehicle routing problem *Research Report* 949-M Universite Joseph Fourier Grenoble France.

[2] Bodin L, Golden B, Assad A and Ball M O 1983 Routing and scheduling of vehicles and crews; the state of the art *Comp. & Op. Res.* **10** 63-211.

[3] Brandão J 2004 A tabu search algorithm for the open vehicle routing problem *European J. of Op. Res.* **157** 552–64.

[4] Clarke G and Wright J W 1964 Scheduling of vehicles from a central depot to a number of delivery points *Op. Res.* **12** 568–81.

[5] Fleszar K, Osman I H and Hindi K S 2009 A variable neighbourhood search algorithm for the open vehicle routing problem *European J. of Op. Res* **195** 803–9.

[6] Fu Z, Eglese R W and Li L Y O 2005 A new tabu search heuristic for the open vehicle routing problem *J. of the Op. Res. Soc* **56** 267–74.

[7] Fu Z, Wright M 1994 Train plan model for British Rail freight services through the Channel Tunnel *J. of the Op. Res. Soc.* **45** 384–91.

[8] Imran A, Luis M and Okdinawati L 2016 A variable neighborhood search for the heterogeneous fixed fleet vehicle routing problem *J. Teknologi* **78** 53-8.

[9] Imran A, Salhi S and Wassan N A 2009 A variable neighborhood-based heuristic for the heterogeneous fleet vehicle routing problem *European J. of Op. Res* **197** 509-18.

[10] Letchford A, Lysgaard J and Eglese R W 2007 A branch and cut algorithm for the capacitated open vehicle routing *J. of the Op. Res. Soc.* **58** 1624-51.

[11] Li L Y O and Fu Z 2002 The school bus routing problem: a case study *J. of the Op. Res. Soc.* **53** 552–8.

[12] Li F, Golden B and Wasil E 2007 The open vehicle routing problem: algorithms, large-scale test problems, and computational results *Comp. & Op. Res.* **34** 2918–30.

[13] MirHassani S A and Abolghasemi N 2011 A particle swarm optimization for the open vehicle routing problem *Expert Syst. with Appl.* **38** 11547–51.

[14] Pichpibul T and Kawtummachai R 2013 A heuristic approach based on Clarke-Wright algorithm for open vehicle routing problem *The Scientific World J.*

[15] Pisinger D and Ropke S 2007 A general heuristic for vehicle routing problems *Comp. & Op. Res.* **34** 2403–35.

[16] Repoussis P P, Paraskevopoulos D C, Zobolas G, Tarantilis C D and Ioannou G 2009 A web-based decision support system for waste lube oils collection and recycling *European J. of Op. Res.* **195** 676–700.

[17] Salari M, Toth P and Tramontani A 2010 An ILP improvement procedure for the open vehicle routing problem *Comp. & Op. Res.* **12** 2106-02.

[18] Salhi S, Imran A and Wassan N A 2014 The multi-depot vehicle routing problem with heterogeneous vehicle fleet: Formulation and a variable neighborhood search implementation *Comp. & Op. Res.* **52** 315-25.

[19] Salhi S and Sari M 1997 A multi-level composite heuristic for the multi-depot vehicle fleet mix problem *European J. of Op. Res.* **103** 95-112.
[20] Sariklis D and Powell S 2000 A heuristic method for the open vehicle routing problem *The J. of the Op. Res. Soc.* **51** 564–73.
[21] Sedighpour M, Ahmadi V, Yousefikhoshbakht M, Didehvar F and Rahmati F 2014 Solving the open vehicle routing problem by a hybrid ant colony optimization *Kuwait J. of Science* **41** 139-62.
[22] Sorensen K, Arnold F and Cuervo D P 2017 A critical analysis of the improved Clarke and Wright savings algorithm *Int. Trans. on Op. Res.DOI*: 10.1111/itor.12443.
[23] Tarantilis C D, Ioannou G, Kiranoudis C T and Prastacos G P 2005 Solving the open vehicle routing problem via a single parameter metaheuristic algorithm *J. of the Op. Res. Soc.* **56** 588-96.

IOP Conf. Series: Materials Science and Engineering **532** (2019) 012007 doi:10.1088/1757-899X/532/1/012007

Effect of adhesive thickness on the strength of steel-composite joints aged in wet environment

S Sugiman[1], S Salman[1], P D Setyawan[1], S Sulardjaka[2]
[1] Department of Mechanical Engineering, Faculty of Engineering, University of Mataram, Jl. Majapahit 62 Mataram, 83125 Indonesia
[2] Department of Mechanical Engineering, Faculty of Engineering, Diponegoro University, Semarang, Indonesia

E-mail: s.sugiman@unram.ac.id

Abstract. This paper presents the effect of adhesive layer thickness on the static strength of adhesively bonded steel-glass fiber-reinforced unsaturated polyester resin (GFRP) lap joints after being aged in water at a temperature of 60°C for 15 days. The adhesive thickness was varied from 0.1 mm to 0.5 mm. It was found that for the dry joints, the static strength increased with the increase of adhesive thickness. However, in wet condition, water really had a significant detrimental effect on the strength of the joints, as most of the strengths of wet joints were much lower than those of the dry joints. In dry condition, mode of failure changed from interfacial (at the adhesive thickness of 0.1 mm) to mix failure; however, for the wet joints; most of failures were at the interface of adhesive/steel.

1. Introduction

Nowadays, adhesively bonded metal-composites joints gain more applications in transportation industry in order to develop lightweight vehicles. Other sectors used adhesively bonded joints technology, mainly for strengthening steel-made bridge structures, concrete and wood structures. The drawback in making use of adhesively bonded joints is the capability of the adhesive to absorb water from the surrounding environment. It is well known that the absorbed water has deleterious effect on the strength of the adhesive joints; therefore, designing an adhesive joint in those applications become attractive researches, considering the effect of adhesive thickness, type of adhesive and substrate, and type of environment.

Theoretically, (i.e. Goland-Reissner's and Volkersen's theories) an adhesively bonded joint strength increased with the increase of adhesive thickness [1], however experimentally, most literature reported that the joint strength decreased with the increase of adhesive thickness [2–8]. To obtain a high joint strength, the adhesive thickness should be thin enough, in the range of 0.1 – 0.5 mm [9]; however, it depended on the adhesive type. For very ductile adhesive, such as polyurethane, the adhesive thickness between 0.5 mm and 1 mm gave the optimum strength for aluminium joints [4]. For many structural adhesives, the optimum adhesive thickness was in the range of 0.1 – 0.2 mm [9].

Water can enter into the adhesive joint through several mechanisms; diffusing into the bulk adhesive, wicking into the interface, capillary action into micro voids and cracks, and diffusing into permeable substrate before reaching the adhesive layer [10]. The absorbed water then degrades the mechanical properties of adhesive and affects the interface of adhesive/substrate. Often during aging, the joint fails prematurely at the interface and is difficult to predict. For simplification (as many factors influences the durability), most of durability studies for metal-metal joints, composite-composite joints and metal composites joints used a constant adhesive thickness assuming the homogeneous water diffusion across the adhesive thickness. In fact, from literature, the optimum adhesive thickness can vary depending on the adhesive type. In moist environment, especially for metal-composite joints, water can diffuse from the adhesive layer and the permeable composites substrate. This can add the complexity of the effect of water on the joint strength. This paper elucidates the effect of water on the steel-glass fiber reinforced unsaturated polyester joints immersed in distilled water at temperature of 60°C at various adhesive thicknesses. Finite element modeling had been undertaken to investigate the water distribution in the adhesive layer as well as in the composites substrate.

2. Experimental methods

2.1. Materials

Low carbon steel was used for substrate with a thickness of 3.96 mm and a width of 30 mm. A room temperature cured Araldite® epoxy was used for adhesive. A composite plate was manufactured by hand lay up using Yukalac BQTN 137 unsaturated polyester resin as matrix and plain weaves glass fiber with an area density of 600 g/m^2 as the reinforcement. Methyl ethyl ketone peroxide (MEKPO) was used as curing agent (hardener), with resin to hardener ratio of 100:1 (by weight). The thickness of composite plate was 4 mm. The composite was cured at room temperature, under pressure approximately of 0.1 MPa. This composite was then called as glass fiber-reinforced unsaturated polyester (GFRP).

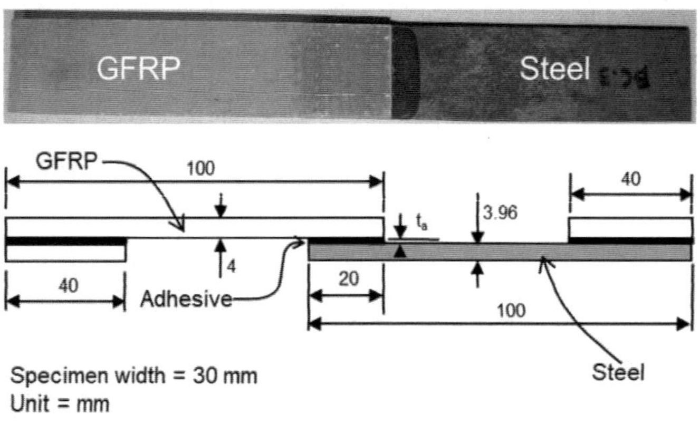

Figure 1. A single lap joint specimen (not to scale).

2.2. Single lap joint manufacturing

Single lap joints (SLJs) were manufactured using low alloy carbon steel and GFRP substrates, and an Araldite adhesive. The dimension of SLJ is shown in figure 1. Before being bonded, the steel surfaces were abraded using a steel brush and an emery paper (grid 80) to remove the smudge and the contaminants, such as grease and oil. After abrading, they were rinsed using detergent and acetone. The surfaces were subsequently etched using sulphuric acid (30% by weight) for 5 min, cleaned using acetone and wiped using a cloth. For GFRP materials, their surfaces were lightly abraded using an 80 grid emery paper and were cleaned using acetone. The substrates were bonded immediately after the surface treatment using Araldite adhesive. The ratio of resin to hardener was 1:1 (by volume). The thickness of adhesive layer (t_a) was varied; 0.1, 0.2, 0.3, 0.4 and 0.5 mm and was maintained by inserting wires in the overlap area, close to both edges of the overlap. Paper clamps were used to hold the joints during curing. The curing was conducted at a temperature of 120°C for 2 hours. The specimens were conditioned in two conditions; dry and wet conditions. In dry condition, the specimens were tested a day after the adhesive was cured. In wet condition, the specimens were immersed in distilled water at a temperature of 60°C for 15 days (wet) before being tested. A container with a temperature controller was used for immersion.

2.3. Water absorption and the water dependent mechanical properties of adhesive and composites

Water absorption was conducted on the adhesive and the GFRP plate. The water uptake and its water-dependent mechanical properties for Araldite adhesive immersed in the same environment have been conducted in previous research [11]. Water uptake into a GFRP plate was carried out using a plate specimen with dimension of (45 × 45 × 3.25) mm^3. The specimens were immersed in distilled water at temperature of 60°C, and periodically taken for weighing. The experimental water uptake (M_t) data were plotted vs. square root of time. Fickian diffusion model, as in equation (1) [12] were fitted with the experimental water uptake data. The diffusion coefficient (D) was obtained from the experimental water uptake data at the linear part of the curve up to M_t/M_∞ of 0.6 using equation (2).

$$\frac{M_t}{M_\infty} = 1 - \frac{8}{\pi^2}\sum_{n=0}^{\infty}\frac{1}{(2n+1)^2}\exp\left[\frac{-(2n+1)^2\pi^2}{h^2}Dt\right] \tag{1}$$

$$D = \frac{\pi}{16}\left(\frac{m \times h}{M_\infty}\right)^2 \tag{2}$$

where m is the slope of the linear part of water uptake vs. root square time curve, t is the time, h is the specimen thickness and M_∞ is the water uptake at equilibrium.

The tensile specimens of composites were immersed in the same environment as for water uptake specimen. After 15 days, the specimens were taken out from the immersion and immediately tested using a Control universal testing machine, with a load cell capacity of 2000 kN.

3. Experimental results and discussion

3.1. Water uptake and water dependent mechanical properties

Figure 2 shows the water uptake of Araldite adhesive and GFRP with square root of time, while the diffusion parameters are shown in table 1. Fickian diffusion model [12] was plotted along with the experimental water uptakes. For GFRP, in the linear part, the experimental and the Fickian model were in good agreement, but after 3 days immersion (at water uptake of 1.4%), the water uptake started to slow down, deviating from linearity and the Fickian fit curve. The equilibrium water uptake was reached after immersion for about 15 days for both the adhesive and the GFRP. It was observed that water diffusion into the GFRP did not fully follow the Fickian diffusion model. The saturation level of composite was lower (about 55%) than that of Araldite adhesive, but the diffusion constant was almost tenfold higher. For heterogeneous materials such as GFRP, the diffusion rate can be faster as the water molecules can wick into the interface of matrix and glass fiber, in addition to the diffusion from the matrix [10].

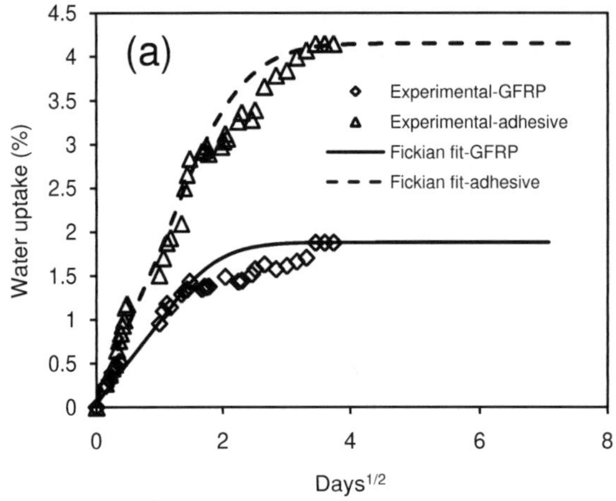

Figure 2. Water uptakes of GFRP and adhesive.

Tabel 1. The diffusion parameters of adhesive and GFRP.

Materials	M_∞ (%)	D (mm^2/day)
Adhesive	4.15	0.072
GFRP	1.88	0.61

After 15 days immersion, the GFRP specimens were tested under tension and the mechanical properties are shown in table 2. Referring to figure 2, after 15 days immersion, the composite has reached the saturation level. In this condition, as seen in table 2, the decrease of elastic modulus and the tensile strength seemed not to be significant as the standard deviation was high. For woven glass fiber-reinforced polymer composites, the tensile strength (X) and the elastic modulus (E) in longitudinal (indicated by 11 subscript) and transverse (indicated by 22 subscript) directions were the same. The adhesive properties in dry and wet (after 15 days immersion) conditions obtained from published literature [11], can be seen in table 3. The adhesive tensile properties degraded considerably after ageing in distilled water at a temperature of 60°C for 15 days. The decrease of elastic modulus and tensile strength was 83% and 71%, respectively.

Tabel 2. The GFRP properties in dry and wet conditions.

Condition	GFRP	
	$E_{11}=E_{22}$ (GPa)	$X_{11}=X_{22}$ (MPa)
Dry	12.39 ± 1.69	301.15 ± 21.2
Wet	11.00 ± 0.64	296.78 ± 5.24

Table 3. The adhesive properties in dry and wet conditions.

Condition	Araldite adhesive [11]	
	E (MPa)	Tensile strength (MPa)
Dry	2500	41.29 ± 4.65
Wet	420	12.06 ± 3.10

IOP Conf. Series: Materials Science and Engineering **532** (2019) 012007 doi:10.1088/1757-899X/532/1/012007

3.2. Residual static strength of aged single lap joint

Figure 3(a) shows the shear strength of steel-GFRP composites joints in dry and wet conditions at various adhesive thicknesses. It is seen that in dry condition, the shear strength tended to increase linearly with the increase of adhesive thickness up to 0.5 mm. At the thickness from 0.2 to 0.5 mm, the variation, shown by the vertical bars in the graph, was high. It might be caused by the different mode of failure of each joint, either interfacial or mix failure. Sugiman et al. [11] also reported the same trend on the shear strength of bonded steel joints with the adhesive thickness using the same adhesive as used in this study. Generally, the shear strength decreased with the increase of adhesive thickness [2–8], as the increase of adhesive thickness tended to have more flaws [1]. The increase of shear strength in this case, may be caused by the increase of plastic deformation in the fracture zone as the adhesive is quite ductile. Campilho et al. [13] found that the critical energy release rate of joints with a ductile polyurethane adhesive increased essentially linear with the increase of adhesive thickness up to 2 mm. This was due to enlargement of adhesive plastic zone in front of crack tip in the joint with the increase of adhesive thickness.

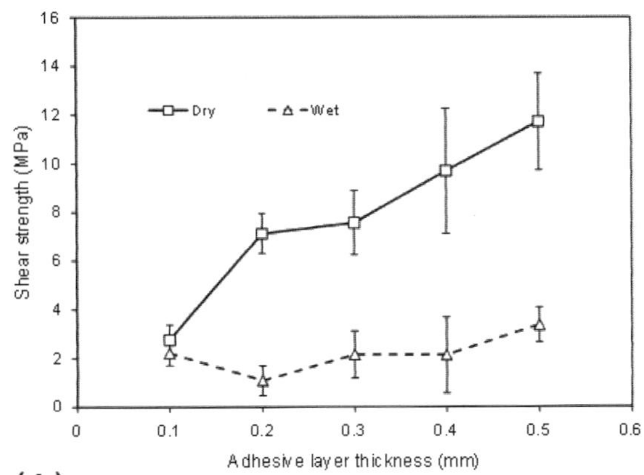

(A)

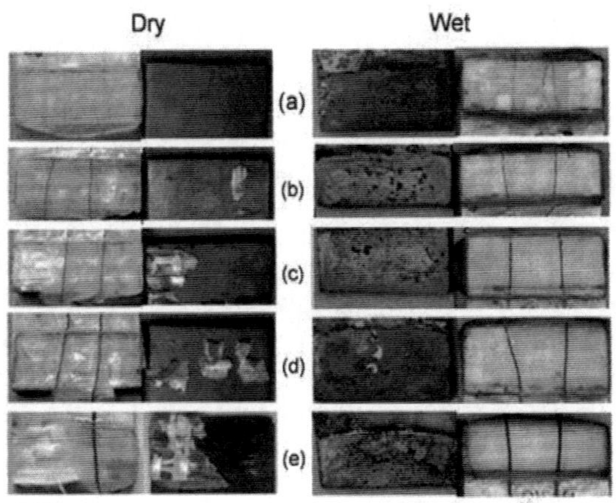

(B)

Figure 3. Showing, (A) the shear strength with adhesive thickness and (B) the failure surfaces of steel-GFRP bonded joints, in dry and wet conditions. In the right figure, symbols a, b, c, d, and e are respectively showing the adhesive thickness of 0.1 mm, 0.2 mm, 0.3 mm, 0.4 mm and 0.5 mm.

In wet condition, the joint strength as seen in figure 3(a), was very low and seemed slightly increasing with the adhesive thickness. This could be due to the absorbed water that influenced the adhesive bulk strength and attacked the interface adhesive/substrate. Interestingly, at the adhesive thickness of 0.1 mm, the dry and the wet joint strengths were about the same, while the strength of other thicknesses in wet condition dropped considerably compared to the dry joints strength. This is interesting as in our previous study [11], in wet condition, on the steel-steel joints, the joint strength still increased with the increase of adhesive thickness up to 0.3 mm, and then dropped steeply afterward.

As seen in figure 3(b), in dry condition, mode of failure changed from interfacial (at 0.1 mm) to mix failure (0.2 mm - 0.5 mm) with the increase of adhesive thickness. The mix failure was indicated by the tearing of adhesive and by the delamination of part of the GFRP, adhering to the steel substrate. The extent of mix failure increased as the size of the delaminated GFRP got bigger and bigger with the increase of adhesive thickness. It is expected as the mix failure gives higher joint strength than the interfacial failure. However, the cause of this kind of failure needs to be further investigated by analysing the stress state of the joints.

IOP Conf. Series: Materials Science and Engineering **532** (2019) 012007 doi:10.1088/1757-899X/532/1/012007

In wet condition, modes of failure were mostly interfacial at the adhesive-steel interface, regardless the adhesive thickness. The interfacial failure occurred at the adhesive-steel interface rather than at the adhesive-GFRP interface, because the adhesive-metal interface was not stable in water environment [14, 15]. Moreover, the steel tended to corrode after being aged in water. The corrosion product might further displace the adhesive from the steel surface, decreasing the joints strength [16]. Some corrosion products were seen at the steel surface after failure. After 15 days immersion, the adhesive and the adhesive-steel and the adhesive-GFRP interfaces had been wetted by water, even though the water content at both interfaces was different. This will be discussed further in the water diffusion modelling section.

4. Finite element modelling of moisture diffusion

To obtain water distribution in the adhesive layer and in the composites substrate, a finite element modelling was carried out using ABAQUS utilizing a mass diffusion analysis. The detail of this method can be found elsewhere in ABAQUS documentation [17]. For Fick's law, the concentration flux, F can be described as in equation (3),

$$F = -sD\left(\frac{\partial \phi}{\partial x}\right) \tag{3}$$

The normalized concentration (ϕ) along the diffusion path x, is defined as c/S, where c is the mass concentration and S is the solubility of diffusing substance in base material. Two parameters are required to model the water diffusion using mass diffusion; diffusivity (D) and solubility (S), corresponding to the diffusion rate and the equilibrium water uptake, respectively.

Finite element modelling was performed in 3-dimensional (3D) with meshing as seen in figure 4. There were four elements across the adhesive thickness, so that the thickness of an element was 0.025, 0.05, 0.075, 0.08, 0.125 mm for the adhesive thickness of 0.1, 0.2, 0.3, 0.4 and 0.5 mm, respectively. Both of the element length and width were 0.5 mm.

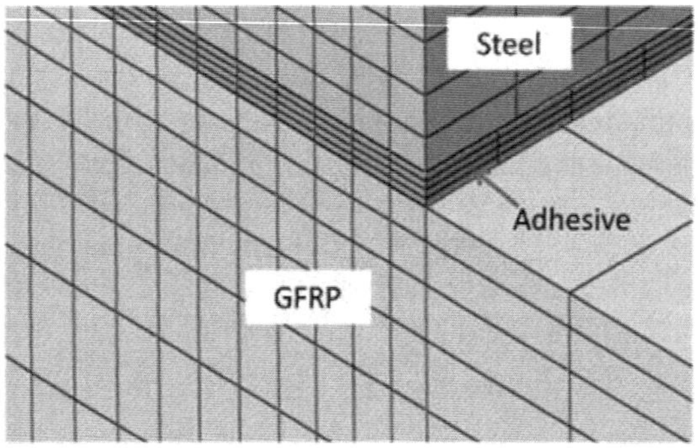

Figure 4. Meshing of finite element modelling for water diffusion in adhesive and GFRP.

The diffusion parameters of GFRP and the adhesive discussed in the previous section were used as the input parameters in the diffusion finite element modelling. Water content was normalised using the equilibrium water uptake and then it was called as the normalised concentration (NNC). Water diffused from four sides of the adhesive layer of the joint and from the thickness direction of the GFRP. Water diffusion from the GFRP width sides was not modelled because in practice water usually came in from the thickness direction.

Figure 5 shows the contour plot of water distribution in the GFRP and the adhesive layer at the various adhesive thicknesses, while the distribution of water both in the adhesive layer along the overlap length and in the through thickness of the joint is shown in figure 6. The contour shows a half of width wise of the overlap region and the water distribution plots shown in figure 6 were taken at the midplane of the overlap length. As seen in figure 6(a), water distribution across the adhesive thickness shown by the normalised concentration, seemed not homogeneous and it was more obvious with the increase of adhesive thickness. The water diffusion from four sides of adhesive layer was not different in all adhesive thicknesses; however, the water diffusion from the through thickness of GFRP across the adhesive layer contributed to the difference of water content at the adhesive-steel and adhesive-GFRP interfaces. As water came in first from the GFRP, the water content at the adhesive-GFRP interface was higher than at the adhesive-steel interface. At the thickness of 0.1 mm, water content was more homogeneous, but as the thickness increased, the different of water content at both interfaces was more pronounced. After 15 days immersion, the saturation in the adhesive layer had not been reached yet. At the middle of the joint, the normalised concentration was around 0.7 and 0.5 for the adhesive thickness of 0.1 and 0.5 mm, respectively.

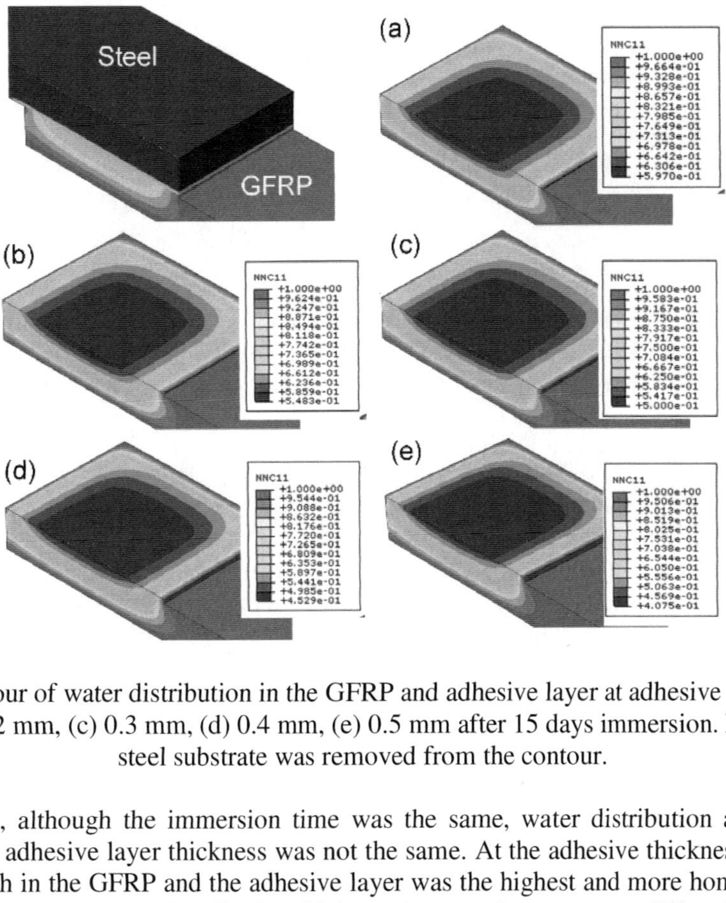

Figure 5. Contour of water distribution in the GFRP and adhesive layer at adhesive thickness of (a) 0.1 mm, (b) 0.2 mm, (c) 0.3 mm, (d) 0.4 mm, (e) 0.5 mm after 15 days immersion. For clarity, the steel substrate was removed from the contour.

In figure 6(b), although the immersion time was the same, water distribution across the GFRP thickness and the adhesive layer thickness was not the same. At the adhesive thickness of 0.1 mm, the water content both in the GFRP and the adhesive layer was the highest and more homogeneous across the adhesive layer; however, as the adhesive thickness increased, more water diffused from GFRP into the adhesive layer, so that at the same frame time, water distribution in GFRP and in adhesive layer shifted to the lower value. The decrease of water content from the adhesive thickness of 0.1 mm to 0.5 mm was about 36%. As the water content at the adhesive thickness of 0.5 mm was lower than that in 0.1 mm, the joint strength at the adhesive thickness of 0.5 mm should be higher. However, the experimental results showed that the joint strength at all the adhesive thicknesses seemed not significantly different. It suggests that there is critical water content at the adhesive-steel interface above which the joint strength tends to be constant. Loh et al. [18] reported that the joint strength decreased sharply up to the water content of 3% and tended to level off afterward. Referring to figure 6(b), the water content was higher than 3%, so that it is consistent with that in [18].

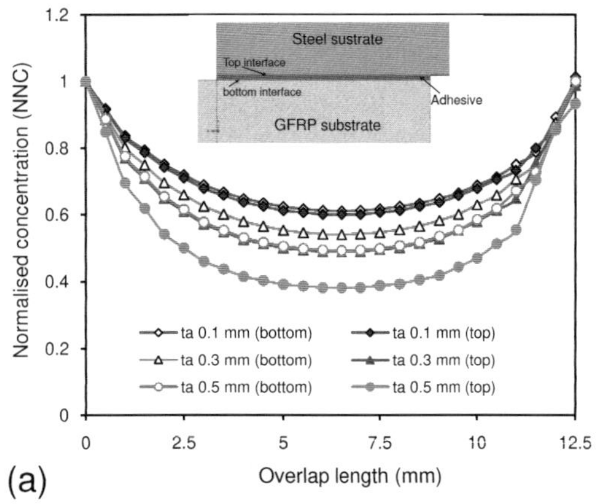

Figure 6. Showing distribution of (a) water across the adhesive layer thickness along the overlap length, (b) water in the GFRP and adhesive layer. The distribution in (a) was taken at the middle of width wise, while the distribution in (b) was taken at the center point of the joint.

5. Conclusions

Experimental studies on water diffusion in the glass fiber reinforced unsaturated polyester (GFRP) and in the Araldite adhesive and its effect on the tensile properties had been undertaken. The investigation of the effect of adhesive thickness on the joints strength had also been carried out after being immersed in distilled water at a temperature of 60°C for 15 days. It was found that the equilibrium water content of the GFRP was lower than that of the Araldite adhesive, but the diffusion rate was much higher. The effect of water on the tensile properties of GFRP seemed not to be significant, whereas its effect on the tensile properties of adhesive was significant. In dry condition, the joint strength tended to increase with the increase of adhesive thickness; however, in wet condition, the effect of adhesive thickness appeared not to be significant. At the thickness of 0.1 mm, the joint strength in dry and in wet conditions was about the same, but the effect of water was more obvious at the thicker adhesive thickness. Finite element modelling on water diffusion had been undertaken utilizing a mass diffusion analysis. Water content across the adhesive thickness was not homogeneous and it was more noticeable at the thicker adhesive layer. The water content at the adhesive-GFRP interface was higher than at the adhesive-steel interface due to the contribution of water from the GFRP.

Acknowledgements

This work was supported by Ministry of Research, Technology, and Higher Education, the Republic of Indonesia with grant number [733/UN18.L1/PP/2018]. Thanks also to I Wayan Januartha for specimens preparation.

References

[1] Adams R D and Peppiatt N A 1974 Stress analysis of adhesive-bonded lap joints *Journal of Strain Analysis* **9** 185–96
[2] Yang S, Xu W, Liang L, Wang T and Wei Y 2014 An experimental study on the dependence of the strength of adhesively bonded joints with thickness and mechanical properties of the adhesives *Journal of Adhesion Science and Technology* **28**(11) 1055–71

[3] da Silva L F M, Rodrigues T N S S, Figueiredo M A V, de Moura M F S F and Chousal J A G 2006 Effect of adhesive type and thickness on the lap shear strength *The Journal of Adhesion* **82**(11) 1091–1115

[4] Boutar Y, Naïmi S, Mezlini S, da Silva L F M, Hamdaoui M and Ali M B S 2016 Effect of adhesive thickness and surface roughness on the shear strength of aluminium one-component polyurethane adhesive single-lap joints for automotive applications *Journal of Adhesion Science and Technology* **30**(17) 1913–29

[5] Adnan A and Sun C T 2008 Effect of adhesive thickness on joint strength: A molecular dynamics perspective *The Journal of Adhesion* **84**(5) 401–20

[6] Castagnetti D, Spaggiari A and Dragoni E 2011 Effect of bondline thickness on the static strength of structural adhesives under nearly-homogeneous shear stresses *The Journal of Adhesion* **87**(7-8) 780–803

[7] van Tooren M J L, Gleich D M and Beukers A 2004 Experimental verification of a stress singularity model to predict the effect of bondline thickness on joint strength *Journal of Adhesion Science and Technology* **18**(4) 395–412

[8] Kahraman R, Sunar M and Yilbas B 2008 Influence of adhesive thickness and filler content on the mechanical performance of aluminum single-lap joints bonded with aluminum powder filled epoxy adhesive *Journal of Materials Processing Technology* **205** 183–9

[9] Gleich D M, Van Tooren M J L and Beukers A 2001 Analysis and evaluation of bondline thickness effects on failure load in adhesively bonded structures *Journal of Adhesion Science and Technology* **15**(9) 1091–1101

[10] Adams R D, Cowap J W, Farquharson G, Margary G M and Vaughn D 2009 The relative merits of the Boeing wedge test and double cantilever beam test for assessing the durability of adhesively bonded joints, with particular reference to the use of fracture mechanics *International Journal of Adhesion and Adhesives* **29** 609–20

[11] Sugiman, Akbar I, Sulistyowati E D and Setyawan P D 2016 Effect of adhesive layer thickness on the shear strength of adhesively bonded steel joints in wet environment *Applied Mechanics and Materials* **836** 78–82

[12] Crank J 1975 *The mathematic of diffusion, 2nd ed.* (London: Oxford University Press)

[13] Campilho R D S G, Moura D C, Banea M D and da Silva L F M 2014 Adhesive thickness effects of a ductile adhesive by optical measurement techniques *International Journal of Adhesion and Adhesives* **57** 125–32

[14] Kinloch A J, Dukes W A and Gledhill R A 1975 Durability in adhesive joints, 2 *Polymer Science and Technology* vol **9B**, ed L H Lee (New York: Plenum Press) pp 597–614

[15] Kinloch A J 1979 Interfacial fracture mechanical aspects of adhesive bonded joints – A review *The Journal of Adhesion* **10**(3) 193 – 219

[16] Zanni-Deffargess M P and Shanahan M E R 1995 Diffusion of water into an epoxy adhesive: comparison between bulk behaviour and adhesive joints *International Journal of Adhesion and Adhesives* **15** 137–42

[17] ABAQUS 2009 *User manual version 6.9.1* (USA: Habbit, Karlsson & Sorensen, Inc.)

[18] Loh W K, Crocombe A D, Wahab M M A and Aschcroft I A 2002 Environmental degradation of the interfacial fracture energy in adhesively bonded joint *Engineering Fracture Mechanics* **69** 2113–28

IOP Conf. Series: Materials Science and Engineering **532** (2019) 012008 doi:10.1088/1757-899X/532/1/012008

Effect of fiber length on the mechanical properties and water absorption of bamboo fiber/polystyrene-modified unsaturated polyester composites

S Sugiman[1], P D Setyawan[1], B Anshari[2]
[1] Department of Mechanical Engineering, Faculty of Engineering, University of Mataram, Jl. Majapahit 62 Mataram, 83125 Indonesia
[2] Department of Civil Engineering, Faculty of Engineering, University of Mataram, Mataram, Jl. Majapahit 62 Mataram, 83125 Indonesia

E-mail: s.sugiman@unram.ac.id

Abstract. The paper presents the effect of fiber length on the mechanical properties and water absorption of bamboo fiber/polystyrene-modified unsaturated polyester composites. The fiber length varied from 2.5 mm to 15 mm at the fiber volume fraction of 0.2, 0.25 and 0.3. For water absorption, the specimens were immersed in distilled water at temperature of 40°C. The results showed that for all volume fractions, the tensile, flexural, and the impact properties tended to increase with the increase of fiber length; however, there was an optimum fiber length for the tensile and flexural properties. The water uptake increased with the increase of both fiber length and volume fraction. In wet condition, for all volume fractions, the flexural properties degraded, whereas, the impact strength did not indicate degradation, even; it significantly increased for all fiber lengths.

1. Introduction

In the recent years, bamboo fiber has been increasingly used for manufacturing green composites as bamboo is abundantly available, and the most importance is that bamboo fiber has high mechanical properties [1]. Among the bamboo species, a *Gigantochloa apus* strives in Indonesia, and is potential to be used as reinforcement of polymer for biodegradable composites. Unsaturated polyester resin (UPR) is low price and has high mechanical properties [2]. This thermosetting polymer combined with glass fiber has been used for manufacturing boats. Furthermore, UPR has low water absorption [3], which can function as water barrier for natural fiber composites exposed in humid environments.

Expandable polystyrene (EPS) or styrofoam is post-consumer product and can be considered as a plastic waste. This waste is non biodegradable and can stand for 500 years [4]. So this pollutes water, landfill and dangerous for life. In the previous research [5], styrofoam waste had been used for modifying UPR. Adding small amount of polystyrene did not affect the mechanical properties of UPR. This paper investigated the use of polystyrene-modified UPR for manufacturing natural fiber composites with bamboo fiber as the reinforcement. The effect of bamboo fiber length at various fiber contents on the tensile, flexural, and the impact properties as well as the water absorption was investigated.

2. Experimental methods

2.1. Materials

Unsaturated polyester resin was obtained from Eternal (Taiwan). Methyl ethyl ketone was used as hardener. Styrofoam was obtained from electronic packaging waste. Bamboo fibers were mechanically extracted from *Gigantochloa apus* species that strives in Lombok island.

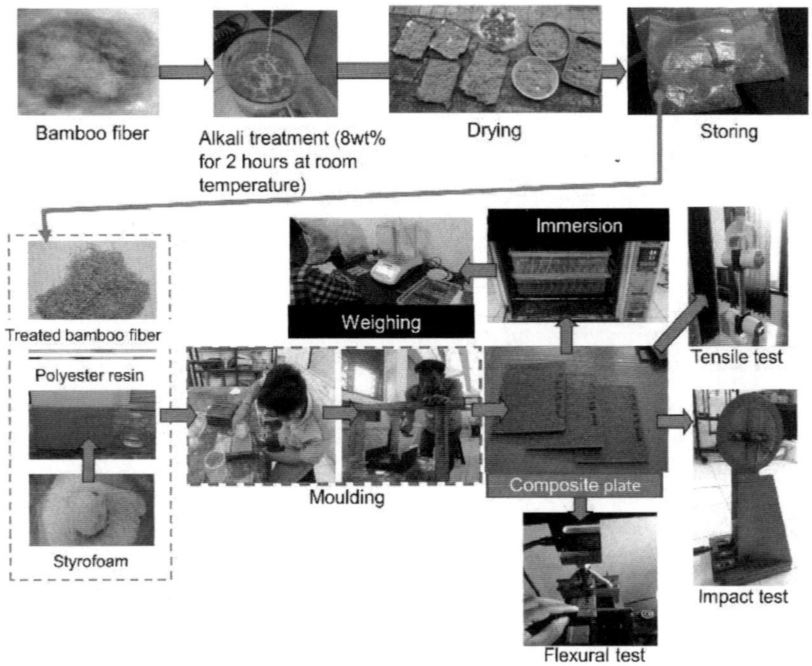

Figure 1. Graphical flow chart of the experimental procedure.

2.2. Manufacturing of composites specimen

Bamboo fibers were cut to obtain fiber length (FL) of 2.5, 5, 10, and 15 mm. The diameter of bamboo fiber was in the range of 0.2 - 0.46 mm. Bamboo fibers was treated using alkali (NaOH) solution at a concentration of 8% for 2 hours [6]. After treatment, the bamboo fibers were rinsed many times using tap water to remove the alkali. After rinsing, the bamboo fibers were dried under the sun for couple of days and then were transferred into an oven for final drying. The oven drying process was carried out at a temperature of 80°C for 24 hours. The dried bamboo fibers were then kept in a sealed plastic bag.

The styrofoam was ground using a blender to obtain a particle size about 1 mm. The UPR and styrofoam particles were mixed, and then subsequently stirred using hand mixer until all styrofoam particles were dissolved. The styrofoam content was 2% (by weight) [5, 6].

Figure 1 shows the graphical flow chart of the bamboo composites manufacturing and testing. Bamboo fibers were laid up randomly on the steel female mould after having covered with PVC sheet film. The volume fraction (VF) of bamboo fiber was 0.2, 0.25, and 0.3. The modified UPR was mixed with hardener, with resin to hardener ratio of 100:1 (by weight). After mixing, the modified UPR was poured into the mould and waited for several minutes to allow the resin to flow. A PVC sheet film was used to cover the top of lay up before being covered with the male mould. Two spacers with thickness of 3.8 mm were inserted between the female and male moulds to maintain the specimen thickness. The layup was pressed and left under pressure, at least for 6 hours to cure.

After the composite was cured, the mould was opened and the bamboo fiber/modified UPR was released from the mould. The composite then was cut to obtain the tensile, flexural and the impact specimens according to ASTM D3039 [7], ASTM D790 [8] and ASTM D256 [9], respectively. The tensile and flexural testing were conducted using a Tensilon universal testing machine with displacement rate of 2 mm/min for both. For the impact test, an Izod testing machine was used. The specimen dimensions are shown in figure 2.

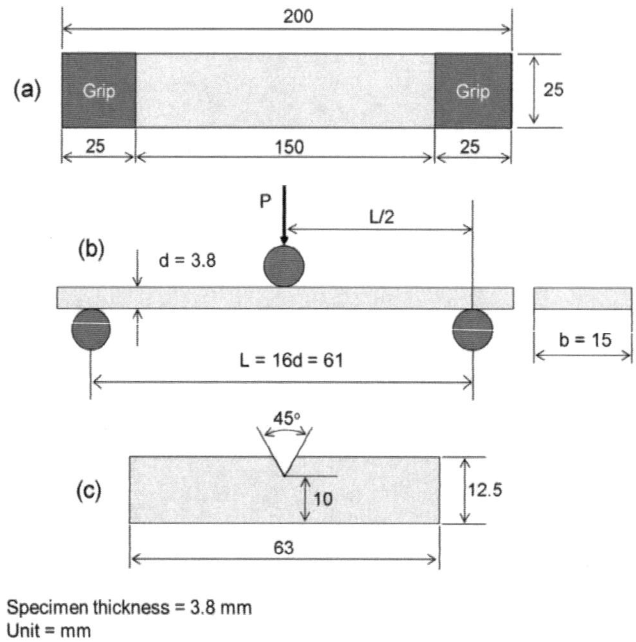

Figure 2. The specimen dimension of (a) tensile, (b) flexural, (c) Izod impact tests.

2.3. Water absorption studies

Water absorption study used specimens with dimension of (60 × 30 × 3.8) mm^3, in triplicate. The specimens were immersed in distilled water at temperature of 40°C. The specimens were periodically taken from the immersion and weighed using a Kenko microbalance. Before weighing, the specimens were wiped out using a tissue to remove water on the surface of the specimens. The absorbed water (M_t) was calculated using equation (1),

$$M_t = \frac{(W_t - W_o)}{W_o} \times 100\% \quad (1)$$

where W_t and W_o are respectively the weight at time t and at the initial. The water uptake data were plotted against square root of time. Fickian diffusion model (equation (2)) was fitted into the experimental water uptake data. The diffusion rate (D) was determined experimentally using the slope of the linear part of water uptake vs. square root of time, according to equation (3) [10].

$$\frac{M_t}{M_\infty} = 1 - \frac{8}{\pi^2}\sum_{n=0}^{\infty}\frac{1}{(2n+1)^2}\exp\left[\frac{-(2n+1)^2\pi^2}{d^2}Dt\right] \quad (2)$$

$$D = \frac{\pi}{16}\left(\frac{m \times d}{M_{\infty}}\right)^2 \quad (3)$$

where t is the immersion time, d is the specimen thickness, m is the slope of the linear part of water uptake vs. square root of time curve, M_{∞} is the water uptake at equilibrium.

After the saturation was reached and after the specimen was weighed, the specimens were cut to obtain the flexural and impact specimens. The flexural and impact testing were conducted immediately after cutting to avoid water desorbed from the specimen. This was intended to obtain the flexural and impact properties in wet condition. Three replications were used in this study.

3. Results and discussion

3.1. Tensile properties

Figure 3 shows the typical tensile stress-strain curves, the tensile strength and the elastic modulus of bamboo fiber/modified UPR composites with fiber length at various fiber volume fractions. It is seen in figure 3(a, b) that the initial stress-strain curves were dominated by the linear curve (elastic portion) and close to the peak load the curves deviated to be nonlinear, which was caused by the damage process. It is observed in figure 3(c, d) that for all volume fractions, the tensile strength and the flexural modulus increased with the increase of fiber length up to 10 mm and then tended to decrease afterward. For example, compared to the fiber length of 2.5 mm, the tensile strength at the fiber length of 10 mm and at the volume fraction of 0.25 increased by 37%. For all the volume fractions, the highest tensile strength was obtained at the volume fraction of 0.25 (except at the fiber length of 2.5 mm), whereas the lowest tensile strength was obtained at the volume fraction of 0.3. At the fiber length of 10 mm, the tensile strength at the volume fraction of 0.25 was about 12% and 6.5% higher than those at the volume fractions of 0.3 and 0.2, respectively.

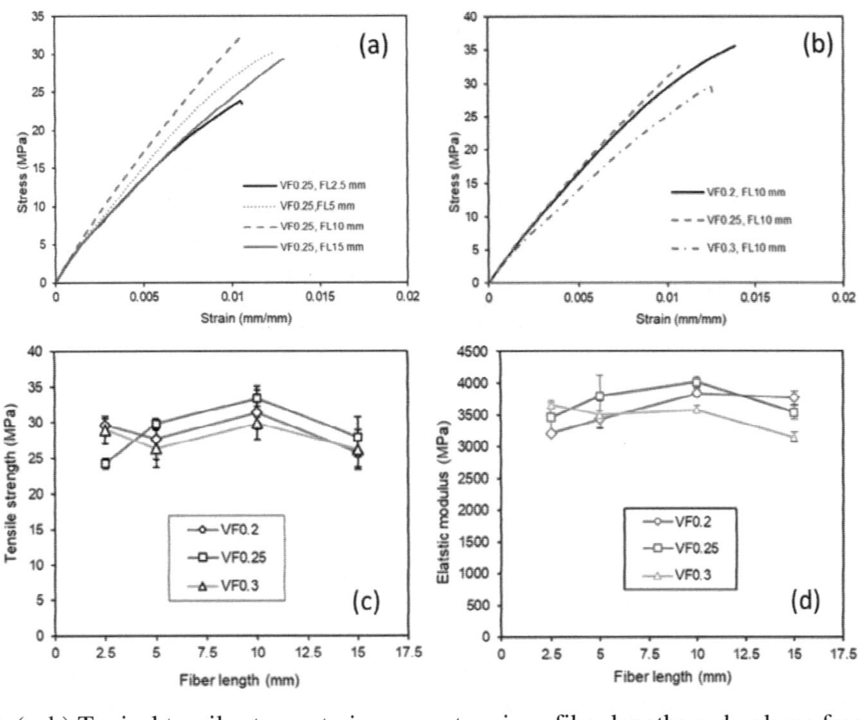

Figure 3. (a, b) Typical tensile stress-strain curve at various fiber lengths and volume fractions, (c) Tensile strength and (d) Elastic modulus of bamboo fiber/modified UPR composites.

Generally, a long fiber length enables to transfer load easily, so that the tensile strength is high. However, for randomly oriented fibers, the fiber could transfer the load effectively when the fiber direction is parallel to the applied load [11]. Furthermore, for randomly oriented fiber, the longer fiber contributed to more voids. This is expected as the density of the composite tended to decrease with the increase of fiber length. For example, at the volume fraction of 0.2, the density of bamboo fiber/modified UPR composites at the fiber lengths of 2.5 and 15 mm was 1.224 g/cm^3 and 1.168 g/cm^3, respectively. Meanwhile, the low tensile properties for higher volume fraction (i.e. volume fraction of 0.3) could be caused by the lack of matrix for bonding the bamboo fibers and more voids content. Figure 4 show the images of fibers in the composite at the different volume fraction. It is seen that qualitatively, the increase of volume fraction (the same fiber length) tended to increase the spaces between fibers, so it tended creating more voids.

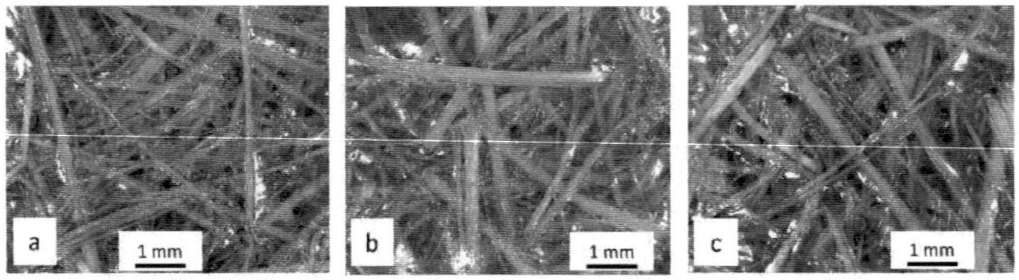

Figure 4. The image of fibers in the bamboo fiber/modified UPR composites at the fiber lengths of 10 mm, at the volume fractions of (a) 0.2, (b) 0.25, and (c) 0.3.

Figure 5. The photograph of tensile fracture of the composite at fiber length of 10 mm, at the volume fractions of (a) 0.25 and (b) 0.3.

Figure 5 shows the images of tensile fracture specimens at the fiber length of 10 mm and at different volume fraction. It is observed that at low volume fraction (0.25), the fiber tended to break, indicating high interfacial strength between the fiber and the matrix. Meanwhile at the volume fraction of 0.3, some fibers tended to be pulled out and some fibers were broken. The former indicated that some fibers could not bond well with the matrix due to lack of matrix, while the latter have good bonding. As the tensile strength and the elastic modulus tended to decrease at the volume fraction of 0.3 and at the fiber length beyond 10 mm, more voids and lack of matrix were considered to be the main cause of the low strength at the longer fiber length and high volume fraction for randomly oriented fibers.

3.2. Flexural properties

Figure 6 shows the typical of load-displacement curves, the flexural strength and the flexural modulus of bamboo fiber/modified UPR composites with fiber length at various volume fractions, in dry and wet conditions. As seen in figure 6(a), the initial load-displacement curves showed linearity and close to the peak load they deviated from the linearity due to damage process. After the peak load, at dry condition, the loads progressively dropped, indicating the brittle failure of the composite. In general, as seen in figure 6(b, c), the trend of flexural strength and flexural modulus with the fiber length was about similar to that of the tensile strength, as expected. The optimum flexural strength and flexural modulus were obtained at the fiber length of 10 mm. The flexural strength of composites having the volume fraction of 0.2 was the highest at the fiber length of 10 mm; however, as the variation at this volume fraction was high so that it might not be the case. The best performance of flexural properties was obtained at the volume fraction of 0.25.

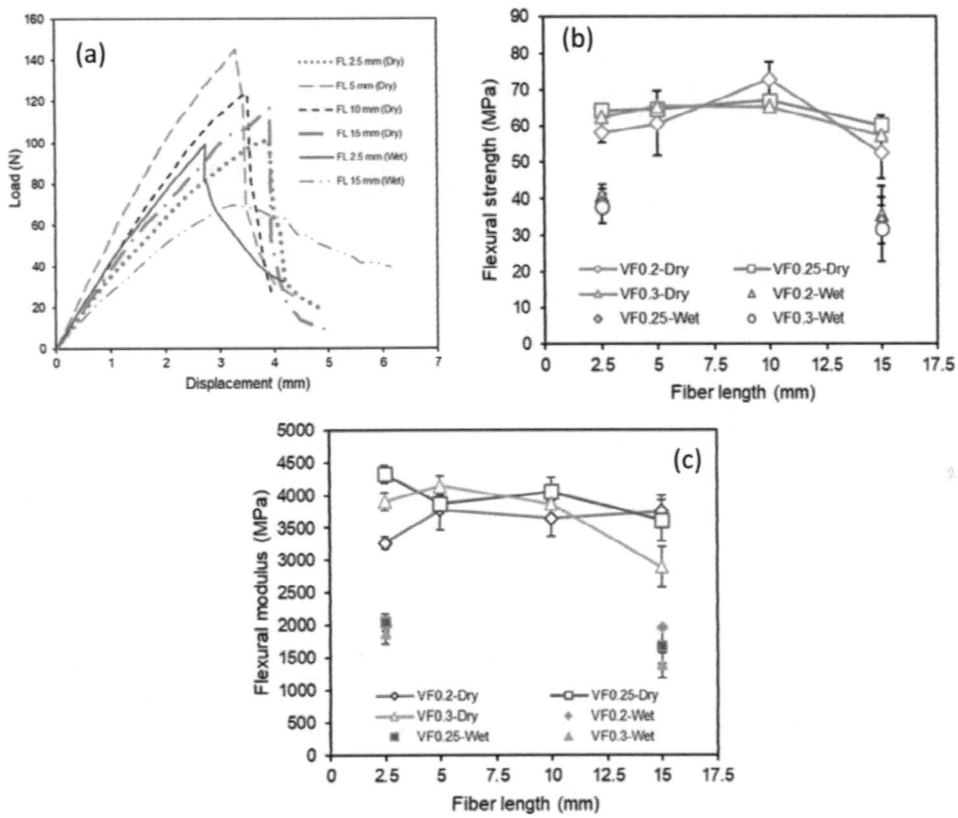

Figure 6. (a) Typical load-displacement curve at the volume fraction of 0.25, (b) Flexural strength and (c) Flexural modulus of bamboo fiber/modified UPR composites.

Figure 7 shows the failed flexural specimens at the volume fraction of 0.25 and at various fiber lengths in dry and wet conditions. It is seen that the failure occurred at the bottom of the specimens, but the crack did not always occur straight under the load applied, where the stress was the highest. The crack initiated at the weakest part of the bottom, where it could not sustain the stress.

IOP Conf. Series: Materials Science and Engineering **532** (2019) 012008 doi:10.1088/1757-899X/532/1/012008

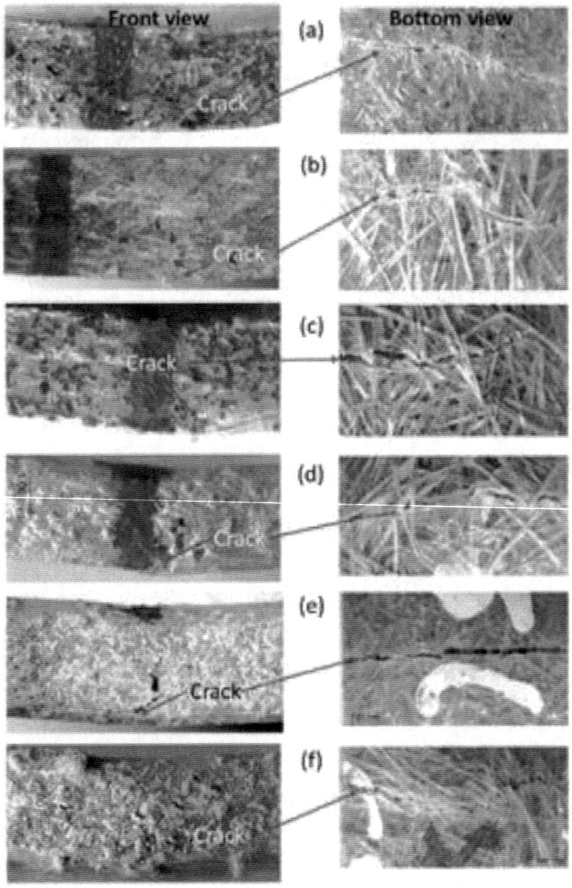

Figure 7. The photograph of the failed flexural specimens at the volume fractions of 0.25 at the fiber lengths of (a) 2.5 mm (dry), (b) 5 mm (dry), (c) 10 mm (dry), (d) 15 mm (dry), (e) 2.5 mm (wet), and (f) 15 mm (wet).

3.3. Impact properties

Figure 8 shows the impact strength of bamboo fiber/modified UPR composites with fiber length at various volume fractions. Different to the tensile and flexural strength, the impact strength tended to linearly increase with the increase of fiber length up to 15 mm. Furthermore, the impact strength also tended to increase with the increase of volume fraction for all fiber lengths. A composite material that has capability to absorb more energy is expected to have high impact strength. The energy absorption for composite was contributed by debonding of the fiber, the pulling out of the fiber, and matrix and fiber fractures. Debonding and fiber pulled out more contributed to the impact strength than the matrix and fiber fractures [12]. Therefore, it is expected that longer fiber length contributed to the high-energy absorption, as it needs high energy for debonding and pulling out of the fiber rather than the shorter fiber. Similarly, high fiber volume fraction also needs more energy for debonding and fiber pulling out. It is observed that the fiber length seems having significant effect in increasing the impact strength. As seen in figure 9, in dry condition the longer fibers pulled out was observed for fiber length of 15 mm (figure 9b) than the fiber length of 2.5 mm (figure 9a).

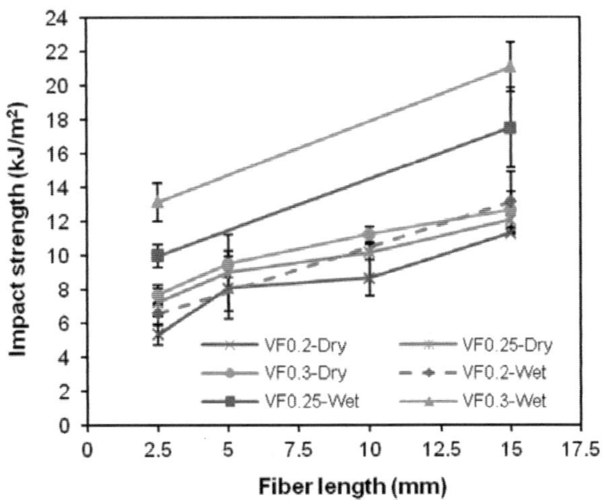

Figure 8. Impact strength of bamboo fiber/modified UPR composites.

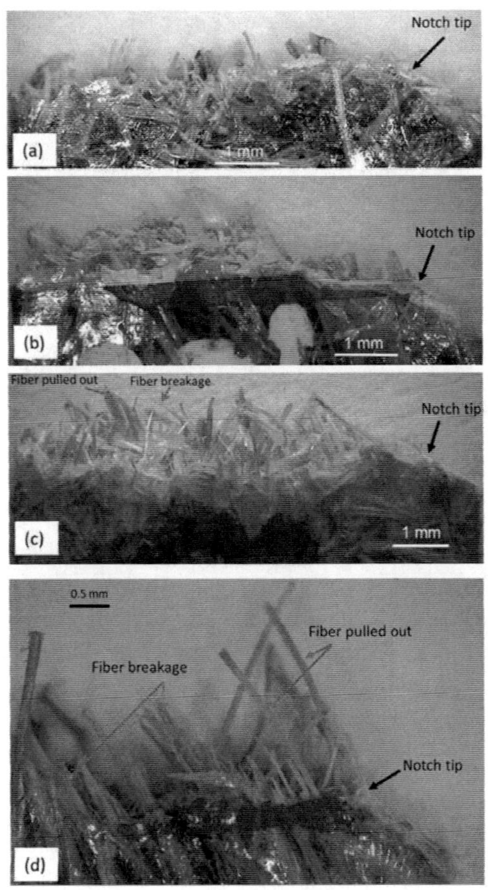

Figure 9. The photographs of the failed impact specimens at volume fraction of 0.3, at the fiber lengths of (a) 2.5 mm (dry), (b) 15 mm (dry), (c) 2.5 mm (wet), (d) 15 mm (wet).

3.4. Water absorption

Figure 10 shows the water uptake vs. square root of time curves of bamboo fiber/modified UPR composites at two fiber lengths and various volume fractions. The water uptake parameters (the equilibrium water uptake and the diffusion rate) are shown in table 1. The Fickian fits corresponding to the studied volume fraction and fiber length are also shown in figure 10. At the beginning, the water uptake increased linearly and then the water absorption slowed down and reached the equilibrium. It is observed that for all volume fractions, the longer fiber (15 mm) absorbed more water than the shorter fiber (2.5 mm). Water absorption also increased with the increase of volume fraction for all fiber lengths. Similarly, the diffusion rate also increased with the increase of fiber length and volume fraction. It is also observed that regardless the volume fraction, at the fiber length of 2.5 mm, the water uptake followed the Fickian diffusion law, but at the fiber length of 15 mm, most water uptake curves deviated from the Fickian diffusion law.

At the same volume fraction, the longer fiber was able to absorb water along the fiber and the interface of fiber/matrix, without interruption. Whereas, for the shorter fiber, the capability of fiber to absorb water was hindered by the matrix, as matrix absorbed less water than the fiber. Sugiman et al. [13] showed that at the volume fraction of 0.2, the 8% alkali treated bamboo fiber/modified UPR composites absorbed water five times higher than the modified UPR matrix. Similar to the effect of fiber length on the absorbed water, the absorbed water also increased with the increase of volume fraction, and this effect was getting higher with the increase of fiber length. The increase of water uptake with the increase of fiber length at the volume fraction of 0.2, 0.25, and 0.3 was 30%, 34%, and 44% respectively. Meanwhile, the increase of water uptake with the volume fraction for the fiber length of 2.5 mm was 35% and 44% for the volume fractions of 0.25 and 0.3, respectively, relative to the volume fraction of 0.2. For the fiber length of 15 mm, the increase of water uptake was 39% and 59% for the volume fractions of 0.25 and 0.3, respectively relative to the volume fraction of 0.2.

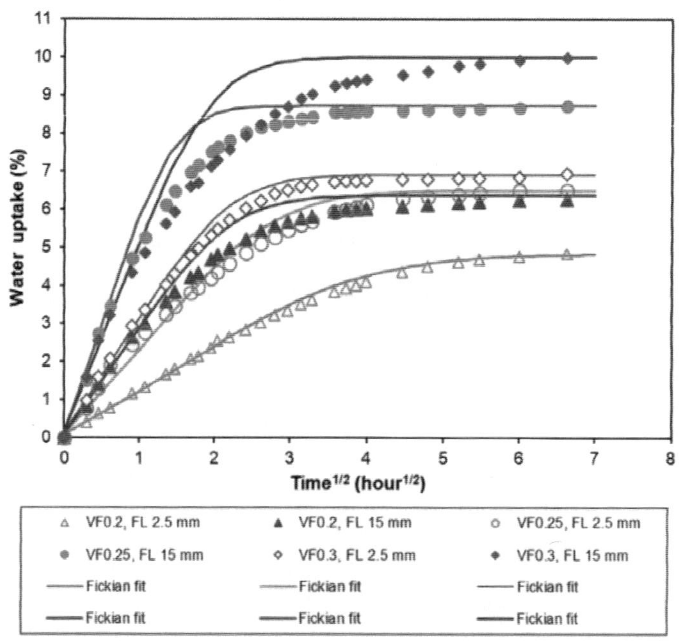

Figure 10. Water uptake vs. square root of time of bamboo fiber/modified UPR composites.

Table 1. Water uptake parameters of bamboo fiber/modified UPR composites.

Volume fraction (VF) and fiber length (FL)	M_∞ (wt%)	D (mm^2/day)
VF0.2, FL2.5 mm	4.83 ± 0.10	0.14 ± 0.03
VF0.25, FL2.5 mm	6.50 ± 0.95	0.33 ± 0.13
VF0.3, FL2.5 mm	6.92 ± 0.52	0.55 ± 0.06
VF0.2, FL15 mm	6.37 ± 0.20	0.52 ± 0.02
VF0.25, FL15 mm	8.72 ± 0.71	0.94 ± 0.38
VF0.3, FL15 mm	9.52 ± 1.03	0.71 ± 0.17

3.5. Effect of water on the flexural and impact properties

The flexural strength and the flexural modulus of bamboo fiber/modified UPR composites after being immersed in distilled water at a temperature of 40°C for approximately 1055 hours can be seen in figure 6. In this condition, the composites had been saturated with water. It is seen that the flexural strength and the flexural modulus significantly reduced, and the reduction is higher for the longer fiber. It is also seen that although the reduction increased with the increase of volume fraction; however, the effect of volume fraction on the reduction seemed not significant. The reduction of flexural strength in wet condition was in the range of 30-38% for the fiber length of 2.5 mm and 39-45% for the fiber length of 15 mm. While the reduction of flexural modulus in wet condition was in the range of 39-52% for the fiber length of 2.5 mm and 52% for the fiber length of 15 mm, regardless the volume fraction.

Most of the absorbed water resided in the bamboo fiber and interface fiber-matrix. The water attacked the interface fiber/matrix and then weakened the interface strength. Water also caused the reduction of fiber and matrix strength. For matrix, water caused plasticization, leading to the tensile strength degradation. The longer fiber has longer interfacial region and able to absorb more water. As a result, the flexural strength and flexural modulus degradation of the longer fiber was higher than the shorter fiber.

The impact strength of bamboo fiber/modified UPR composites in wet condition can be seen in figure 8. Similar to the flexural specimen, the impact specimens had reached the saturation before being tested. The impact strength in wet condition increased significantly compared to those in dry condition both for the volume fraction and the fiber length. Compared to dry condition, the significant increase was seen for the volume fractions of 0.25 and 0.3, which was about 38% and 71% for the fiber length of 2.5 mm, respectively, and 45% and 66% for the fiber length of 15 mm, respectively. In wet condition, the increase of impact strength with the fiber length from 2.5 mm to 15 mm was 100%, 75%, and 60% for the volume fractions of 0.2, 0.25, and 0.3, respectively.

Again, the absorbed water attacked the fiber, matrix, and the interface fiber/matrix. The most important factors contributing the impact strength is the interface strength, as the interface provides high-energy absorption by debonding and fiber pulling out. As seen in figure 9(c-d), water that attacked the interface degraded the interface strength and promoted the failure by debonding and fiber pulled out rather that the fiber fracture. This leaded to high impact strength. The effect of fiber length (at the same volume fraction) seemed higher than the effect of volume fraction (at the same fiber length). This is expected, as the longer fiber needed high energy to debond and then to pull it out.

4. Conclusions

Investigation on the effect of fiber length on the mechanical properties and water absorption of bamboo fiber/polystyrene-modified unsaturated polyester composites had been undertaken. It was found that the tensile properties had the similar trend to the flexural properties, in which the tensile and flexural properties tended to increase with the increase of fiber length, and reached the optimum at the fiber length of 10 mm, for all fiber volume fractions. Meanwhile, the impact strength tended to increase with the fiber length up to 15 mm. The water absorption tended to increase with the fiber length and fiber volume fraction. In wet condition, the flexural properties degraded and the degradation was higher for the longer fiber. In contrast to the flexural properties, in wet condition, the impact strength significantly increased, compared to dry condition for all fiber lengths and volume fractions. The effect of fiber length was more dominant than the volume fraction in increasing the impact strength.

Acknowledgements
This work was supported by the Ministry of Research, Technology, and Higher Education, the Republic of Indonesia with grant number [808/UN18.L1/PP/2018]. Authors would like thank Mr. B. Ulum for preparing the specimens.

References

[1] Amada S, Ichikawa Y, Munekata T, Nagase Y and Shimizu K 1997 Fiber texture and mechanical graded structure of bamboo *Composites Part B* **28** 13–20
[2] Gay D, Hoa S V, Tsai S W 2003 *Composite materials design and applications* (Florida: CRC Press)
[3] Sugiman S, Gozali M H and Setyawan P D 2017 Hygrothermal effects of glass fiber reinforced unsaturated polyester resin composites aged in steady and fluctuating conditions *Advanced Composite Materials* **In Press** DOI: 10.1080/09243046.2017.1405597
[4] Gaggino R 2012 Water-resistant panels made from recycled plastics and resin *Construction and Building Materials* **35** 468–82
[5] Setyawan P D and Sugiman 2012 Pengaruh penambahan styrofoam dan partikel karet terhadap sifat mekanik resin polyester tak jenuh *Proceeding Seminar Nasional Tahunan Teknik Mesin XI (SNTTM XI) & Thermofluid IV* (Yogyakarta: Universitas Gadjah Mada)
[6] Sugiman S, Setyawan P D and Anshari B 2018 Effect of alkali treatment on the flexural strength of bamboo fibers reinforced styrofoam-modified polyester resin *AIP Conference Proceedings* **1983** 050005 doi: 10.1063/1.5046278
[7] ASTM D3039 2002 *Standard test method for tensile properties of polymer matrix composite materials* (PA: ASTM International)
[8] ASTM D790 2003 *Standard test methods for flexural properties of unreinforced and reinforced plastics and electrical insulating materials* (PA: ASTM International)
[9] ASTM D256 2002 *Standard test methods for determining the izod pendulum impact resistance of plastics* (PA: ASTM International)
[10] Sugiman S, Putra I K P and Setyawan P D 2016 Effects of the media and ageing condition on the tensile properties and fracture toughness of epoxy resin *Polymer Degradation and Stability* **134** 311–21
[11] Manalo A C, Wani E, Zukarnain N A, Karunasena W and Lau K T 2015 Effects of alkali treatment and elevated temperature on the mechanical properties of bamboo fibre-polyester composites *Composites Part B* **80** 73–83
[12] Öztürk S 2005 The effect of fibre content on the mechanical properties of hemp and basalt fibre reinforced phenol formaldehyde composites *Journal of Materials Science* **40**(17) 4585–92
[13] Sugiman S, Setyawan P D and Anshari B 2018 Water absorption and impact strength of bamboo fiber/polystyrene-modified unsaturated polyester composites: Effects of alkali treatment *IJASEIT,* Submitted for publication

Single stage dimethyl ether plant model based on gasification of palm empty fruit bunch

R Heryadi[1*], A S Uyun[1], E Yandri[1,2], S M Nur[1,2], K Abdullah[2]

[1] Graduate School of Renewable Energy, Darma Persada University, Jalan Taman Malaka Selatan, Pondok Kelapa, Jakarta 13450, Indonesia
[2] Center of Renewable Energy Studies, Darma Persada University, Jalan Taman Malaka Selatan, Pondok Kelapa, Jakarta 13450, Indonesia

[*]E-mail: rudy.heryadi@live.com

Abstract. Biomass from empty fruit bunch (EFB) is considered as potential renewable energy sources to be developed in Indonesia. EFB can be efficiently converted into valuable and useful fuel products through gasification. Gasification is a process of biomass conversion into syngas. Syngas is a raw material for most other chemical products, such as Methanol, Ammonia, and DME. One of the biomasses that has big potential to be utilized is palm EFB waste. Research about DME synthesis from various biomass has been done, though little research about DME production based on EFB gasification. The aim of this research is to simulate single stage DME Plant Model based on gasification of EFB by assuming that the gasification reaction and DME synthesis reaction are under equilibrium condition. Method encompasses step to perform simulation of the model and DME plant design. The model predicts yield of DME product in agreement with published real DME plant operation. The result shows flow rate of DME produced is 50% of EFB feed flowrate and 73% energy efficiency obtained for pre-treated EFB. Electric power requirement has been investigated at 4096 kW and can be fulfilled by DME Plant through steam generation cycle and internal combustion engine fuelled by DME product. The model developed will be used as a basis for techno-economic analysis of DME plant based on EFB gasification.

Keywords: Empty Fruit Bunch, Dimethyl Ether, Gasification, Model, Simulation

1. Introduction

A The need for renewable energy in Indonesia is a necessity. From various renewable energy alternatives available, renewable energy from biomass has great potential to be developed in Indonesia. The biomass potential in Indonesia is about 32 GWe, and recent utilization was only around 1,740.40 GWe or 5.4% of the total potential [1]. It was estimated that biomass potential in Indonesia at 146.7 million tons per year equivalent to 470 GJ/y [2].

Utilization of biomass as one of energy source can be done through gasification process. Gasification is a thermochemical process that converts biomass into a gas called a producer gas or synthetic gas (syngas). Syngas is a fuel-based mixture consisting mainly of Hydrogen (H_2), Carbon Monoxide (CO), and Methane (CH_4) [3]. Syngas is a raw material for most other chemical products, such as Methanol, Ammonia, and Dimethyl Ether (DME). One of the biomass sources which have big potential to be utilized is palm empty fruit bunch (EFB) waste. The total EFB waste generated in Indonesia was 17.43 and 30.6 million tons for the year 2014 [4] and 2015 [5], respectively.

Some research on DME production using biomass through gasification process has been widely performed, such as research conducted by [6]–[10]. In addition to research on the process of producing DME from biomass, research done by [7] also examines the technical analysis and feasibility of DME plant investment by utilizing wood pellets that have undergone a torrefaction process. The wood pellets have similar characteristics to coal; therefore, the gasifier used was the same with gasifier built for coal gasification.

Specific study on simulation of gasification process based on EFB gasification to DME had been done by [11] which discussed the parametric analysis of DME production. Another research on DME plant modelling was conducted by [12], which one of the paper was discussing about zero-dimension DME plant model based on torrefied biomass gasification and low CO_2. From those studies which already mentioned above, yet no model developed for DME plant based on gasification of EFB. Hence, this study originated to provide a model of DME plant based on EFB gasification and simulate DME Plant model developed by using Aspen Hysys. In this paper, the model developed was simulated to obtain DME yield, energy efficiency, power consumption, and power generation.

2. Theory of gasification and DME synthesis

2.1. Gasification

Gasification is a process that converts biomass into a gaseous fuel called producer gas (CO, H_2, CH_4, etc.) using a little air or oxygen/steam. The reaction is occurred as incomplete combustion. According to the reference [13], gasification reactions involve several series of chemical reactions such as: drying, pyrolysis, combustion, gasification, and auxiliary processes such as gas and water phase displacement reactions.

2.2. DME synthesis

The process of producing DME using biomass as raw material was divided into 4 stages: gasification, water-gas shift, gas purification, and DME synthesis [11]. In general, two-stage DME is produced by methanol dehydration reaction in (1).

$$2\,CH_3OH \rightarrow (CH_3)_2O + H_2O \tag{1}$$

DME can be produced by single-stage (SS-DME) synthesis and two-stage (TS-DME) synthesis process. The TS-DME two-stage process passes through two different process units, one process with a special catalyst to produce Methanol, and another one for the methanol dehydration reaction as main reaction to produce DME [14]. The basic reaction for single-stage DME synthesis consists of several reaction steps shown in Table 1.

According to [15], there are two routes of the overall reaction of SS-DME synthesis from syngas, those are reactions (a) and (b) as shown in Table 1. The reaction (a) which is the overall reaction consists of 3 reaction steps that is, methanol synthesis reaction (c), dehydration reaction (d), and water-gas shift (WGS) reaction (e). The overall reaction (b) consists of reaction step (c) and (d). The single-stage DME synthesis reaction of JFE technology follows the overall reaction (a), while the Haldor Topsoe A/S technology and other technologies follow the overall reaction (b).

In this study, the simulation was performed using SS-DME synthesis process route. SS-DME synthesis process is considered more economical. In addition, it can overcome the problems associated with equilibrium on synthesis of methanol when synthesis is done through a two-stages synthesis process [11].

The technology used for simulation purposes on the model developed was adopting JFE technology combined with research results done by [7] about the techno-economic analysis of torrefied biomass and process configuration described on [16], which described the potential of DME as an alternative to conventional fossil fuel. Reference [16] was made by EPCM consultant based on South Africa. The description of the process and reaction between [7][16] were quite similar with reference from [15] with topic about pilot plant 5 tons/day DME production through single-stage synthesis route. Those references based on the technology developed by JFE, and transparently provides detail information about the process.

Table 1. Reaction of single stage DME synthesis

Reaction	Heat of reaction (kJ/mol)
(a) $3CO+3H_2 \rightarrow CH_3OCH_3+CO_2$	-246
(b) $2CO+4H_2 \rightarrow CH_3OCH_3+H_2O$	-205
(c) $2CO+4H_2 \rightarrow 2CH_3OH$	-182
(d) $2CH_3OH \rightarrow CH_3OCH_3+H_2O$	-23
(e) $CO+H_2O \rightarrow CO_2+H_2$	-41

3. Method and DME plant design

The following steps are undertaken to perform the modelling.

i) Collect information on existing or model of DME plant that was available.

ii) Perform the simulation by utilizing process engineering simulation tools. In this study, Aspen Hysys was utilized.

iii) Prepare material stream and energy stream to find DME yield, process efficiency, and energy recovery.

The Process Flow Diagram (PFD) of DME synthesis was developed and shown on Fig.1, included some important parameters. DME plant design process was described in 10 steps as follow.

3.1 Simulation tools

The simulation was done by using Aspen Hysys process simulator. Aspen Hysys has been used in several studies to simulate biomass gasification [12][18] and DME production [11]. The various components that comprise Aspen Hysys provide an extremely powerful approach to steady state process modelling. The user describes the process in terms of pieces of equipment interconnected by process stream, and the program solves all the mass/energy/equilibrium equations, taking into consideration the specified design for the units [19]. Since feedstock EFB is not a default component in Aspen Hysys, hence it must be put manually as Solid Hypothetical component. The input of EFB as hypothetical solid component was based on ultimate analysis of EFB measured feedstock. Base for simulation is dry ash free (daf), from weight percentage of components C, H, O, N, S

3.2 Biomass feedstock

Feedstock used was EFB. The composition of the EFB was based on dry ash free base (daf) taken from [20]. Based on ultimate analysis chemical formula can be determined [18]. In 100 grams of biomass, there is 51.67 grams of carbon. Mass in grams divided by carbon molecular weight (12 g/mol) will give result of 4.306 mol of carbon. The same is applied to other elements which lead to the following chemical formula for EFB biomass: $C_{4.306}\,H_{6.176}\,O_{2.587}\,N_{0.048}\,S_{0.003}$, and this chemical formula is used as an input to Aspen Hysys as hypothetical solid component. The input of EFB is at 6.16 tonnes per hour, based on [21].

3.3 Biomass pre-treatment.

Before the gasification process, raw biomass in any form needs to go through pre-treatment process. The pre-treatment process was consisting of several parts; such as granulator, dryer, hammermill, and pelletizer. As pellets, EFB has a greater energy density, reducing the bridging problem in gasifiers that use biomass with non-uniform sizes [20]. According to [22], with pelletizing, bulk density of EFB increases almost three times after the pellet-making process. Energy required for granulator, hammermill, and pelletizer were 1.1 %, 2.2% and 2.5% of LHV (low heating value), respectively [20].

3.4 Gasification.

The process of gasification occurs in Circulating Fluidized Bed (CFB) gasifier. Gasifier used Steam–Oxygen as an oxidant. The Oxygen was supplied by cryogenic air separation process. The purity of oxygen was assumed at 100%. The electricity consumption of air separation unit was based on [7] at 1.0 MWe/ kg O_2/s. The composition of the syngas as result of gasification was calculated based on equilibrium reaction at 900 °C. The process was assumed isothermal during the gasification reaction [23]. The waste heat from reactor was used as heat for steam generation, since reactor was cooled using water, and water will turn into steam (steam generator I) for gasification oxidant supply. From the CFB gasifier, syngas outlet were cooled to 400 °C, before entering Water Gas Shift (WGS) reactor to adjust for the H_2 to CO ratio (H_2/CO) equal 1 for DME synthesis requirement [24]. Raw syngas outlet was filtered using bag house filter to remove particulates.

3.5 Water gas shift.

Water gas shift reaction takes place to adjust for the composition of H_2/CO equal 1. Temperature of the reaction is at 400 °C and pressure 1000 kPa based on [18]. Prior to acid gas removal process, syngas was cooled to 25 °C [12]. Gibbs reactor was used to simulate the reaction and 130 kg/h steam was supplied to WGS reactor to get H2/CO = 1. Type of WGS used is high temperature WGS, because it favors WGS reaction with a greater reaction rate [18].

3.6 Water removal and syngas cleaning.

Water removal was modelled using separator as a dryer and acid gas was removed by MEA scrubber. The removal process was modeled by water separator that removes water content less than 1 % mol as acceptable level [18]. After water removed, syngas entering Acid removal modelled as component splitter. Assumed, CO_2 and H_2S were removed at 90% and 100% [18].

3.7 DME synthesis.

Sweet syngas is compressed into 5000 kPa, according to the operating pressure of JFK process [15]. The reactor used is Gibbs reactor to simulate the synthesis of DME and reactions involved is equilibrium reaction based on [25]. The operating temperature of the reactor was based on JFK process [15] at 260 °C. The temperature was maintained at 260 °C throughout the reaction. Water jacketed cooler is used to maintain temperature of DME reactor, and water as a cooler will turn into saturated steam by absorbing heat of DME synthesis reaction. Steam produced was a part of steam integration at steam generator II. Product gas was cooled to 15 °C [7], then separated by gas–liquid separator where unconverted syngas recycled back to DME synthesis reactor. On this simulation, 95% of unconverted is recycled and 5% is sent to steam boiler to generate electricity [7]. Liquid outlet of gas-liquid separator flows to purification section which consist of three distillation towers

3.8 Distillation.

The distillation process occurred in the first tower was mainly to separate CO_2 and other light gasses with DME, Methanol (MeOH) and water. The tower has 9 stages and feed stage located on stage number 1 from condenser at the top. The gas as a top product mixed with the outlet gas from Gas – Liquid separator and sent as a fuel for steam as part of steam generator II system. Bottom product contains DME, methanol, and water entering DME Tower. The distillation in DME tower is to separate DME as top product with Methanol and water as bottom product. DME recovered from DME tower is 99.6% purity according to [14, 15].

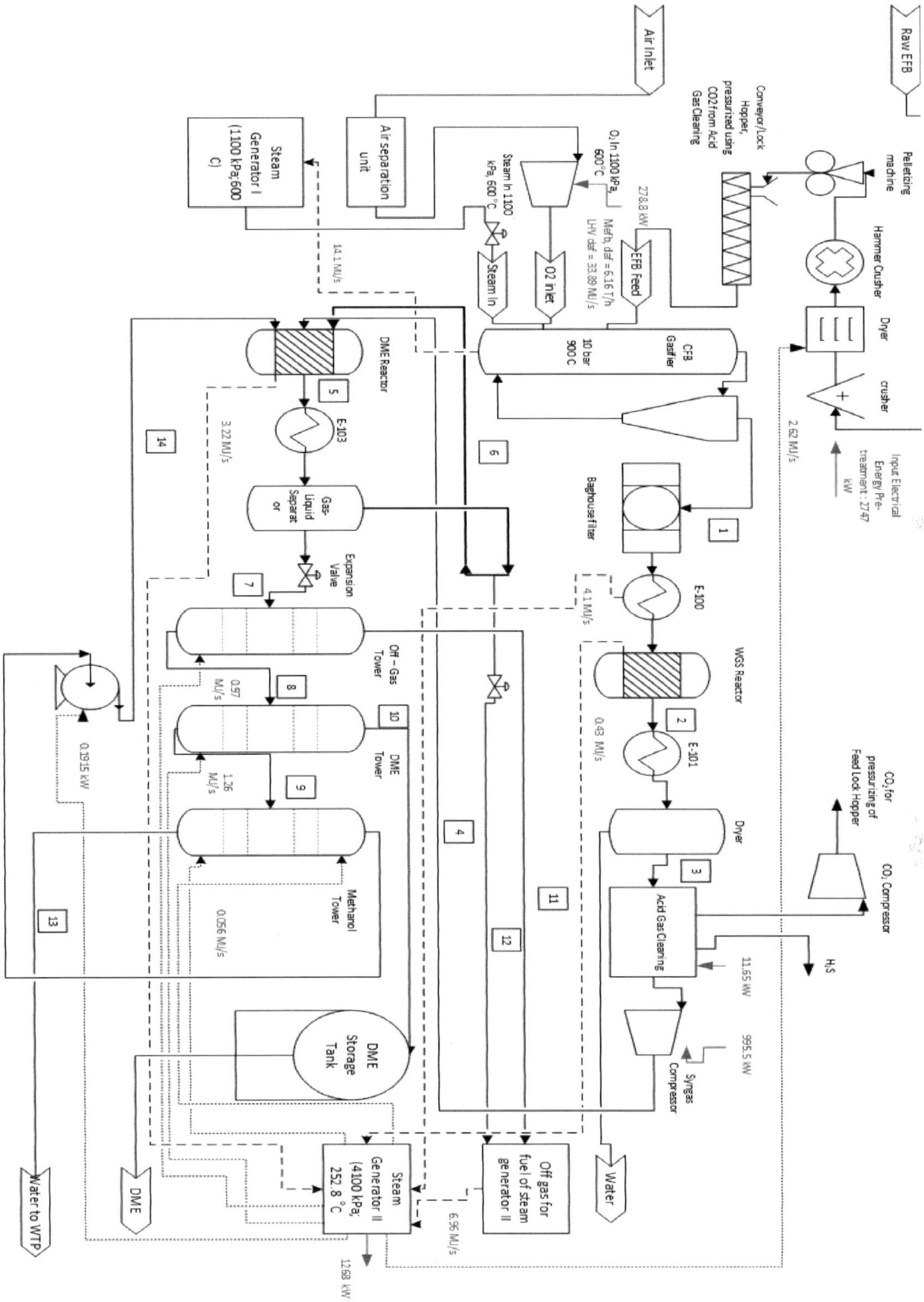

Fig. 1. Process flow diagram DME production based on EFB gasification

DME tower has 10 stages and feed stage is feed number 3 from the top. The bottom product of DME Tower which were methanol and water was further separated in methanol tower. Methanol tower has 12 stages, the top product of the distillation in methanol tower was methanol and the bottom product were water. Methanol was recycled, while water delivered to water treatment facility before discharged or re-utilize.

3.9 Steam and power generation.

Steam generation for the process and power generation was from the steam boiler fueled by waste heat from the plant combined with the utilization of off-gas from plant. There were two steam generators available. First, steam was generated by waste heat from CFB gasifier and was named as steam generator I as shown on Fig.1. Second, combination of off-gas boiler with waste heat boiler from WGS reactor, DME reactor, and E-100, all were combined as steam generator II as shown on Fig.1. Power for DME plant was generated by using one steam turbine at steam generator II system.

3.10 Efficiency Calculation

DME $\eta_{product}$ (product efficiency) was calculated by using (2). Formula was adopted from [23].

$$\eta_{product} = \frac{\dot{m}_{product}.LHV_{product}}{\dot{m}_{biomass}.HHV_{biomass}} \qquad (2)$$

Where $\dot{m}_{product}$ is flow rate of product DME in kg/h, and $\dot{m}_{biomass}$ is flow rate of EFB Biomass in kg/h. $LHV_{product}$ and $LHV_{biomass}$, are the low heating value (in kJ/kg) of the product and biomass respectively.

4. Result and discussion

4.1 Process simulation result

The simulation result on every stream number shown on Fig.1 was tabulated on Table 2. Results showed the total mass and molar flow, operating temperatures, operating pressures, heat flow, and composition of each substance produced. Ratio of H_2 to CO on stream No. 2,3, 4 was equal to 1, and this is required according to JFE technology [15] for direct synthesis of DME (single-stage). Raw DME product was represented by stream No.5. The stream further cooled and separated in gas-liquid separator, where gas product consists mostly CO_2 and unreacted syngas (H_2 and CO), CH_4, and CO_2. The gas product recycled at 95% mol and 5% mol of it delivered and mixed with top product of off-gas tower, further was utilized as fuel of steam boiler for power generation and process consumption.

The liquid product of gas-liquid separator was expanded from 5000 kPa to 1000 kPa, to follow operating condition of DME purification section [12]. Methanol produced at stream No.9 was recycled into DME reactor. Stream No.10 was DME product with purity 99.9 %, since DME has similarity with LPG in term of physical properties, its handling and storage are similar to LPG [26]. Yield of DME over input EFB for the model developed was 50%. Table 3 shows mass of EFB input and DME output, LHV of EFB input and LHVof DME output, and DME product energy efficiency.

Table 2. Material stream and stream composition of DME plant

Stream No.		1	2	3	4	5	6	7	8	9	10
Vapour Fraction		0.9932	1	1	1	1	1	0.2532	0	0	0
Temperature	*C*	900	400	25	25	260	15	-14	49	150	43
Pressure	*kPa*	1,000	997	997	5000	5000	5000	1000	999	999	950
Molar Flow	*kgmole/h*	602	605	601	528	1,027	819	156	76	9	67
Mass Flow	*kg/h*	11,400	11,478	11,404	8,200	33,970	25,684	6,672	3,325	239	3,086
Heat Flow	*MJ/h*	-41,080	-54,627	-60,855	-28,918	-218,000	-176,900	-44,300	-15,620	-2,112	-13,520
Mole Frac (%)											
H_2		35.00	41.51	41.79	47.58	9.68	11.29	0.47	0.00	0.00	0.00
CO		48.47	41.62	41.91	47.71	17.26	20.12	1.70	0.00	0.00	0.00
CO_2		6.66	13.22	13.31	1.52	41.60	41.65	41.19	0.00	0.00	0.00
H_2O		6.46	1.02	0.34	0.39	0.31	0.05	1.80	3.71	31.95	0.05
CH_4		2.20	2.19	2.21	2.51	14.04	15.99	3.20	0.00	0.00	0.00
H_2S		0.19	0.19	0.19	0.00	0.00	0.00	0.00	0.00	0.00	0.00
N_2		0.26	0.25	0.26	0.29	2.00	2.32	0.21	0.00	0.00	0.00
CH_3OH		0.00	0.00	0.00	0.00	0.62	0.06	3.78	7.81	68.02	0.01
CH_3OCH_3		0.00	0.00	0.00	0.00	14.48	8.53	47.64	88.48	0.04	99.94

4.2 Power consumption and generation

Total electrical power consumption of main unit/section in DME plant was shown on Table 4. Power generated by steam turbine was 1268 kW. According to Table 4, there was deficit 2717.19 kW on the electricity for the DME plant to run. Steam cycle converts 13519 kW waste heat into 1268 kW electricity. This is corresponding to 9.4% electrical efficiency. To overcome deficit on electricity, generator coupled with internal combustion engine (ICE) should be added for power generation. Fuel used for ICE is DME itself as a product from the DME plant. Type of ICE utilized is diesel engine and the fuel used is DME, since DME can be used as replacement for diesel fuel [26].

DME most prominent characteristic is high cetane number that makes DME can be considered as a replacement of Diesel fuel [26], with modification on fuel tank.

Table 3. Energy eefficiency of DME plant

Parameters	Unit	Value
$\dot{m}_{biomass}$	Kg/h	6160
$LHV_{biomass}$	MJ/kg	19.81
$\dot{m}_{product}$	Kg/h	3080
$LHV_{product}$	MJ/kg	28.8
$\eta_{product}$	%	73

Table 4. Power cconsumption of DME plant

Plant Section/Unit	Power Consumption (kW)
Recycle pump	0.1915
Air Separation Unit	1214
Pre-treatment EFB	2747
Acid Gas Cleaning	11.65
CO2 compressor	12.35
Power Deficit	**-2717.19**

By selecting diesel engine as genset prime mover, high thermal efficiency can be achieved, in this study the thermal efficiency of the ICE was 45 % [27]. Compressors in DME plant needs huge power to run, and this is supplied by ICE as prime mover of compressor. Table 5 shows the power consumption of compressors prime mover. Power required for refrigeration and compression of syngas consumed huge amount of power and supplied by ICE as prime mover. With 45% efficiency, total DME required to be supplied to power generation ICE is 754 kg/h, in order to generate deficit power at 2717.19 kW. While DME required by compressors is 1098 kg/h. From total 3086 kg/h DME produced, almost two-third of it is used as ICE fuel.

Table 5. Power consumption of compressor prime mover

Prime Mover ICE	Power Consumption (kW)
Oxygen compressor prime mover	279
Syngas Compressor prime mover	956
Refrigeration compressor prime mover	2725

Based on the result, conversion of CO on the simulation was 64% at 5000 kPa and 260 °C. It was tally with the experiment result in [15], where CO conversion was higher than 50% at 5 MPa and 260°C. The efficiency of this model ($\eta_{product}$) is 73% for pre-treated EFB, and this is slightly different with the result given by JFE process at 69.4% when using natural gas as the feed. The efficiency for torrefied biomass as reported by [12] was 67%, while untreated biomass only 60%. This study revealed that $\dot{m}_{product}$ was 50% of $\dot{m}_{biomass}$, and energy efficiency from EFB to DME was 73%. Conversion of torrefied biomass model into DME as studied by [7] was 54%, and it is quite close with the result obtained in this study. To run the plant electricity is required, and it can be fulfilled by the waste heat utilization through steam generation and genset, where genset fuel was using DME as final product.

5. Conclusion

The model of DME plant was designed and simulated by using Aspen Hysys. Simulation of the model was based on EFB gasification, conducted in CFB gasifier. The results of the simulation were in good agreement with reference data, especially for DME yield and CO conversion. The electric power for DME plant based on EFB gasification was provided by itself, where all the power supplied by steam cycle and ICE fueled by DME product. Total electric power required is 4096 kW. Total power required for compressor prime mover is 3960 kW. DME can be used to replace LPG and Diesel fuel, because of its physical properties and high cetane number. $\dot{m}_{product}$ produced at 50% of $\dot{m}_{biomass}$ and $\eta_{product}$ from simulation was 73% for pre-treated EFB. In the future study, DME plant model developed will be used as a basis for techno-economic analysis of bio-DME plant based on gasification of EFB.

References

[1] Hutrindo E 2015 *Modul Pengenalan Bioenergi, Kesatu* (East Jakarta: Pusdiklat Ketenagalistrikan, Energi Baru, Terbarukan dan Konservasi Energi)

[2] Abdullah K 2001 *Biomass Energy Potentials And Utilization In Indonesia* (Bogor: Dept. of Agri Eng IPB and IRES)

[3] Basu P 2010 *Biomass Gasification and Pyrolysis. Burlington* (Massachusets: Elsevier)

[4] Peryoga Y, Dewi Solikhah M, and Agus Raksodewanto A 2014 Production cost assessment of palm empty fruit bunch conversion to bio-oil via fast pyrolysis *Int. J. Adv. Sci. Eng. Inf. Technol.* **4** 394

[5] Hambali E and Rivai M 2017 *IOP Conf. Ser. Earth Enviromental Sci.* **65** 012050

[6] Ju F *et al* 2009 Process simulation of single-step dimethyl ether production via biomass gasification *J. Biotechnol. Adv.* **27** 599–605

[7] Clausen L R, Elmegaard B, and Houbak N 2010 Technoeconomic analysis of a low CO2 emission dimethyl ether (DME) plant based on gasification of torrefied biomass *J. Energy* **35** 4831–4842

[8] Do T X and Lim Il Y 2016 Techno-economic comparison of three energy conversion pathways from empty fruit bunches *J. Renew. Energy* **90** 307–318.

[9] Wang T, Li Y, Ma L, and Wu C 2011 Biomass to dimethyl ether by gasification/synthesis technology-an alternative biofuel production route *J. Front. Energy Power Eng. China* **5** 330–339

[10] Yuping L *et al 2015* Demonstration of pilot-scale bio-dimethyl ether synthesis via oxygen-and steam-enriched gasification of wood chips *J. Energy Procedia* **75** 202–207

[11] Inayat A, Ghenai C, Naqvi M, Ammar M, Ayoub M, and Hussin M N B 2017 Parametric study for production of dimethyl ether (DME) as a fuel from palm wastes *J. Energy Procedia*, **105** 1242–1249

[12] Clausen, L R N, Elmegaard B, and Houbak N 2011 *Design of novel DME/methanol synthesis plants based on gasification of biomass (*Denmark: DTU)

[13] Pradana Y S, and Budiman A 2015 Bio-syngas derived from Indonesian oil palm empty fruit bunch (EFB) using middle-scale gasification *J. Eng. Sci. Technol.***10** 1–8

[14] Inokoshi O, Ohno Y, Ogawa T, Inoue N, and Tokoeda N 2005 a new DME production technology 100tons / day DME direct synthesis demonstration plant project *The Fifteenth International Offshore and Polar Engineering Conference* vol 8 (Seoul: *Intl. Soc. of offshore and polar engineers*) pp. 49–53,

[15] Ogawa T, Inoue N, Shikada T, and Ohno Y 2003 direct dimethyl ether synthesis *J. Nat. Gas Chem.***12** 219–227

[16] Direct DME Synthesis From Natural Gas - EPCM Consultants [Online]. Available: http://epcmconsultants.co.za/direct-dme-synthesis-from-natural-gas/. [Accessed: 01-Jul-2018].

[17] Bassyouni M *et al* 2014 Date palm waste gasification in downdraft gasifier and simulation using ASPEN HYSYS *J. Energy Convers. Manag.* **88** 693–699

[18] Florence Hand Bour A 2016 *Modelling and Optimization of a Process from Biomass to Liquid Fuels via Fischer-Tropsch Synthesis* (Norway: NTNU)

[19] 2013 *HYSYS Design Tutorial For CHEE 332* (Ontario: Queen ' s University Department of Chemical Engineering) p 1–36

[20] Salomon M, Gómez M F, Erlich C, and Martin A 2013 Pelletization: an alternative for polygeneration in the palm oil industry *J. Biomass Convers. Biorefinery* **3** 213–229

[21] Salomon M, Gómez M F, Erlich C, and Martin A 2013 Technical polygeneration potential in palm oil mills in Colombia: a case study *J. Sustain. Energy Technol. Assessments* **3** 40–52

[22] S. Kerdsuwan and K. Laohalidanond 2011 Renewable energy from palm oil empty fruit bunch in *Renewable Energy - Trends and Applications*, Nayeripour N (Bangkok: InTech) p 29

[23] Doherty W, Reynolds A, and Kennedy D 2008 Simulation of a circulating fluidised bed biomass gasifier using ASPEN plus: a performance analysis (Dublin: Dublin Inst. Technol.) pp 1241–1248

[24] Parvez A M, Wu T, Li S, Miles N, and Mujtaba I M 2018 Bio-DME production based on conventional and CO 2 -enhanced gasification of biomass: A comparative study on exergy and environmental impacts *J. Biomass and Bioenergy* **110** 105–113

[25] Tavan Y, and Nikou M R K 2013 A novel distillation design for vapor phase dimethyl ether production process *ASIA_PACIFIC J. Chem. Eng.* **1776** 258–261

[26] Park S H, and Lee C S 2014 Applicability of dimethyl ether (DME) in a compression ignition engine as an alternative fuel *J. Energy Convers. Manag.***86** 848–863

[27] Easley W L, Kapic A, and Milam D M 2005 The Path to a 50 % Thermal Efficient Engine *DEER conf.* (Chicago: Caterpillar) pp 1–28

IOP Conf. Series: Materials Science and Engineering **532** (2019) 012010 doi:10.1088/1757-899X/532/1/012010

Noise reduction in near infrared-based glucose concentration measurement sensor

K Prawiroredjo, E S Julian, G Tjahjadi
Electrical Engineering Department, Faculty of Industrial Technology
Trisakti University, Kampus A, Jl. Kyai Tapa No. 1 Jakarta 11440, Indonesia

E-mail: kiki.prawiroredjo@trisakti.ac.id, eshintadewij@trisakti.ac.id,

Abstract. This paper describes the method of measuring glucose concentrations in solution using near infrared light (NIR) and photodiode sensor. A sensitive and accurate non invasive blood glucose level sensor is our main concern in designing this device. The sensor consists of a transimpedance amplifier, a notch filter and a low pass filter to reduce high frequency noise and 50 Hertz noise from the power line induction. Light Emitting Diode (LED) with 1450 nm wavelength is used as a light source and a photodiode is used as a sensor to sense this wavelength. The NIR light received by the photodiode after passing glucose solution is converted to voltage signal and filtered to determine the output voltage. The measurement results show that the filters reduced the AC voltage or noise signal under 100 mV and the DC voltage tended to decrease as glucose concentration increase until 700 mg/dL.

1. Introduction

Diabetes Melitus or simply diabetes, is a chronic disease that occurs when the pancreas is no longer able to make insulin, or when the body cannot make good use of the insulin it produces. World Health Organization, Global Report on Diabetes reported that over time diabetes can damage the heart, blood vessels, eyes, kidneys and nerves and increase the risk of heart disease and stroke [1]. According to IDF Atlas 2015, 415 million adults live with diabetes and there were 5 million deaths caused by diabetes. In Indonesia approximately 10 million adults (20 to 79 years old) live with diabetes and as many as 6 million adults live with undiagnosed diabetes, making Indonesia's diabetes population the seventh largest in the world [2].

The causes of diabetes in humans are not yet fully understood, but the widely accepted hypothesis is that it may be genetic and may be caused by high sugar intake as part of daily meal serving. Diabetics are insisted to regulate their blood glucose levels through proper diet and by injecting insulin. Effective treatment for diabetics is to measure blood glucose levels periodically. The current method available is invasive one that needs blood samples to measure blood glucose level. This is the main disadvantage of the invasive method because it requires pricking the finger that gives pain and discomfort due to frequent finger pricks and increase risk of infection. Figure 1. shows the invasive method for measuring blood glucose level.

Developing non invasive technique would be much more useful and user friendly from the user point of view. The main advantage of non-invasive method is the relief from pain and discomfort due to frequent finger pricks. This method is proper for continues monitoring of blood glucose level and reduces the healthcare cost [3].

One of popular optical method for non invasive blood glucose is near infrared (NIR) spectroscopy since NIR light is found to penetrade a great depth into biological tissue. This concept is based on the transmission of NIR light band through a vascular area of the body (finger, ear, tongue, etc), and the glucose concentration in vivo is calculated from the spectral information obtained at the reception [4].

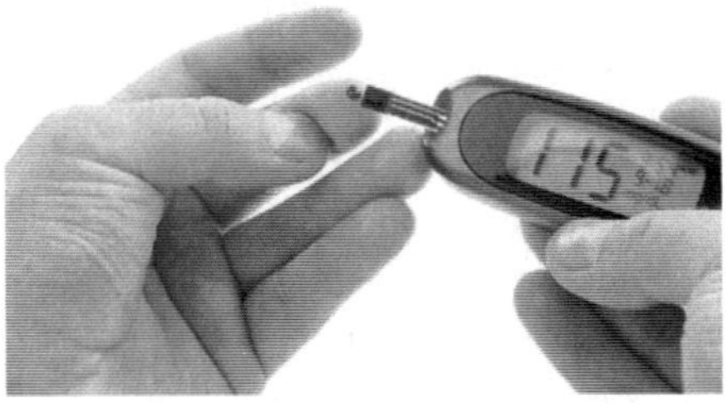

Figure 1. Invasive blood glucose measurement.

For the last two decades, NIR sensor for blood glucose measurement has been widely explored. Menon et al. proposed a method that makes use of the variation in the intensity of the signal that received from the NIR sensor to detect whether a patient is diabetic or not [3]. Chua et al. used 8 pairs of LEDs with different colors and tested for sensitivity to different glucose concentrations [5]. Narkhede et al. used NIR LED of 940 nm wavelength through fingertip and the reflected signals were detected by phototransistor placed beside the LED [6]. Bobade et al. proposed a technique used a near infrared sensor for transmission and reception of rays from forearm. By analyzing intensity variation in received signal by using photo detector at another side of forearm, level of blood glucose can be predicted [7].

In our preliminary research, we designed and developed a non invasive sensor using a pair of NIR LED at 1450 nm because the wavelength is in the overtone band of glucose absorption and a photodiode that sensitive to 800 to 1700 nm wavelength to determine glucose concentration in glucose solution. The results showed that higher glucose concentration yield lower sensor output voltage. This result is concistent with Beer Lambert's law that attenuation of light is proportional to the level of glucose solution, thickness of the sample and molar extinction coefficient. Equation (1) shows the attenuation A according to Beer Lamberts law [8].

$$A = log\left(\frac{I_0}{I}\right) = \mathcal{E}Cl \quad (1)$$

where I_o = transmitted light intensity, I = incident light intensity, $\mathcal{E}$ = extinction coefficient, C = concentration of glucose solution, l = length of light path through solution being observed.

The linier trend line of sensor output voltage as function of different glucose concentration showed good fit with those recorded data that measured in our preliminary research. The value of correlation coefficient is -0.99 which indicated strong relationship between the sensor output voltages and glucose concentrations [8]. In our second year of research, a sensitive and accurate non invasive blood glucose level sensor is our main concern in designing this device so we analyzed and reduced the noise signals that occured.

IOP Conf. Series: Materials Science and Engineering **532** (2019) 012010 doi:10.1088/1757-899X/532/1/012010

2. Research method

2.1. Block Diagram of the System

In this proposed design we complete the system with a transimpedance amplifier and noise reduction circuits to get an accurate small signal that the photodiode senses for a high accuracy reading of glucose concentration solution. Figure 2. shows the block diagram of the system.

An NIR LED of 1450 nm wavelength from ThorLabs is the light source as the input section of the system. An infrared photodiode from the same factory is used as the detector of the IR light that senses the 800 to 1700 nm light wavelength. The amount of IR light passing through glucose solution is received by an infrared photodiode. This attenuated signal is amplified and filtered.

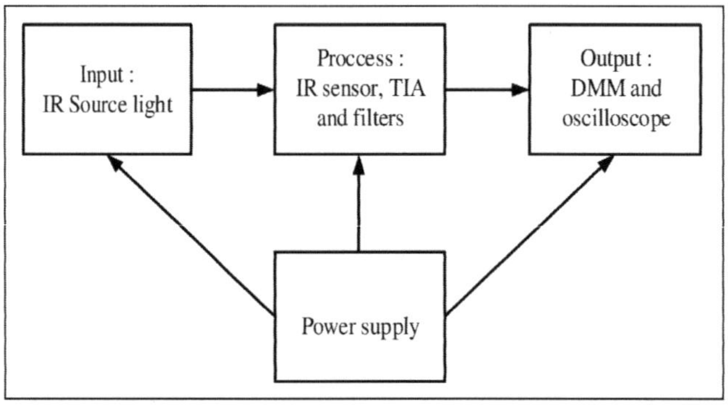

Figure 2. System block diagram.

2.2 The Amplifier and Noise Reduction Circuits

The IR sensor photodiode is connected to a transimpedance amplifier to convert the current from photodiode sensor to voltage as seen in Figure 3. Due to its low output impedance, this circuit allows viewing the output signal with an oscilloscope to analyze the noises that occurs.

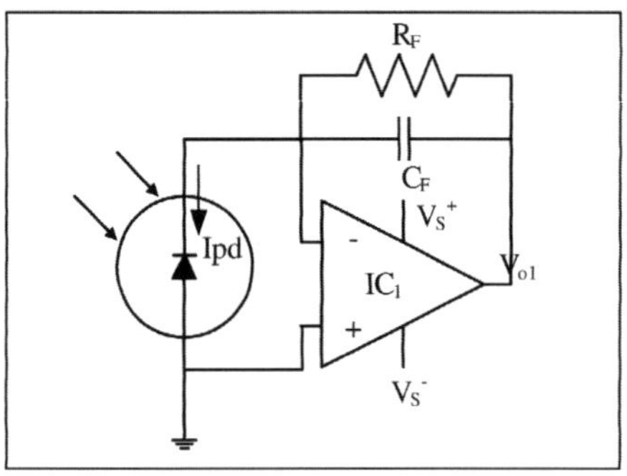

Figure 3. Transimpedance amplifier.

The circuit output voltage is:

$$V_{01} = I_{pd} \, . R_F \tag{2}$$

I_{pd} is the reverse current from the photodiode sensor. The TLC274 op amp is used for this transimpedance amplifier due to its low input offset voltage and low input bias current, the output offset voltage can be neglected.

The output signal from the transimpedance amplifier is filtered by a 50 Hz notch filter and a 50 Hz low pass filter because it is the most disturbing frequency noise from the power line besides the high frequency noise. Capacitor C_F in feedback path in transimpedance amplifier reduces the high frequency noise. Figure 4. shows the 50 Hz second order notch filter and Figure 5. is the frequency response of the filter.

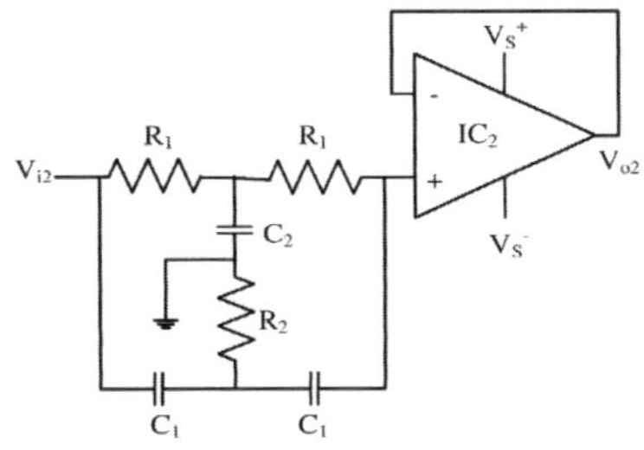

Figure 4. Notch filter.

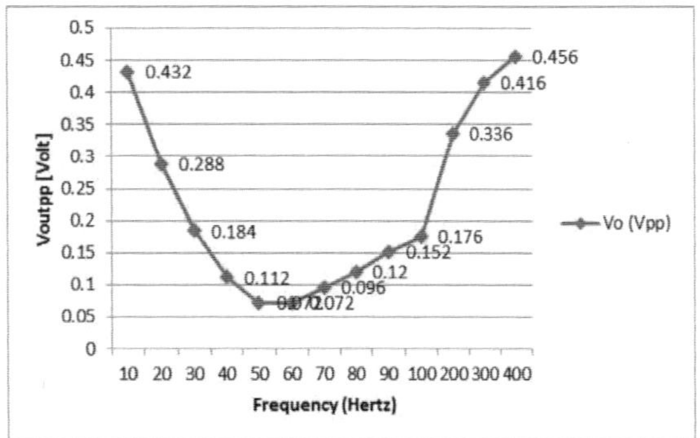

Figure 5. Notch filter frequency response.

The notch frequency of the filter is:

$$f_n = \frac{1}{2\pi R_1 C_1} \quad (3)$$

We measured the output voltage by giving 1 volt peak to peak sine wave from 10 Hz to 400 Hz to the input of the notch filter. As it is shown in Figure 5., the 50 Hz output voltage gets the highest attenuation. The 50 Hz low pass filter is cascaded to the notch filter to give more attenuation to the 50 Hz noise and it is shown in Figure 6. and Figure 7. is the frequency response of the second order low pass filter.

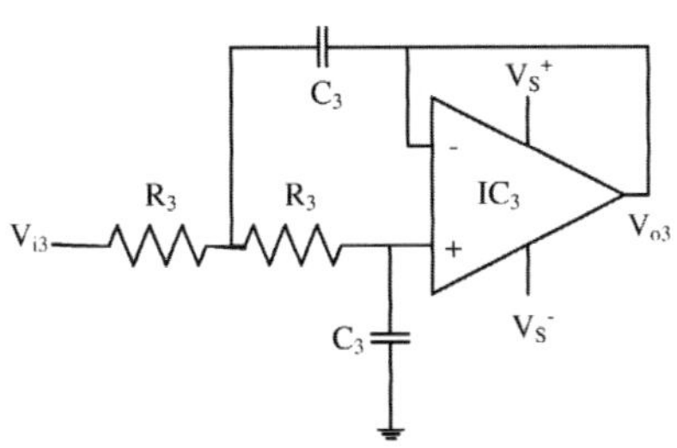

Figure 6. Second order low pass filter.

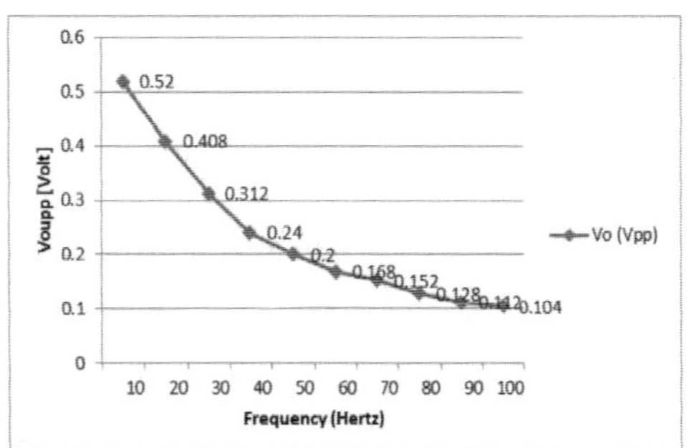

Figure 7. Low pass filter frequency response.

The cut off frequency of the low pass filter is:

$$f_{CT} = \frac{1}{2\pi R_3 C_3} \quad (4)$$

We measured the output voltage of the low pass filter by giving 1 volt peak to peak sine wave from 10 Hz to 100 Hz to the input of the low pass filter.

3. Measurement method

In this experiment we put some glucose solution concentrations in an acrylic box with 1 cm thickness, 5 cm height and 5 cm width. Figure 8 shows the experiment diagram.

IOP Conf. Series: Materials Science and Engineering **532** (2019) 012010 doi:10.1088/1757-899X/532/1/012010

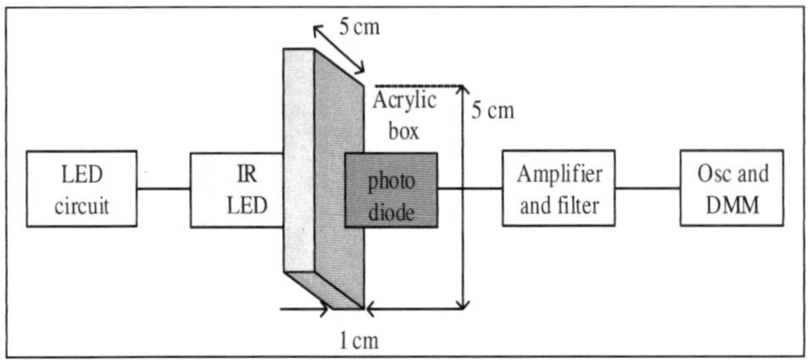

Figure 8. The experiment diagram.

The glucose solutions were made by mixing 5000 mg/dL infuse glucose solution with water to get 50, 100, 200, 300, 400, 500, 600, 700, 800, 900, and 1000 mg/dL glucose concentrations. The environmental conditions, such as room temperature and supply voltage were maintained constant in a laboratory. The infrared light was emitted through the glucose solution and the incident light was received by the photodiode on the other side of the acrylic box. The infrared light was then converted to voltage by a transimpedance amplifier and filtered by the notch filter and low pass filter to reduce the 50 Hz and high frequency noises. We measured the photodiode output voltage and the filter output voltage with a Metrix digital voltmeter and a digital oscilloscope to see the noises.

4. Result and discussions

The LED was driven by a DC voltage so the photodiode received a DC voltage too. The oscilloscope showed there were a 50 Hz sine wave noise voltage and high frequency noise besides the DC input voltage from the LED. The transimpedance amplifier output voltage and the filter output voltage for water and various glucose concentrations measurement result is shown in Table 1. The DC and rms AC output voltage signal from the transimpedance amplifier is shown in Figure 9.

Table 1. Transimpedance amplifier dan filter output voltage.

Glucose conc	Transimp Vout		Filter Vout	
(mg/dl)	V_{DC}(V)	V_{AC}(Vrms)	V_{DC}(V)	V_{AC}(Vrms)
Water	0,69	0,75	0,71	0,09
50	0,79	0,74	0,72	0,01
100	0,72	0,79	0,75	0,03
200	0,53	0,68	0,72	0,01
300	0,62	0,67	0,70	0,01
400	0,69	0,74	0,70	0,01
500	0,75	0,70	0,69	0,01
600	0,56	0,70	0,56	0,01
700	0,71	0,77	0,56	0,01
800	0,67	0,72	0,73	0,01
900	0,65	0,68	0,70	0,01
1000	0,66	0,71	0,69	0,01

As seen in Figure 9. almost all the rms AC voltage is higher than the DC voltage, it means that the noise voltages are higher than the information signal voltage.

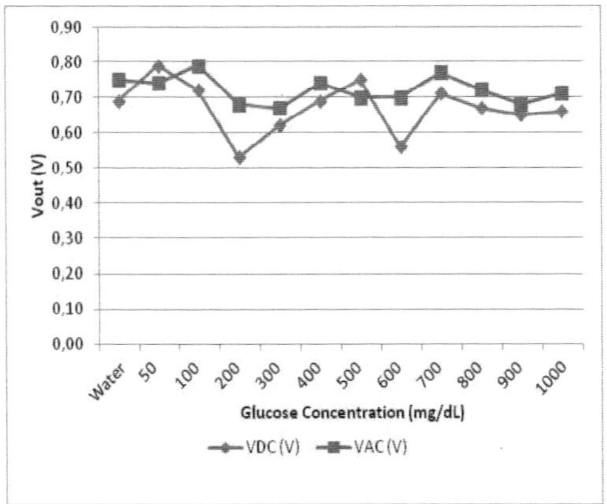

Figure 9. The transimpedance amplifier output voltage.

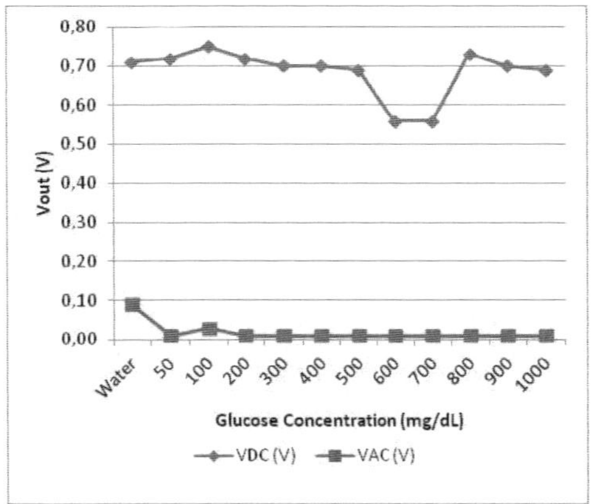

Figure 10. The filter output voltage.

The blue line or the DC signal voltages do not reduce linierly as the glucose concentrations increase. The DC voltages were randomly changed as the glucose concentrations increase. According to Beer Lamberts law, as the glucose concentration increases the lower the sensor output voltage. The output voltage from the filter in Figure10. can be seen that the rms AC voltages are reduced lower than 100 mV after it passed the filters. The DC voltages tend to decrease as the glucose concentrations increase until 700 mg/dL and it is in the range of human blood glucose. The DC voltages increase again after the glucose concentration is more than 700 mg/dL.

5. Conclusion

In this research we complete the system with a transimpedance amplifier, a notch filter and a low pass filter to get a high accuracy reading of blood glucose level. The filter circuits in this system reduced noise significantly and the DC voltage signals from the photodiode sensor tend to decrease as the glucose concentrations increase. This system is still not sensitive enough to distinguish one glucose concentration to the other concentration because the voltage difference is not proportional between concentrations so we still have to improve the system.

Acknowledgement

We would like to express our gratitude to Trisakti University and DRPM, and Ministry of Research, Technology and Higher Education of the Republic of Indonesia, Grant Number 097/A/LPT/USAKTI/III/2018, that funded this research.

References

[1] World Health Organization 2016 *Global Report on Diabetes* ed Angela Burton (Geneva, Switzerland: WHO Press) p 13
[2] International Diabetes Federation 215 *Diabetes Atlas* IDF 7th ed p17
[3] Menon K A U, Hemachandran D and Kunnath A T 2013 A Survey on Non-Invasive Blood Glucose Monitoring Using NIR 2013 *Proc. in Int. Conf. on Communication and Signal Processing (India, April 2013)* p 1068.
[4] Ciudin A, Hernández C and Simó R 2012 Non-Invasive Methods of Glucose Measurement: Current Status and Future Perspectives *Current Diabetes Reviews* **8** pp 48-54
[5] Chua C D, Gonzales I M, Manzano E and Manzano M C 2014 Design and Fabrication of a Non-Invasive Blood Glucometer Using Paired Photo-Emitter and Detector Near-Infrared LED *Proc. DLSU Research Congress (Manila, Philippines, 6-8 March 2014)* pp 1-7

IOP Conf. Series: Materials Science and Engineering **532** (2019) 012010 doi:10.1088/1757-899X/532/1/012010

[6] Narkhede P, Dhalwar S and Karthikeyan B 2016 NIR Based Non-Invasive Blood Glucose Measurement *Indian Journal of Science and Technology* **9** 1-7

[7] Bobade C D and S M. Patil S M 2016 Non-invasive Blood Glucose Level Monitoring System for DiabeticPatients using Near-Infrared Spectroscopy *American Journal of Computer Science and Information Technology* **4** 1-8

[8] Julian E S, Prawiroredjo K, Tjahjadi G 2017 The Model of Near Infrared Sensor Output Voltage as a Function of Glucose Concentration in Solution *Proc. The 15th Quality in Research (QIR) (Bali, Indonesia, 24 - 27 July 2017)*

IOP Conf. Series: Materials Science and Engineering **532** (2019) 012011 doi:10.1088/1757-899X/532/1/012011

Design and Mechanism Analysis of Paddlewheel Aerator with Movable Blades

Samsul Bahri[1], **Radite Praeko Agus Setiawan**[2], **Wawan Hermawan**[2] **and Muhammad Zairin Junior**[3]

[1] Jurusan Teknik Mesin, Politeknik Negeri Lhokseumawe, Lhokseumawe, Indonesia
[2] Departemen Teknik Mesin dan Biosistem, Institut Pertanian Bogor, Bogor, Indonesia
[3] Departemen Budidaya Perairan, Institut Pertanian Bogor, Bogor, Indonesia

E-mail: samsul@pnl.ac.id

Abstract. The successful of intensive aquaculture is strongly influenced by the ability of the farmers to overcome the deterioration of water quality. The problem is low dissolved oxygen through aeration process. The aerator device which widely used in pond farming is paddle wheel aerator because it is the best aerator in aeration mechanism and usable driven power. However, this aerator still has a low performance of aeration, so that the cost of aerator operational for aquaculture is still high. Up to now, the effort to improve the performance of aeration was done by two-dimensional blade design. Obviously, it does not provide an optimum result due to the power requirements for aeration is directly proportional to the increase of aeration rate. The development of movable blade is based on fact that power is required only when blade of paddle wheel aerator entering water and in contrary action of aeration effect only when the blade is about leaving the water. This study was carrier out to design paddle wheel aerator with movable blade which will open when entering water and close when leaving water. Wheel closed at quadrant I to IV (entering water surface) and was about to open at quadrant III to II (leaving water surface). The blade was designed referring to commonly used Taiwan wheel model. The size of the blade was 15 cm of width, 20 cm of length, trapezoid-shape with 15° of bottom side and 30° of top side, had 40 holes with diameter of 1.6 cm. The component of mobable blade mechanism consisted of cam and shaft, rim, rim cap, blade holder, follower, spring and bearing. Follower was able rotate with angle of rotation was 125^0, rotational displacement was 50 mm, maximum velocity was 0.55 m/s and acceleration was 6.09 m/s^2. The largest angle of pressure occurred between cam and follower was 40.12^0 and the maximum torque required to rotate movable wheel was 80.09 Nm. The result shows that the testing without a load at 115 rpm, the torque that occured is 43.05 N·m and the electric power consumption is 511.72 Watt. However, the electrical consumption is still higher comparing with Taiwan paddlewheel aerator model due to the friction occured on the follower and cam.

1. Introduction

Aeration is a mechanism of adding some amount of oxygen into water to provide sufficient amount of oxygen. Aeration is carried out by increasing water and air contact using aerator device. One type of aerator device which widely used in pond farming is paddle wheel aerator [1]. Paddle wheel aerator is considered as the most appropriate aerator device due to aeration mechanism and wide usable driven power [2].

Some of parameters including water and air surface contact, differential oxygen concentration, film surface coefficient and turbulence influence aeration rate [3]. Aeration performance was influenced by geometry, size and wheel velocity [4]. Higher size tends to have higher aeration which simultaneously followed by higher driven power needs due to higher drag force. This condition creates certain problem in utilizing paddle wheel aerator as it may increase operational cost including electrical and fuel consumption.

[1] To whom any correspondence should be addressed.

Various models of paddle wheel aerator are offered in market. Aerator based on Taiwan design is widely used by consumers due to affordable price, light in weight and corrosion-resistant but has low efficiency as shown Figure 1 [1][5]. Aerator that was designed and fabricated by Taiwan has *SAE* (standard aeration efficiency) value of 1.063 kg O_2 kW h^{-1} [6]. Bhuyar *et al.* designed aerator with SAE value 2.269 kg O_2 kWh^{-1} [7]. The most appropriate paddle wheel aerator was designed by Moore and Boyd with *SAE* value 2.54 kg O_2 kWh^{-1}. Some of fabrications use aerator design with specification 2.25-7.5 kW and *SOTR* 17.4- 23.2 kg O_2 h^{-1} and average value of SAE was 2.2 kg O_2 kW h^{-1} [8].

Figure 1. Paddlwheel aeator based of Taiwan design

The development of movable blade is based on fact that power is required only when blade of paddle wheel aerator entering water and in contrary action of aeration effect only when the blade is about leaving the water. This study was carrier out to design paddle wheel aerator with movable blade which will open when entering water and close when leaving water, also the torque and power consumption assessment its done.

2. Materials and methods

2.1. Materials and functional design

The size of the blade was 15 cm of width, 20 cm of length, trapezoid-shape with 15° of bottom side and 30° of top side, had 40 holes with diameter of 1.6 cm. The wheel was designed to rotate clockwise with movable blade that enabled to open and close. The blade was about to close at quadrant I to IV (entering water surface) and open at quadrant III to II (leaving water surface). Blade opened to 45° from its close position which parallel to rim. Wheel dimension was designed similarly with commonly used wheel size i.e. 20 cm width, 30 cm rim diameter and 60 cm total dimension.

2.2. Measurment methods

2.2.1. Torque measurement. Torque measurement was done by using a strain gauge mounted on the wheel shaft. The sensor was connected to the strain amplifier (DAS-406B DC Strain Amp) through the slip ring and the bridge box were recorded by using data loggers (minilab 1008) and stored in the computer (Figure 2). Measurement data in the form of voltage (mVolt) was converted to strain (μst) and torque measurement values (N·m) with calibration values that have been done before.

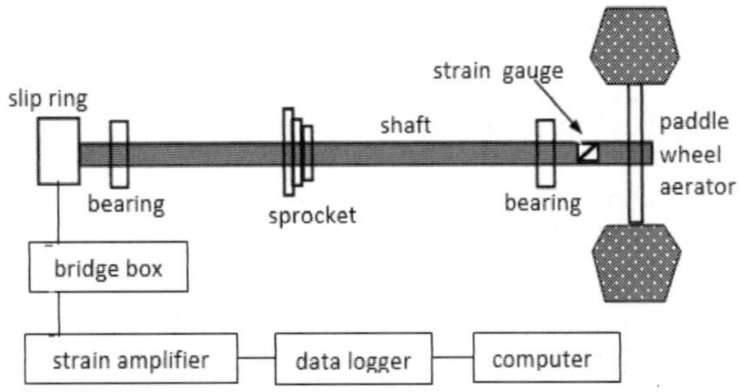

Figure 2. Torque measurement scheme

2.2.2. Power measurement. Paddlewheel power measurement was done by measuring the electrical power consumption of electric motors using Ammeter (DO2A) which was connected to an electrical outlet. Reading the power measurement value (Watt) was done by using a digital video camera recording on display Ammeter. Rated power was taken on the average value that often was showed from the reading video playback recording for each treatment testing.

3. Results and discussion

The blade was designed referring to commonly used Taiwan wheel model. The structural design based of functional design. The design paddle wheel aerator with movable blade which will open when entering water and close when leaving water. Wheel closed at quadrant I to IV (entering water surface) and was about to open at quadrant III to II (leaving water surface). The blade was designed referring to aerator based on Taiwan design.

3.1. Structural design

The wheel structure consisted of two main components i.e.stationary and rotary component as shown in Figure 3.

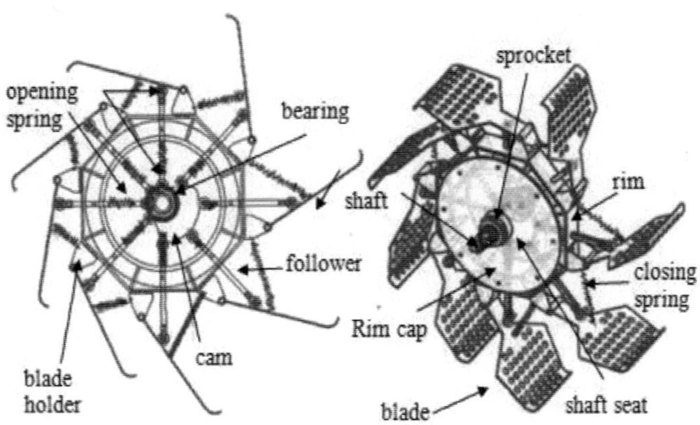

Figure 3. Wheel structure with movable blade

Stationary component consisted of cam and shaft. The longest and shortest radius of cam were 680 mm and 17.5 mm, respectively. Cam was mounted to shaft with diameter of 25 mm and attached to machine frame.

Rotary component consisted of the rim, rim cap, blade holder, follower, bearing and spring. The rim was octagonal-shape encircling tube with diameter of 218 mm and height of 30 mm. One side of the tube was enclosed with metal sheet, shaft seat and bearing with diameter of 25 mm. Outside the shaft seat,

IOP Conf. Series: Materials Science and Engineering **532** (2019) 012011 doi:10.1088/1757-899X/532/1/012011

sprocket that engage onto chain was attached for transmission purpose. The rim cap was a shaft seat made from metal sheet and similar bearing with rim tube which mounted to rim tube using bolt. Blade was used to directly bursting up water. Blades formed 30° of angle towards rim with radius of curvature was 40 cm. The size of the blade was 15 cm of width, 20 cm of length, trapezoid-shape with 15° of bottom side and 30° of top side, had 40 holes with diameter of 1.6 cm.
Blade holder was used to place blade with shaft of 8 mm and height of 25 mm and bolted at the end side of rim. The follower stem was used to push blade to open and close adjusting to cam profile. The follower stem was 150 mm of height and bearing with 19 mm of external diameter was attached on the two end-sides. Spring consisted of opening blade and closing blade. The opening spring was inserted to follower stem with diameter of the spring was 10.5 mm, length was 60 mm, wire diameter was 1 mm and spring constanta was 0.35 Nm. The closing spring of blade was attached on the front blade holder with diameter of 10 mm, length of 45 mm, wire diameter of 1 mm and spring constanta of 0.5 Nm.

3.2. Movable component mechanism

3.2.1. Motion mechanism. Movable blade were driven using cam mechanism. The cam is a simply mechanism that can provide almost all types of follower movement. The movement analysis of cam mechanism is shown in Figure 4.

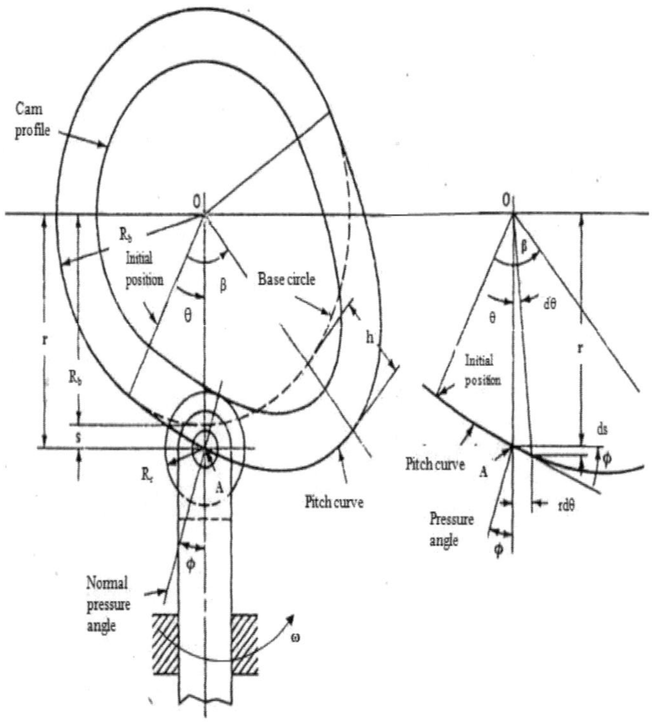

Figure 4. Profile analysis of cam-follower

Diagram of displacement, velocity and acceleration of cam is important factors in determining cam design [9]. Equation of cam displacement is written as follows:

for $\frac{\theta}{\beta} \leq 0.5$ $\qquad s = 2h\frac{\theta^2}{\beta^2}$

for $\frac{\theta}{\beta} \geq 0.5$ $s = h\left[1 - 2\left(1 - \frac{\theta}{\beta}\right)^2\right]$ (1)

Equation of cam velocity is written as follows:

for $\frac{\theta}{\beta} \leq 0.5$ $\frac{ds}{dt} = \frac{4h\omega\theta}{\beta^2}$

for $\frac{\theta}{\beta} \geq 0.5$ $\frac{ds}{dt} = \frac{4h\omega}{\beta}\left(1 - \frac{\theta}{\beta}\right)$ (2)

Equation of cam acceleration is written as follows:

for $\frac{\theta}{\beta} \leq 0.5$ $\frac{d^2s}{dt^2} = \frac{4h\omega^2}{\beta^2}$

for $\frac{\theta}{\beta} \geq 0.5$ $\frac{d^2s}{dt^2} = -\frac{4h\omega^2}{\beta^2}$ (3)

The maximum displacement of follower for one rotation 50 mm with angle of rotation 125° is shown in Figure 5.

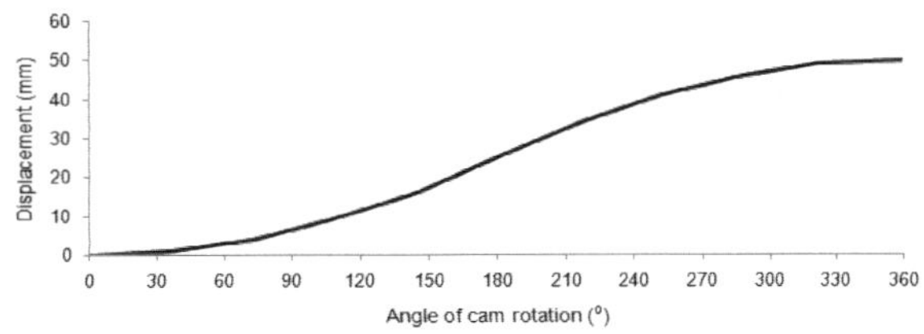

Figure 5. Displacement of follower

The maximum velocity of follower was 45.84 m/s as shown in Figure 6.

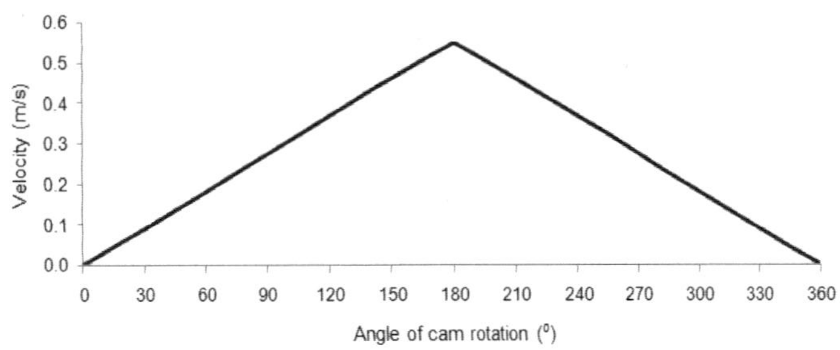

Figure 6. Velocity of follower

The constant acceleration was 6.09 m/s^2 as shown in Figure 7.

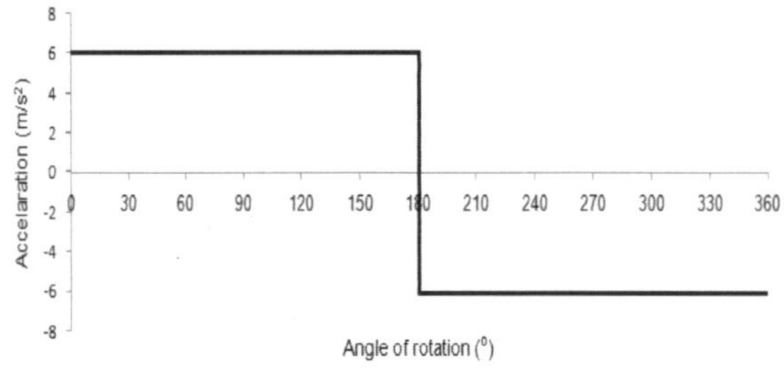

Figure 7. Acceleration of follower

3.2.2. Angle of pressure. Angle of pressure determines the smoothness of cam movement. The analysis of angle of pressure was illustrated in Figure 3. Angle of pressure (Ø) for every angular position was equated as follows:

$$r = R_b + s \qquad \tan\phi = \frac{ds}{r\,d\theta} \tag{4}$$

The magnitude of pressure angle for every angle of rotation is shown in Table 1.

Table 1. Pressure angle of follower

θ (deg)	r=R$_b$+s	tanθ=ds/r(dθ)	ds/dθ	Ø (deg)
0	27	0.00	0.00	0.00
36	28	9.17	0.33	18.13
72	31	18.33	0.59	30.60
108	36	27.50	0.76	37.38
144	43	36.67	0.85	40.46
180	52	45.84	0.88	41.40
180	52	45.84	0.88	41.40
216	61	36.67	0.60	31.01
252	68	27.50	0.40	22.02
288	73	18.33	0.25	14.10
324	76	9.17	0.12	6.88
360	77	0.00	0.00	0.00

The largest angle of pressure between cam and follower was 41.40°. This magnitude was too large and not necessary for cam-follower mechanism as it required high force and caused mechanism failure that led to machine damage.

3.2.3. Spring mechanism. Each blade had two types of spring i.e. blade-closing spring and blade-opening spring. Blade-closing spring (s1) worked against drag force (Fd) and gravifty of the blade (w), while blade-opening spring (s2) worked against force of blade-closing spring from cam pressure due to wheel rotation. Analysis of spring force is shown in Figure 8.

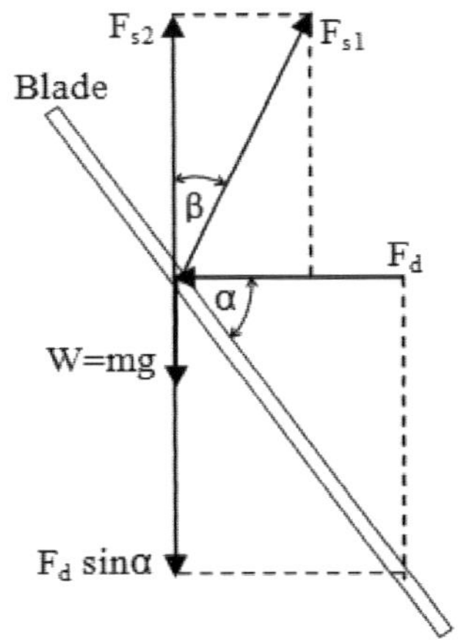

Figure 8. Spring force analysis

Based on the calculation, some of spring data were collected, including installed length, operating length, operating force, springs material, wire diameter, average diameter, inside diameter, outside diameter, free length, number of coils and allowable shear stress for blade-opening springs. The magnitude was 75 mm, 25 mm, 2.45 N, 49.05 N, chromium-vanadium A231, 2 mm, 2 mm, 10.5 mm, 14.5 mm, 80 mm, 12 coils and 922.74 MPa, respectively. The magnitude for blade-closing spring was 124 mm, 38 mm, 2.45 N, 264.50 N, chromium-vanadium A231, 2 mm, 2 mm, 8 mm, 125 mm, 130 mm, 20 coils and 815.75 MPa, respectively.

3.2.4. Innertia and torque. Innertia and torque analysis are shown in Figure 9. The influencing parameters consisted of force acting on follower (P), inertia force of follower (f), force of gravity on follower (W), shear stress acting on follower (F), normal force on rim towards follower (F_1,F_2), normal force of cam toward follower (N), follower overhang (a), distance between bearing surface (b), diameter of follower stem (d), pressure angle (Ø) and friction coefficient between follower (μ).

$$F = P + f + W + F_s \tag{5}$$

Total vertical force was:

$$N \cos\phi = F + \mu\left(F_1 + F_2\right)_s \tag{6}$$

Total horizontal force was:

$$F_1 = F_2 + N \sin\phi_s \tag{7}$$

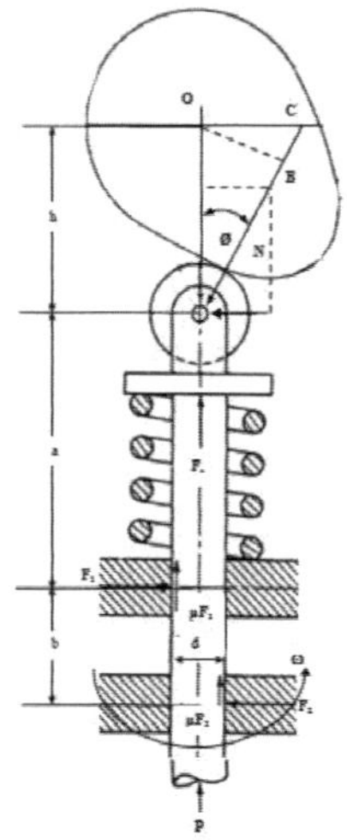

Figure 9. Free body diagram for analysis of angle of pressure

Summing moments to a point where F_1 works, gave:

$$F_2(b-\mu d)=N_a \sin\phi+\frac{d}{2}(F-N\cos\phi) \quad (8)$$

By neglecting F_1 and F_2 at the last 3 equations, normal force of cam acting on follower was:

$$N=\frac{F_b}{b\cos\phi-(2\mu a+b\mu-\mu^2 d)\sin\phi} \quad (9)$$

Torque required to rotate paddle wheel was calculated as follow:

$$T=N(OB) \quad (10)$$

The results of normal force, vertical force and horizontal force are shown in Table 2. The maximum torque required to activate blade mechanism was 80.09 N·m.

Table 2. Normal force cam-follower

θ (deg)	2πθ/β (deg)	Ø (deg)	N (N)
0	0	0	239.64
12.5	36	18.13	-380.96
25	72	30.60	-146.87
37.5	108	37.38	-120.09
50	144	40.46	-122.97
62.5	180	41.40	-144.87
62.5	180	41.40	-144.87
75	216	31.01	-324.70
87.5	252	22.02	-3141.54
100	288	14.10	612.74
112.5	324	6.88	328.35
125	360	0	239.64

3.3. Torque and power consumption

Testing without a load at 115 rpm showed the torque that occured 43.05 N·m and the electric power used 511.72 Watt. Torque occuring reduction compared with calculation results due to modification follower. Follower rod are to be bent to minimize the contact angle that happened. That torque still high and not proportional to torque reduction of simulation result effect of using moveable blade 31.51% (Bahri *et al.* 2015). Thus, the electric power consumption was still high compared with Taiwan model paddlewheel aerator was 451 Watt. However, the results showed that the friction force still high occured between the cam and the follower that are caused by a cam profile which must follow the functional design of the padllewheel aerator.

4. Conclusions

The blade of aerator is able to open when entering water and close when leaving water. Structure of the wheel consisted of two main components i.e. stationary and rotary component. Stationary component consisted of cam and shaft. Rotary component consisted of a rim, a rim cap, blade holders, followers, bearings and springs. The follower was able to rotate with angle of rotation was 125°, rotational displacement was 50 mm, maximum velocity was 0.55 m/s and acceleration was 6.09 m/s^2. The follower had constant acceleration. Testing without a load at 115 rpm shows the torque that occured 43.05 N·m and the electric power used 511.72 Watt. These test results showed the friction force still high occured between the cam and the follower that are caused by a cam profile which must follow the functional design of the padllewheel aerator. It is suggested that to overcome the friction by increasing the dimension of mechanism contruction and reduce the angle contact between cam and follower.

References

[1] Laksitanonta S, Singh S and Singh GA 2003 A review of aerators and aeration practices in Thai Aquaculture. *Agricultural Machanization in Asia, Africa and Latin America* **34** 64-71.
[2] Romaire RP and Merry GE 2007 Effect of paddlewheel aeration on water quality in crawfish pond *Applied Aquaculture* **19** 61-75.
[3] Boyd CE 1988 Pond water aeration systems. *Aquaculture Engineering* **18** 9-40.
[4] Moulick S, Mal BC and Bandyopadhyay 2002 Prediction of aeration performance of paddlewheel aerators *Aquaculture Engineering* **25** 217-237.
[5] Wyban JA, Pruder GD and Leber KM 1989 Paddle wheel effect on shrimp growth, production and crop value in commercial earthen ponds. *Journal of the World Aquaculture Society* **20** 18-23.
[6] Peterson EL and Walker MB 2002 Effect of speed on Taiwanese paddelwheel aeration. *Aquaculture Engineering* **26** 129-147.
[7] Bhuyar LB, Thakre SB and Ingole NW 2009 Design characteristics of curved blade aerator w.r.t. aeration efficiency and overall oxygen transfer coefficient and comparison with CFD modeling. *International Journal of Engineering,Science and Technology* **1** 1-15.
[8] Moore JM and Boyd CE 1992 Design of small paddle wheel aerators *Aquaculture Engineering* **11** 55-69.
[9] Martin GH 1982 *Kinematics and dynamics of machines* USA:McGraw-Hill,Ltd. 194-382.

IOP Conf. Series: Materials Science and Engineering **532** (2019) 012012 doi:10.1088/1757-899X/532/1/012012

The effect of time and number of balls on shaker milling process in hydroxyapatite powder synthesis

A Fadli, S R Yenti, R Rasyidin, M Sari

Department of Chemical Engineering, Engineering Faculty, University of Riau Kampus Bina Widya HR Subrantas Km 12,5, Panam, Riau Indonesia

fadliunri@yahoo.com

Abstract. The synthesis of hydroxyapatite (HA) from the shells of blood clams has been successfully performed by the low temperature hydrothermal method. To obtain uniform and nano-HA powder, size of obtained HA was reduced using shaker milling process. The purpose of this research was to determine the effect of time and the number of balls on the shaker mill process on hydroxyapatite particle size. The shells of blood clams were calcined at 1000 °C for 24 hours to obtain CaO powder. Then 55.63 gram $(NH_4)_2HPO_4$ dissolved in 600 mL of distilled water then added 45,91 gram of CaO at 90 °C to form slurry. The slurry was dried at 120 °C for 15 hours, and milled using shaker milling process with variation of time 1, 2 and 4 hours and variation of number 10, 20, and 30 balls. The peak intensities show the high percentage of HA with Tricalcium phosphate (TCP) presence in X-ray diffraction (XRD) analysis. Diameter of crystal produced is 67 nm, with time variation for 2 hours and number of 20 balls. Breuner-Emmet-Teller (BET) analysis result, with time ball milling for 1 hour and the number of 10 balls surface area of HA is 36,974 m^2/g. The longer of milling process, the temperature will also increase causing clumps of particles resulting in an increase in particle size. The more the ball is used then the surface area of obtained hydroxyapatite will be smaller.

1. Introduction

Heavy metals generally refer to its density greater than 5 g/cm^3 with biological toxicity. Such as silver (Ag), copper (Cu), lead (Pb), zinc (Zn), nickel (Ni), tin (Sn), cadmium (Cd), chromium (Cr), mercury (Hg) and metalloid such as arsenic (As). Without appropriate treatment, industrial wastewater, urban sewage and agricultural production wastewater bring the most heavy metal pollution into the natural water bodies and caused one of the most serious globe environmental problems [1]. The heavy metal can be accumulated into the food chain and cause serious problems, not only for the ecosystems but also for human health [2]

Several methods have been used in the removal of heavy metal involve physical (filtration, coagulation, adsorption), chemical (ion exchange, adsorption) and biological processes. Adsorption is the technology most commonly used to reduce heavy metals from waters. This method is often used because the price is relatively cheap, environmentally friendly and simple to do. The most important part in an adsorption process is the type of adsorbent that will used [3]. HA can act as an ion removal material due to its excellent reactivity and low water solubility [4]. HA can be synthesized through various methods, including hydrothermal [5], precipitation [6], sol-gel [7] and mechanochemical methods [8].

Hydroxyapatite ($Ca_{10}(PO_4)_6(OH)_2$), a major inorganic constituent of vertebrate hard tissues such as bone and tooth, due to its excellent bioactivity, biocompatibility and osteoconductivity. HA also considered as one of the most promising mineral phases for remediation of polluted waters and soils, sorption of organic molecules and heavy metal ion removal. HA has strong affinity to their surface for cationic species [9].

In the application of HA as an adsorbent, the morphology and particle size become important parameters. Micro-HA has small surface area and has a strong crystal bond so it is good to be used as an adsorbent. To get small HA particle size, several studies have been conducted using ball milling technique. Wu et al. [10] used eggshell as the source of Ca, was combined with Dicalcium phosphate dehydrate (DCPD) powder. The powder mixture was ball-milled for 1, 5 and 10 h followed by heat treatment at 1000°C for 1 h. The formation of HA phase can be initiated by sintering the 1 h milled at 1000°C for 1 h, while pure HA phase can be obtained upon sintering the 10 h milled sample. Hamidi et al. [11] used eggshell as raw material to synthesize eggshell-derived HA (EHA) via mechanochemical method. HA powder was successfully synthesized with crystallite and particle sizes in the range of 8-47 nm and 250-550 nm respectively. Gergely et al. [12] synthesized HA used seashell and eggshell, the crushed and milled in ball mill or an attritor mill. The characteristic morphology is preserved even after firing at high temperature (900°C). After heat treatment HA, calcium oxide and calcium phosphate phases appeared. In this research, ball milling technique was used to obtain smaller HA size, with variations in the number of ball and grinding time.

2. Materials and Method

2.1. Materials

The clam blood shells were collected from the local market at Pekanbaru, Riau Indonesia. Ammonium dihydrogen phosphate as the source of phosphate and distilled water was used.

2.2. Methodology

The clam shells were cleaned with water, dried and crushed into powder. The powders were then calcined in a furnace at 1000°C for 24 h. The calcination aimed to decompose calcium carbonate ($CaCO_3$) into calcium oxide (CaO) which will be used as Ca precursors. The results were then tested using Atomic Absorption Spectrometer (AAS) to measure calcium content in a sample. The calcium data from the AAS test were used for stoichiometric calculations to find the mass of CaO will be used in the HA synthesis process. This process performed is carried out using a low temperature hydrothermal method which refers to the research of Alqap and Sopyan [13].

Calcinated shells 45.91 gr were mixed with 55.63 gr of ammonium dihydrogen phosphate and distilled water 600 ml. The solution was heated at 90°C and stirring at 300 rpm for 5 h. The slurry were then dried at 120°C for 15 h. The dried samples were then milled to get a smaller size using a shaker ball mill with variation of the number of grinding balls 10, 20 and 30 balls and grinding times of 1, 2 and 4 h. Samples then calcined at 900°C for 2 h at a rate 10°C/min. CaO levels were analyzed using AAS. The crystalline phase of clam blood shells before and after the calcination were analyzed using powder X-ray diffraction (XRD). The microstructure and morphology of the powders were observed using scanning electron microscope (SEM). Particle Size Analyzer (PSA) and Brunauer-Emmet-Teller (BET) were used to determine the distribution of sample particle size and the sample surface area.

3. Results and Discussion

HA was synthesized from blood clam shell as source of calcium. Calcium derived from Calcium oxide (CaO) produced by calcination of clam shells at 1000°C for 24 h. The percentage of CaO levels obtained was 76.2 based on dry weight. The XRD analysis showed that the clam shells before the calcination were $CaCO_3$ phases, which had the highest peak and the calcined process turned clam shells into CaO phases. Calcium oxide can be obtained as follows:

$$CaCO_3 \rightarrow CaO + CO2 \quad (1)$$

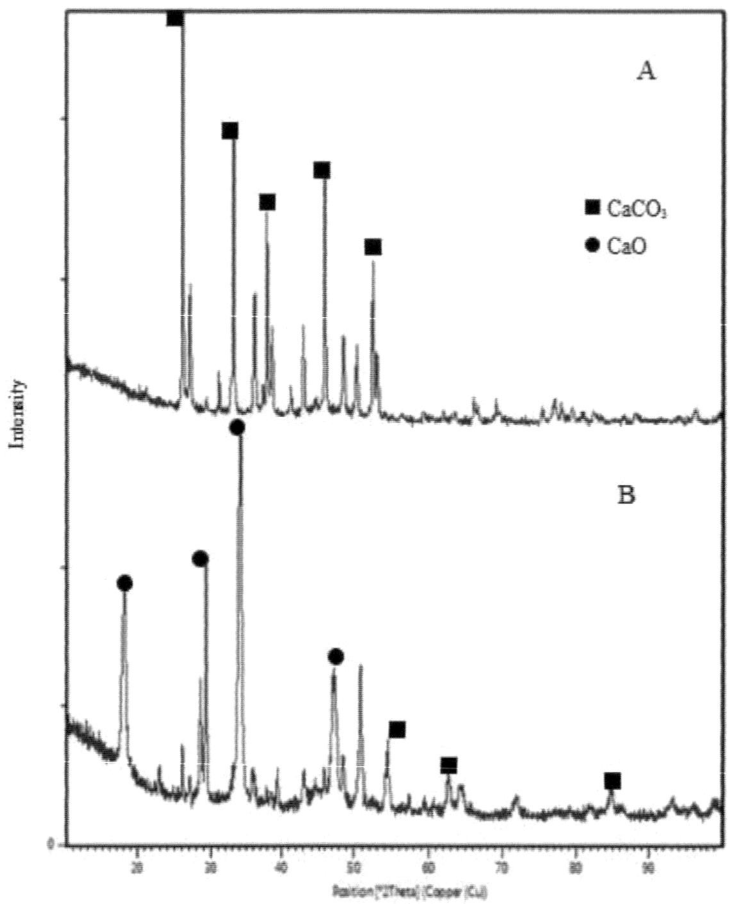

Figure 1. X-ray diffraction of clam blood shells (a) before calcination (b) after calcined at 1000°C

The XRD analysis showed that the clam shells before calcination had the highest peak of $CaCO_3$. The calcined shells turned into phases of CaO which was the highest peak at angle 2θ 35°. From the graph of XRD analysis for the sample after calcination showed crystal peaks or head of CaO as shown in Figure 1.

Based on the XRD analysis results of ball milling samples, HA phase were formed with hexagonal crystalline structure, besides that a Tricalcium phosphate (TCP) phase was formed with rhombohedral crystal form as shown in Figure 2. The HA peaks for the sample (ball milling 4 h 30 grind balls) at 2θ i.e 25.877°, 32, 162°, 43.845°, 53.205° and 61.6653°. The XRD patterns showed the similarity with the JCPDS file no. 09-0432.

Ball milling samples were calcined at 900°C, there was no CaO phase at this calcination temperature. In the sample, the composition contained was HA and TCP. CaO which was still present in the sample cause the formation of TCP [13].

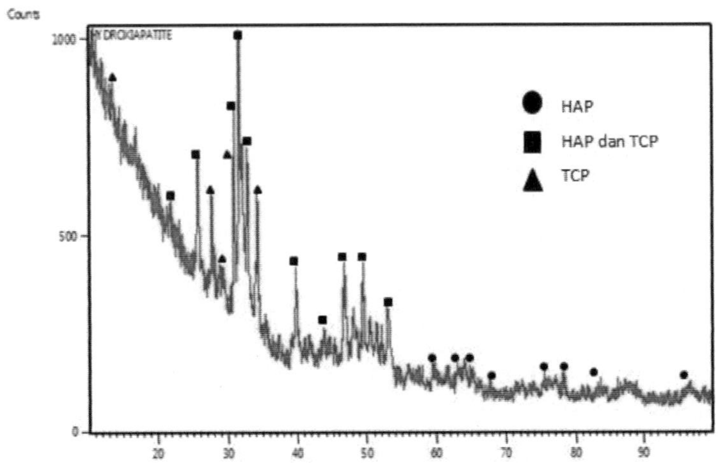

Figure 2. XRD analysis of HA powder at 900°C calcination temperature

The sample also characterized by SEM analysis, which is used to determine the morphology of the sample. The SEM images showed that the powder is smooth and uniform, there is no agglomeration and form the same pore.

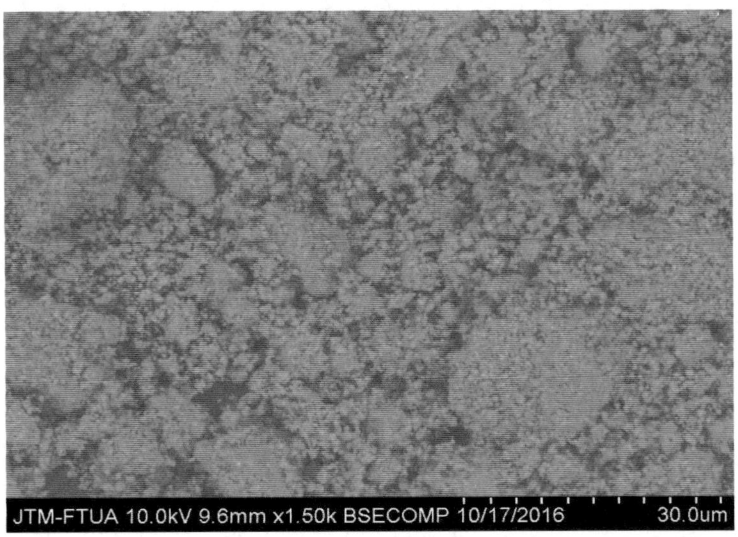

Figure 3. SEM analysis of HA powder after calcination at 900°C

3.1. Effect of Ball Milling Time

During the ball milling process, HA particles would be repeatedly milled. When the balls were milled, the powder will be trapped between them. The load given by the milled balls to the powder will make the plastic deformed so that it undergoes hardening and finally grinded.

IOP Conf. Series: Materials Science and Engineering **532** (2019) 012012 doi:10.1088/1757-899X/532/1/012012

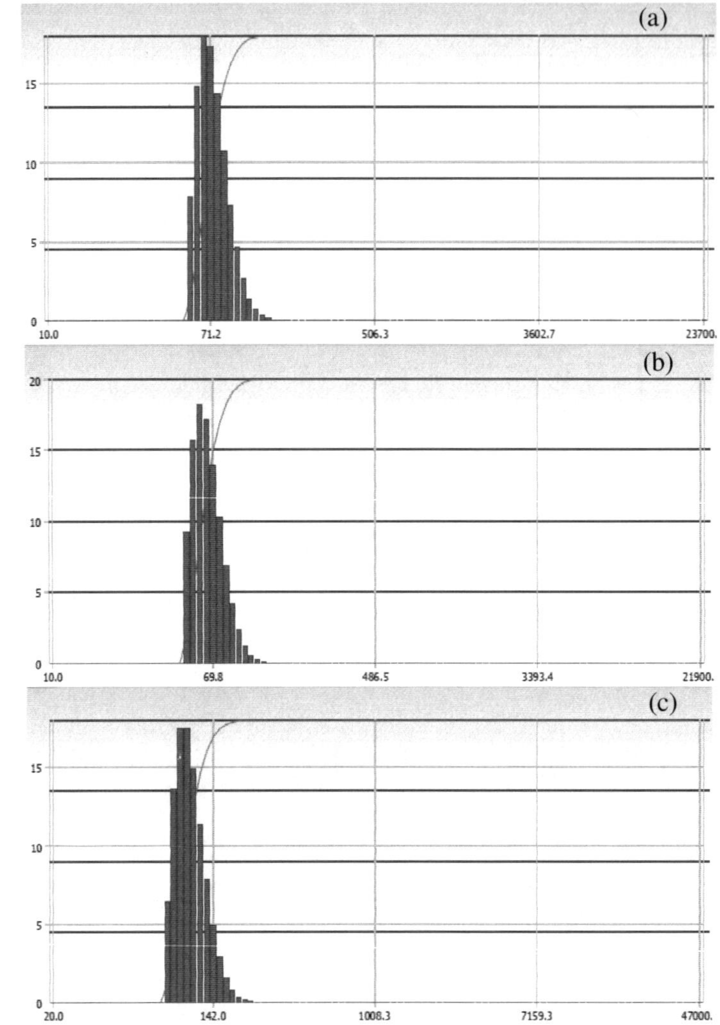

Figure 4. PSA graph particle size distribution of HA using 20 balls with different time (a) 1h (b) 2 h and (c) 4 h

From the results obtained by using the same number of grinding balls at 1, 2 and 4 h it can be seen that 2 h of grinding time, the size of HA particles has the smallest size. HA particles size using 20 ball mill at 1 hour was 74.3 nm, while at 2 h the size of HA particles obtained was 67 nm and at 4 h HA particle size was 110.3 nm. The increasing of milling time also increases the milling temperature so the contaminant can occur in HA powder. The crystalline size of HA from clam blood shells was calculated using the Scherrer's equation and shown in Table 1.

Table 1. HA size based on PSA analysis

Time (h)	Number of balls		
	10	20	30
1	187 nm	74.3 nm	117.4 nm
2	166 nm	67 nm	83.4 nm
4	438.9 nm	110.3 nm	217.1 nm

IOP Conf. Series: Materials Science and Engineering **532** (2019) 012012 doi:10.1088/1757-899X/532/1/012012

After 2 h, the enlargement of particle size occurs. This occur as a result of the cold-welding rate was higher compared to rate of grinding by ball mill. After 2 h, the cold-welding rate and the rate of grinding of the powder has reached the equilibrium limit. To overcome this, a process control agent can be added to the grinding process and the powder will be covered and reduced the influence the rate of cold welding which results of powder particles agglomeration [14].

3.2. Effect of number of ball mill

The effect of the number of mill balls on the particle size obtained is seen at each milling time. At each milling time, the smallest particle size was obtained when using the number of 20 mill balls. Particle size reduction using ball milling will be maximal when the ball milling is filled with ball and the sample to be milled is not more than 50% of the volume of the shaker ball mill.

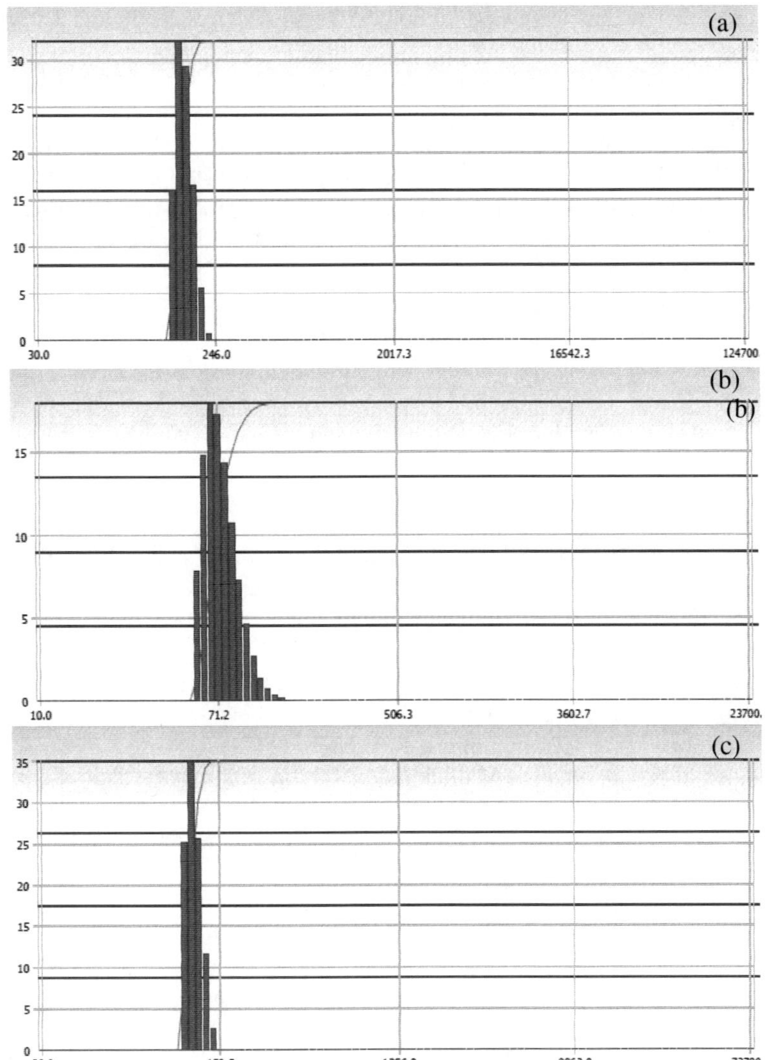

Figure 5. PSA graph of HA at 1 h milling time (a) 10 mill balls (b) 20 mill balls (c) 30 mill balls

Table 2. Effect of time and number of mill balls on HA surface area

Time (h)	Number of balls	*Surface* Area (m^2/g)
1	10	36,974
1	30	1,359
4	30	3,071

BET analysis was also used in this study. This analysis was conducted to determine the sample surface area. Based on table, with the same ball milling time of 1 h using 10 and 30 grinding ball obtained smaller surface with the increasing number of balls, which were obtained by milling of HA powders with 10 and 30 balls were 36.974 and 1.359 m^2/g. Whereas ball milling using the same number of grinding balls as 30 balls at 1 and 4 h of grinding time, surface are increase with increasing time. The surface area of HA powders using 30 balls at 1 and 4 h were 1.359 and 3.071 m^2/g.

4. Conclusion

Ball milling is simple and efficient method to produce micro or nano powder particles. The time and number of milling balls that produce the smallest HA particle size is 2 h and 20 mill ball with HA particles size used 20 ball mill at 1, 2 and 4 h were 74.3, 67 and 110.3 nm. The increasing of milling time and number of milling ball, the reduction of particle size of HA also increasing. The surface area HA will increase with the smaller size of HA. Milling process of HA powders with 10 and 30 balls, the surface area were 36.974 and 1.359 m^2/g. The ball milling used 30 balls with different time 1 and 4 h, the surface area were 1.359 and 3.071 m^2/g. However, the use of time and number of balls in ball milling process must be adjusted to the efficiency of the ball mill used.

Acknowledgements

The authors are grateful for the financial supports from Riau University for the financing of this research through 2018 Competitive Grant scheme.

References

[1] Hao C, Anbin X and Shaohong Y 2018, IOP Conf. Series: Mater. Sci. Eng **301** 012160
[2] Abbasi Z and Aghababaei M 2014. Universal Journal of Engineering Science **2** 124
[3] Mourabet M, El Rhilassi A, El Boujaady H, Bennani-Ziatni M, El Hamri R, Taitai A 2012 Applied Surface Science. **10** 4402
[4] Avram A, Frentiu T, Horovitz O, Mocanu A, Goga F, Tomoaia-Cotisel M 2017. Studia Universitatis Babes-Bolyai, Chemia. **1**62.
[5] Sopyan I and Kaur J 2009. Ceramic International **35** 3161
[6] Cox SC, Jamshidi P, Grover LM, Mallick KK 2014. Journal of Materials Science: Materials in Medicine **25** 37
[7] Bezzi G, Celotti G, Landi E, and La TMG 2002. Materials Chemistry and Physic **78** 816
[8] Rhee, S 2002. Biomaterials **23** 1147
[9] Yang H, Masse S, Rouelle M, Aubry E, Li Y., Roux C, Journaux Y, Li L. and Coradin T 2015. International journal of environmental science and technology, **4**1173
[10] Wu SC, Hsu SH, Hsu SK, Chang YC, and Ho WF 2015. Journal of Asian Ceramic Societies
[11] Hamidi AA, Salimi MN, and Yusoff AHM 2017. Advanced Materials Engineering and Technology V
[12] Gergely G, Weber F, Lukacs I, Illes L, Toth AL, Horvath ZE, Mihaly J, and Balazsi C 2010. Central European Journal of Chemistry **8** 375
[13] Alqap SF, and Sopyan I 2009. Indian Journal of Chemistry **48** 1592
[14] Suryanarayana, C 2001. Progress in Materials Science **46** 1

Extraction of ElectroEncephaloGraph (EEG) Signal Using the Subband Coefficient of Wavelet Transform on Cursor Moves

H Hindarto[1], A Muntasa[2]

[1]Departement of Informatics Engineering,Universitas Muhammadiyah Sidoarjo - East Java - Indonesia
[2]Departement of Informatics Engineering UTM - East Java - Indonesia

Email : hindarto@umsida.ac.id

Abstract. This research explains the application of ElectroEncephaloGraph (EEG) signal waves used to move the up cursor and down cursor. In each sub band of the waveform, Electro Encephalo Graph (EEG) will produce the average and standard deviation to be used as a feature of the EEG. Artificial Neural Network Backpropagation as the basis for determining whether the cursor moves up or the cursor moves down. The data used in this study is EEG data derived from BCI Competition 2003 (BCI Competition 2003). Decision-making is done in two stages. In the first stage, the mean and standard deviation values on each wavelet subband as a feature extraction of EEG data. This feature is an input to the Backpropagation Neural Network. In the second stage of the process of inserting into two classes (class 0 and class 1) EEG data files, there are 260 EEG and 293 file training data files from EEG file test data files, totaling to 553 EEG data files. The results obtained for the classification of EEG results were 79.2%.

Keywords: EEG, Mean, Device Standart, Wavelet, BackPropagation

1. Introduction

To move a cursor on the computer screen, usually someone needs a keyboard or mouse to run it. This is not possible with someone who does not have a hand or someone who cannot move his hand. Initially it may be just wishful thinking but the creative and revolutionary ideas of researchers both from home and abroad to be able to move the cursor without using the hands.

The human brain without opening the skull when recorded can generate a weak electrical current, the recording format is named brain Electroencephalography (EEG). This has been announced by a psychiatrist named Germany Hans Berger there year 1929. So to connect between the brain and the object that will be controlled by the thought digunkannya tool called Brain Computer Interface (BCI). BCI is a system that acquires and analyzes neural signals with the aim of creating a direct communication channel between the brain (EEG signal) and computer. BCI is also a communication system that does not require muscle activity [1]. The BCI system allows subjects to send commands to electronic equipment simply by using brain activity [2]. The BCI system can also be used to play simple games on mobile devices [3].

Brain Computer Interface (BCI) is a communication system that translates the direct action of the user's brain activity into EEG signals and control commands. BCI can be used for spelling, browsing the Internet, controlling robot devices, or performing other tasks with just thoughts [2-4]. Existing BCI is often used to present information in the signal to assess the state of the subject's brain with different EEG signal categories.

Accurate features and identification methods that are user-specific and specific application requirements may be a problem for EEG-based communication to be efficient. Appropriate input design, feature extraction, and matching classifier. Different attributes of EEG signals have been used as inputs for BCI, such as your rhythms (8-12 Hz) and beta rhythms (18-25 Hz), event-related potentials, such as, for example, established P300-evoked response, or slow cortical potential (SCP) [2-4].

The EEG signals in a person, generally composed of wave components distinguished by their frequency region, ie alpha waves (8 - 13 Hz), often appear in the conscious state, closed eyes and relaxed conditions; beta waves (14 - 30 Hz), often appear when a person is in a state of thinking; teta waves (4 - 7 Hz), commonly occurring in someone who is light sleep, drowsiness or emotional stress; delta waves (0.5 - 3 Hz), often present in someone who is in a deep sleep state. Therefore, the representation of EEG signals into many frequency domains is done in research relating to the analysis of EEG signals.

Representations in the frequency domain include for the identification of waves on EEG signals using Fourier Transform and Neural Network methods to distinguish normal people and those of epilepsy [5]. Other research by taking data from BCI Competition 2003 among others the research used two channel (channel 4 and channel 6),four features by combining slow cortical potentials (SCPs) and wavelet packet transform and Neural Networks for the classification process. The result of the classification process is 91.47% [6].

EEG feature extraction based on wavelet packet decomposition for brain computer interface. This study also uses BCI 2003 data competition using six channels (channel 1, channel 2, channel 3, channel 4, channel 5 and channel 6) and takes 17 (seventeen) features with neural network as a classification process. The result of the classification process is 90.80% [7]. In this study, EEG signal processing to perform to move the cursor up or down on the computer screen while his recorded using SCP method of Wavelet transform. To detect the movement of the cursor this research method using Neural Network by taking a set of statistical features of Wavelet subband Transform is used as input.

2. Materials and Methods

2.1. Material

The EEG signal dataset taken from the BCI 2003 competition data comes from Dr. Birbaumer and his team at the University of Tuebingen, Germany (Blankertz 2004) [13]. Six EEG channels were recorded from a healthy subject and the sampling rate of 256 Hz and a 3.5 second recording time. The result of each experiment of each channel is 896 samples. Subjects were asked to imagine moving the cursor up or down on the computer screen when the SCP was recorded. Subjects receive visual feedback from SCPs (feedback phases). The dataset is divided into training (268 experiments) and trials (293 experiments), according to the BCI 2003 description [8].

2.2. Discrete Wavelet Transformation

In pattern recognition, feature extraction is one specific form to reduce dimensions. Representation of the features needed to transform the input data that is too large to be processed and allegedly very many redundant input data, but does not contain the needed information. Extraction Feature is a way to transform the input data to the feature set. Extraction feature must be selected appropriately and carefully so that the data that diekstrasi can be used as expected. Ekatrasi features are used to extract the relevant information from the input data correspond to the desired task and can reduce the representation, not the input data with full size. Extraction of feature against signal and the results that have been achieved by using the wavelet transform can be found in the following Study [9-10].

The frequency domain of the EEG signal provides useful information rather than time domain representation. The wavelet transform provides a multi-resolution description of the nonstationary signals. EEG is a non-stationary signal then wavelet is suitable for EEG signal [11]. At high frequencies will represent a good time resolution and for low frequencies will also represent better frequency resolution. the multi-scale feature of Wavelet allows the decomposition of a signal into a number of scales, each scale representing a particular roughness of the signal under study. The multi-axis decomposition procedure of the x [n] signal is schematically shown in Figure1.

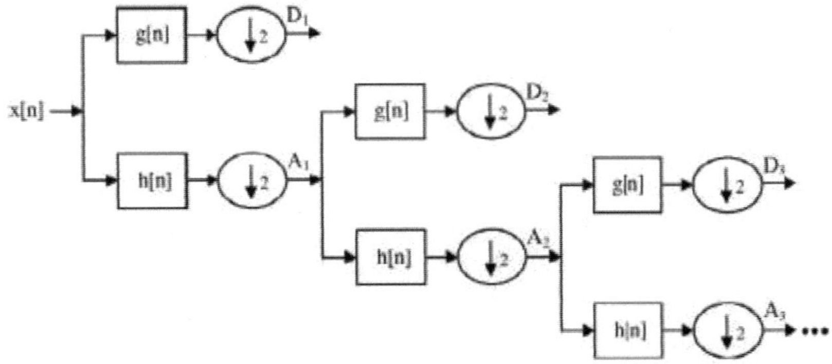

Figure 1. Sub band Decomposition of discrete wavelet transform

Changes in the EEG signal is detected using daubechies wavelet sequence 2 daubechies wavelet, because the order of 2 to feature interiors more appropriate. EEG signal is decomposed into details of D5 to be used in this research.

A wavelet coefficient in the extraction of compact representations that show the distribution of signal energy of EEG in time and frequency bands shown with a good representation on wavelet coefficients that have been in the extraction. Therefore, the feature vectors are represented can be obtained from EEG signal detail that calculated and estimated coefficients of wavelet of EEG signals. Retrieving the value of the statistics of wavelet coefficients can be used to reduce the dimensions of the vector features. The following statistics feature is used to represent the time frequency distribution of EEG signals, i.e. the value of the average and standard deviation of wavelet coefficient on each sub band

2.3 Backpropagation Neural Network

Backpropagation Neural Network training algorithm is social interactions which have many layers. Backpropagation Neural Network using the error output to change the values of the weights does it weigh-in the direction of retrograde (backward). To get this error, the propagation of phase forward (forward propagation) should be done in advance. The terms of the activation function in the Backpropagation Neural Network is a continuous, terdifferensial easily, and is a function that does not go down.

The activation function which can meet the third condition is logsig, tansig, and purelin. The method of recognition is the process of initializing the data to be further processed by Backpropagation Neural Network. The data to be identified are presented in the form of a vector. Each target has data presented also in the form of a vector. Target or output reference is a character map that shows the location of the input vector. While the method of training is the process of storing data and recognize the exercise of knowledge or information obtained into the weighting-weighting. There are 3 phases in the Backpropagation Neural Network training, i.e., phase forward (feed forward), a phase of retreat (back propagation), and a modified phase weighting. In phase feed forward, advanced input pattern is calculated starting from the input layer to output layer. In the phase of back propagation, each output unit receives the target pattern is related to the input pattern to calculated the value of the error.

The error will dipropagasikan withdraw. While the phase weighting modification aimed at lowering the error that occurred. The third phase is repeated continuously until a termination condition is met [12]. At the end of the 80-90s of the 20th century introduced the rule of generalization-delta algorithm, as highlighted by other [13]. Generally, backpropagation algorithms of functions as follows. On the output layer, vector output compared with the desired output. The error is calculated from the rule of the delta and redistributed back through the network to adjust the weights in the interest of minimizing the difference between the output and the desired output is NN. Such a network can be learned by using the activation function is differentiable functions.

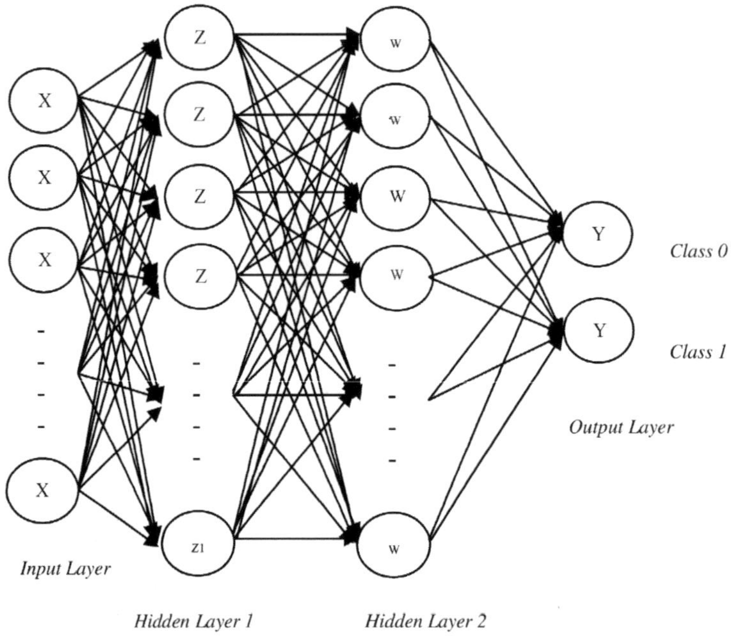

Figure 2. Backpropagation Neural Network Architecture with 2 hidden layers

In this study, the data classification process is done by separating the EEG signals into two parts, ie data for the training process as much as 268 vector data and data for the data testing process used as many as 293 data. This network has an input of 8 nodes (x 1, x2, ..., x8) derived from DWT fiture, 15 nodes for hidden layer 1 (z1, z2, ... z15), 30 nodes for hidden layer 2 (w1, w2, ... w30), and Binary type output for condition identification (y1, y2). Network architecture in research can be seen in Figure 2. Output pattern with 2 target output in binary form. These types of patterns can be seen in Table 1.

Table 1. Output Vector Patterns

No	Data Classification	Output Patterns
1.	Up Cursor Movement	0
2.	Down Cursor Movement	1

3. Results and Discussion

The data used is using data from BCI comptetion 2003 Data set Ia. This data set consists of 6 channels (electrode mounted on the scalp of 6 pieces of electrode sensor, resulting in 6 pieces of signal EEG channel). Data set It consists of Trainning data and Testing data.

A large amount of data, will cause the old computing process caused by a lot of data is processed, so that the presence of features that result in a little quick computing process. In this study, an EEG signal is only taken by the mean and standard deviation of each subband of the DWT process to be a feature extraction for the identification process.

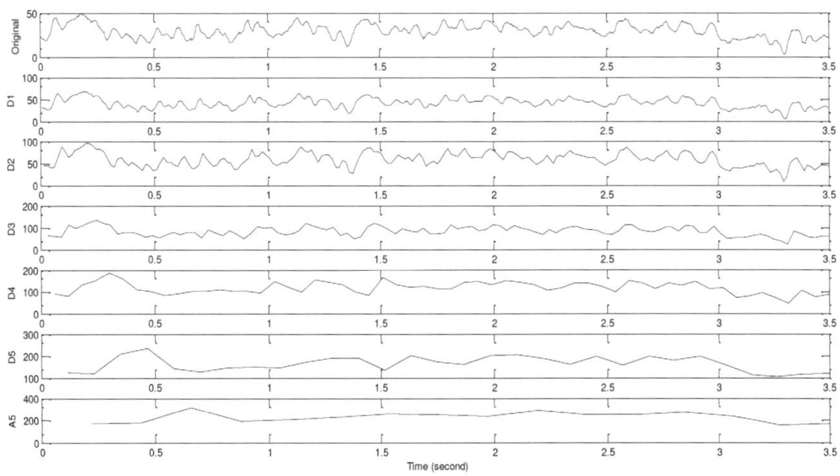

Figure 3. Approximate and detailed coefficients of EEG signal taken from a healthy subject

The EEG record is divided into sub-band frequencies such as the wavelet coefficients A5, D5, D4, and D3 using DWT shown in Figure 3. Then a feature set is extracted from the wavelet sub-band frequency (0-4 Hz), (4-8 Hz), (8-16 Hz) and (16-32 Hz). After normalization, the EEG signal is decomposed using DWT and features extracted from the sub-band.

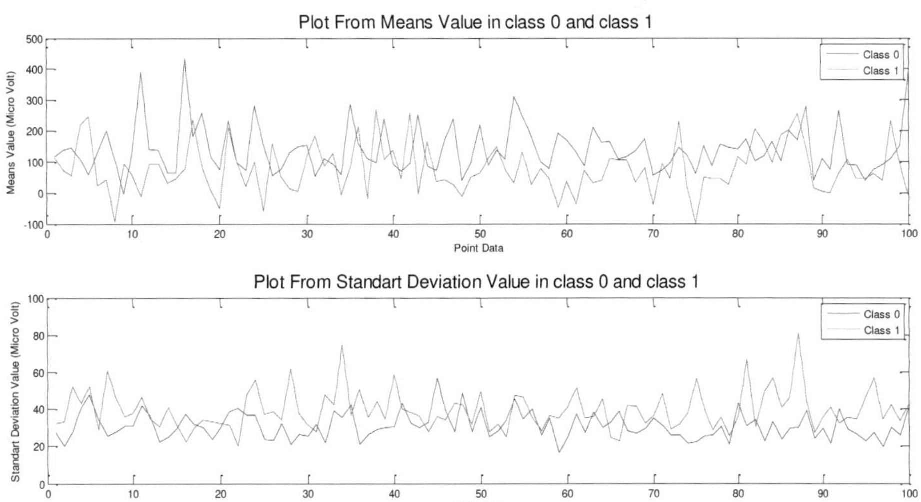

Figure 4. Mean and standard deviation of the wavelet coefficients

Figure 4 shows that for mean and standard deviation on each subband for class 0 and class 1 has a difference value. Different scores indicate that the classification rate by taking the mean and standard deviation is quite good.

Classification using Artificial Neural Network Backpropagation is implemented using the mean value feature and standard deviation of the DWT process as input. In this study, 260 data samples (from normal subjects) for channel 1 were used for training and 293 samples of data (from normal subjects) for each channel were used for testing. The distribution of sample classes in the training and validation datasheets is summarized in Table 2. To improve the capability of backpropagation, the training and test are shaped by data obtained from different subjects. The training data sets are used to train the backpropagatioan, while the test data set is used to verify the accuracy and effectiveness of trained Backpropagation to detect the up cursor and down cursor.

Tabel 2. Class distribution of the samples in the training and test data sets

Class	Training set	Test set
Up Cursor (class 0)	130 x 6 Channel	293 x 6 channel (mix)
Down Cursor (class 1)	130 x 6 Channel	

The result of feature extraction is used for input for neural network, this research uses backpropagation method (8-15-30-2) which is 8 inputs derived from characteristic of EEG signal and 2 hidden layer that is 15 units at hidden layer 1 and 30 units at hidden layer 2 as well as 2 targets (upward cursor movement for class 0 and downward curot movement for class 1). In addition to using the 8-15-30-2 network architecture, for this experimental study also uses additional and reductions for hidden layer.

In the process of identification with the neural network process first done is a training process that is searching the best weight value with the acquisition of the smallest error value of the desired output target. In the process of mapping is done identification of EEG signal from movement of up cursor and downward cursor movement based on weight value which have been got in training process.

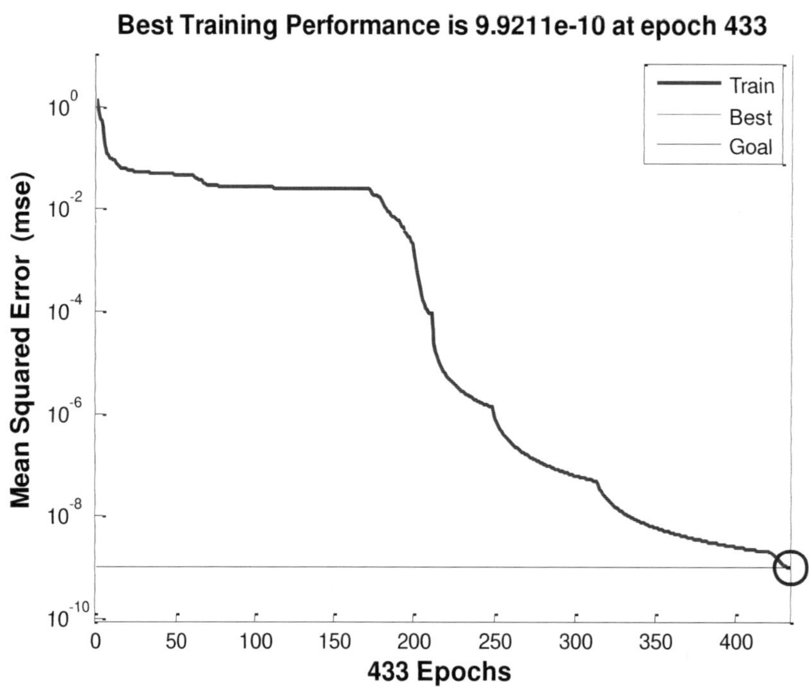

Figure 5. Training Performance of artificial neural networks using 2 hidden layers

260 training data from channel 1 in 433 training period and step size for adaptation parameter has initial value of 9,92.10-10. Performence Backpropagation using 2 hidden layers is able to perform the training process by passing the minimum error limit, so that 100% has the accuracy of the training process.

Table 3. Backpropagation accuracy results with 3 hidden layers for all channels

	Channel 1	Channel 2	Channel 3	Channel 4	Channel 5	Channel 6
Accuracy	61,0 %	64,2 %	79,2 %	72,2 %	77,2 %	74,0 %

Table 3 shows that channel 3 occupies a good level of accuracy compared to other channels with an accuracy of 79.2%.

Table 4. Neural network performance against Hidden Layer numbers different

	MSE (1 Hidden Layer)	**MSE (2 Hidden Layer)**	**MSE (3 Hidden Layer)**
Time	35 second	59 second	152 second
Iteration	1000	433	831
MSE	$2{,}22.10^{-2}$	$9{,}92.10^{-10}$	$9{,}25.10^{-8}$
accuracy	73,0 %	79,2 %	75,4 %

From table 4 it can be seen that by using 2 hidden layers in backpropagation it can achieve the accuracy value of 79.2% of the testing process

4. Conclusions

In this paper, researchers introduced the Discrete Wavelet to extract the feature by taking the mean and standard deviation values on each subband. The process of classifying EEG signals is divided into two classes: class 0 and class 1. This study uses 553 EEG signal data files for training and testing. Backpropagation classification accuracy reached 79.2% for test data. Future research work, will examine the search techniques suitable for feature extraction and EEG signal classification, so the accuracy level for the command move the cursor would be better. The results obtained will be compared with the methods already studied.

Acknowledgement

The authors are grateful to the Chairman of Muhammadiyah University of Sidoarjo who gave time for the research that the researcher and the Directorate of Research and Community Service, Directorate General of Research, Research and Development of the Ministry of Research, Technology and Higher Education of the Republic of Indonesia supported the fund for this research.

References

[1] J. R. Wolpaw, N. Birbaumer, D. J. McFarland, G. Pfurtscheller, and T. M. Vaughan, "Brain-computer interfaces for communication and control.," *Clin. Neurophysiol.*, vol. 113, no. 6, pp. 767–91, Jun. 2002.

[2] T. M. Vaughan, "Guest editorial. Brain-Computer interface technology: a review of the second international meeting," *IEEE Trans. Neural Syst. Rehabil. Eng.*, vol. 11, no. 2, pp. 94–109, 2000.

[3] P. A. Pour, T. Gulrez, O. AlZoubi, G. Gargiulo, and R. a. Calvo, "Brain-computer interface: Next generation thought controlled distributed video game development platform," *2008 IEEE Symp. Comput. Intell. Games*, pp. 251–257, Dec. 2008.

[4] J. R. Wolpaw, N. Birbaumer, D. J. McFarland, G. Pfurtscheller, and T. M. Vaughan, "Brain-computer interfaces for communication and control.," *Clin. Neurophysiol.*, vol. 113, no. 6, pp. 767–91, Jun. 2002.

[5] Puloma Pramanick, 2003. "Classification Of Electroencephalogram (Eeg) Signal Based On Fourier Transform And Neural Network". A Thesis submitted in partial fulfillment of the

requirements for the degree of Bachelor of Technology in "Electrical Engineering". Department of Electrical Engineering National Institute of Technology Rourkela-769008 (ODISHA).
[6] Baojun Wang; Liu Jun; Jing Bai; Le Peng; Yan Li; Guang Li, 2005. "*EEG Recognition Based on Multiple Types of Information by Using Wavelet Packet Transform and Neural Networks* ".Engineering in Medicine and Biology Society, 2005. IEEE-EMBS 2005. 27th Annual International Conference of the Issue Date: 17-18 Jan. 2006 On page(s): 5377 – 5380
[7] Wu Ting, Yan Guo-zheng, Yang Bang-hua, Sun Hong, "*EEG feature extraction based on wavelet packet decomposition for brain computer interface",* 0263-2241/$ - see front matter _ 2007 Elsevier Ltd. All rights reserved. doi: 10.1016/ j.measurement.2007.07.007.
[8] http://bbci.de/competition/ii/
[9] Dipti Upadhyay "Classification of EEG Signals under Different Mental Tasks Using Wavelet Transform and Neural Network with One Step Secant Algorithm" International Journal of Scientific Engineering and Technology, Volume 2 Issue 4, pp : 256-259
[10] Pravin A. Kharat, Sanjay V. Dudul "Daubechies Wavelet Neural Network Classifier for the Diagnosis of Epilepsy" ,wseas transactions on biology and biomedicine.
[11] M. Rajya Lakshmi, Dr. T. V. Prasad, Dr. V. Chandra Prakash "Survey on EEG Signal Processing Methods" International Journal of Advanced Research in Computer Science and Software Engineering Volume 4, Issue 1, January 2014.
[12] D. Rumelhart, G. Hinton, R. Williams, Learning internal representations by error propagation, in: J. Anderson, E. Rosenfeld (Eds.), Neurocomputing, MIT Press, Cambridge, MA, 1988, pp. 675–695.
[13] P.J. Werbos, Beyond regression: new tools for prediction and analysis in the behavioral sciences, PhD Thesis, Harvard University, Cambridge, MA, USA, 1974.

IOP Conf. Series: Materials Science and Engineering **532** (2019) 012014 doi:10.1088/1757-899X/532/1/012014

Friction welding of AZ31-SS316L for partially-degradable orthopaedic pins

A K Nasution[1,*], P Nawangsari[2], A Junaidi[1], H Hermawan[3]
[1]Dept. of Mechanical Engineering, Faculty of Engineering, Muhammadiyah University of Riau, Pekanbaru, Indonesia
[2]Dept. of Mechanical Engineering, Faculty of Engineering, Riau University, Pekanbaru, Indonesia

[3]Dept. Mining, Metallurgical and Materials Engineering & CHU de Québec Research Center, Laval University, Canada

[*]ahmadkafrawi@umri.ac.id

Abstract. Joining dissimilar metals with friction welding is very challenging due to the differences in physical, mechanical and metallurgical properties of the two parent metals. A careful design of welding parameter is necessary especially for joining two metals with a large difference in melting temperature. In this study, friction welding has been viewed as an attractive process to join degradable metal with corrosion resistant alloy. AZ31 type magnesium alloy was friction welded at a low heat input to 316L type stainless steel to form a partially-degradable pin for temporary bone implants. Meanwhile, 6103 type aluminium alloy was also joined with 202 type stainless steel for comparison. A modified lathe machine at a spindle speed of 1600 RPM was used to perform the friction welding. The strength of the weld joints was evaluated by tensile test and their hardness distribution was measured using Vickers micro hardness tester. The results showed that the optimum parameter for the AZ31-316L joint was obtained at the friction time = 78s, forging pressure = 40kPa and the burn-off = 25mm. Meanwhile, for the 6103-202 joint the optimum parameters were: friction time = 50s, forging pressure = 9.3kPa and burn-off = 12mm. Fracture surface analysis showed a spiral defect on the 6103-202 joint, while the fracture surface of AZ31-316L showed an oxide layer formation that prevent a good interfacial bond. In conclusion welding parameters affect the strength of the joint as a result of elemental diffusion between the two metals. The oxide layer disrupts diffusion in the interface of AZ31 and 316L. Vacuum or inert condition during friction is proposed to avoid the oxide layer formation at the interface of AZ31-316L.

Keywords: Friction welding, orthopaedic implant, pin, spiral defect.

1. Introduction

The connection of dissimilar materials is generally more challenging than the connection of similar materials because it has differences in the physical, mechanical and metallurgical properties of the parent material to be joined [1]. Connecting dissimilar materials using conventional welding (fusion welding) is difficult, because of the different physical properties such as the liquid point of the parent materials [2-4]. Another effect arising from the welding process is the shifting of metallurgical properties towards the worse, as in stainless steels there will be a decrease in corrosion resistance at the grain boundaries [5, 6], coarsening grains in weld zone and HAZ areas and decreased toughness and ductility [1].

Currently, the process of joining dissimilar materials is mostly done by the friction welding process. American Welding Society (AWS) classifies the friction welding process as a solid-state welding process (solid-state welding process) [7]. The result of this welding joint is caused by the friction of two solid materials that generate heat. This heat, known as heat input and heat input also affects elements of both parent material to diffuse to form metal bonds (intermetallic bonding) [4]. The heat input in the friction welding process depends on the welding parameters and the material properties [4, 8]. The main parameters in direct drive friction welding are friction pressure, forging pressure, friction time, forging time and the rotational speed of a spindle [4-6]. The parameters proportionately one to the other are interrelated [4]. Ochi, et al. (2001) says that there are two types of friction that affect the heat input; namely friction heat input which has a function of rotation and deformation heat input which has a function of torque [9]. Whereas according to Sahin (2009), practically friction welding is classified in two ways; namely continuous-drive friction welding and inertia-friction welding [10]. In the continuous-drive friction method (Fig 1. (a)), where one material rotates and the other material is stationary. Fig. 1. (b) shows the approach between parameters in continuous friction welding process [7, 10, 11].

Different authors such as James, J. A., and Sudhish, R., (2016) have observed the various effects of SS 304 and AISI 1040, which have circular or non-circular cross-sections, have different thermal and mechanical properties, can be easily incorporated by method friction welding [12]. B.M. Darras, B. and Kishta, E (2013) conducted welding of the commercial magnesium alloy AZ31B-H24, said friction welding produces grain structure into finer and more homogenized grain structure [13]. While D. Wang et al., (2014) said related to optimal welding parameters with the mechanical properties of the welding joints and microstructure characteristics of stainless steel metals [14].

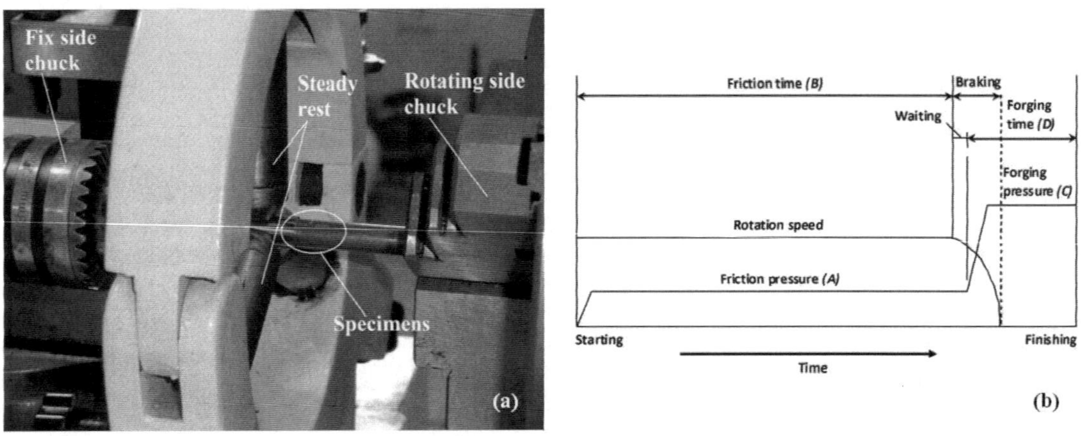

Figure 1. (a) General view of friction welding machine used in the present study. (b) Relation between continuous friction welding parameters.

There are several advantages obtained using friction welding processes such as low heat input, bonds formed under the melting point of the welded material, resulting in higher joint properties and lower distortion [15]. Other advantages are good microstructure, energy efficient process, environmentally friendly and low UV radiation [15, 16]. Therefore, friction welding is done with low heat input, using materials with very different melting points and different applications. A joining between of materials AZ31 and SS316L for biomedical implant applications, while joint of materials AA6103 with SS202 is used for transport applications as a comparison. The result of the welded joint is characterized in the weld zone, tensile test and hardness test.

2. Meterial

The parent material used in this joining process is a magnesium alloy AZ31 with SS316L and an aluminum alloy AA6103 with SS202 for comparison. The magnesium alloy AZ31 with SS316L is a cylindrical stem with a diameter of 10 mm and a length of 70 mm. As for the aluminum alloy series AA6103 with stainless steel series SS202 shaped cylindrical diameter 12.7 mm and 75 mm long. The chemical composition of the parent metal was obtained from the compositional test results using EDS (Hitachi TM3000, Japan) in Table 1. The mechanical properties of the parent metal were obtained from the tensile test results using Universal Tensile Strength (Controlab-TN 20 MD, France) in Table 2.

Table 1. Chemical composition of parent metals (wt.%).

Materials	Al	C	Mn	Si	Cu	Mg	Zn	Cr	Ni	Mo	Fe
AZ31	2.74	-	0.29	0.02		Bal.	0.75	-	-	-	-
AA6103	Bal.	-	0.06	0.93	0.22	1.11	0.12	-	-	-	0.60
SS316L	-	0.03	2.30	0.60	-	-	-	17.80	10.50	2.74	Bal.
SS202	-	0.15	8.50	0.66	0.45	-	-	17.09	3.27	0.12	Bal.

Table 2. Mechanical properties of parent metals.

Property	σ_U (MPa)	σ_Y (MPa)	ε (%)
AZ31	290	180	15
AA6103	339	286	16
SS316L	718	637	45
SS202	1038	768	60

3. Experimental

Friction welding experiments was carried out by modifying lathe machine the type of Knuht DM 1000A-5.7 kW with a maximum spindle rotation speed of 1600 RPM. The material position scheme is shown in Fig. 1. (a). Prior to welding process, the parent metal surface was cleaned in order not to be contaminated, especially with dirt such as grease and the oxide film which would reduce the quality of the connection. This cleansing is an important part to do considering the materials used (magnesium alloys and aluminum alloys) have a high affinity for oxygen [15]. Cleaning was done by sandpaper with size 1000 and then cleaned and wiped using a dry cloth. Furthermore, the welded specimen was disposed of using a lathe machine to be a tensile specimen with a length of 30.0 mm and a diameter of 6.0 mm for AZ31 / SS316L and a gauge length of 45.0 mm and a diameter of 9.0 mm for AA6103 / SS202 following the standard [17].

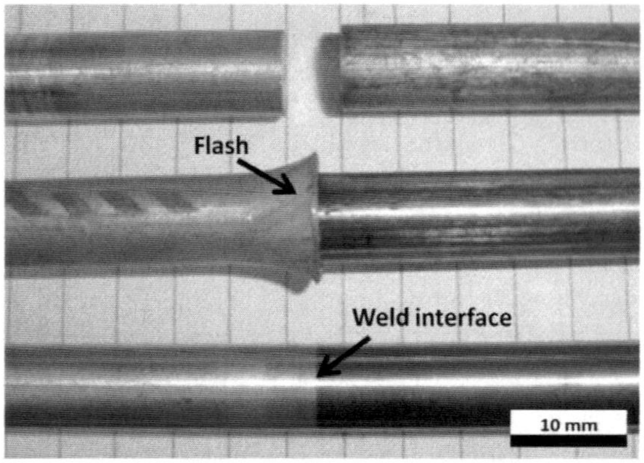

Figure 2. Photograph of a specimen before welding until tensile specimen before testing untuk AA6103/SS202.

The strength of the welded joints was evaluated by tensile testing using a universal tensile testing machine (Controlab-TN 20 MD, France). For microstructure and hardness test, the specimen is cross-cut using electric discharge machining (EDM). The specimen grinding was using abrasive paper grit 1200. Hardness testing was performed using a micro hard test machine (Matsuzawa DVK II, Japan) with a load of 5 kg with a time of 15s and following the ASTM E384 standard. For microstructure, the grinding material was followed by a polish with 0.05 μm diamond paste for etching and analyzed using an optical microscope (Leica DM2500M, UK).

4. Results and discussion

4.1. Tensile tests

The effect of friction welding parameters is noted in the matrix as shown in table 3 (a). In order to obtain the mechanical properties of the result of the connection, a tensile test at room temperature was used to evaluate the mechanical properties of each of the connections shown in table 3 (b).

Table 3. (a) The friction welding parameters, (b) Results of the tensile test of the as-welded for different combination of friction welding parameters.

(a)

Parameters	AA6103/SS202					AZ31/SS316L					
	T1	T2	T3	T4	T5	T1	T2	T3	T4	T5	T6
Friction pressure (kPa)	3.1	3.1	3.1	3.1	3.1	5	5	10	5	10	5
Friction time (sec)	50	50	50	35	70	50	50	50	140	78	154
Forging pressure (kPa)	9.3	22.9	18.6	17	18.6	15	25	40	40	40	50
Burn-off (mm)	12	20	17	16	17	12	15	25	25	25	25

(b)

Parameters	AA6103/SS202					AZ31/SS316L					
	T1	T2	T3	T4	T5	T1	T2	T3	T4	T5	T6
UTS (MPa)	233	214	208	211	211	122	123	122	121	138	79
0.2% YS (MPa)	173	148	146	158	166	-	-	-	-	-	-
ε (%)	8	9	12	7.5	10	-	-	-	-	-	-
Location of failure	**10	**10.4	**10.1	**10.5	**11.2	*WI	*WI	*WI	*WI	*WI	*WI

*WI=Weld interface
** mm from WI
T=Trial

Test results in Fig. 3. (a) and (b) show that the tensile strength of the weld samples is almost all very close to each other. There is a small increase in tensile strength by allowing a friction time of 78 s for the AZ31/SS316L connection. As a results, connection AA6103/SS202 is seen by providing a friction pressure and burn-off a smaller yield higher tensile strength. From this observation, it can be seen that the increase in tensile strength is related to the amount of heat input. While for the joint the AZ31/SS316L, all tensile test results have broken in the welded joint. The fracture shape of the tensile test results shown in Fig. 4.

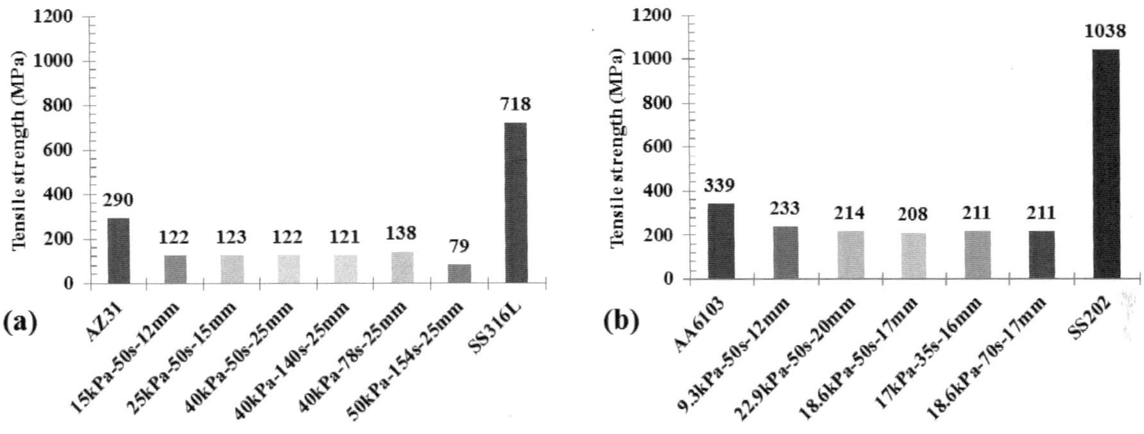

Figure 3. Tensile strength of welded specimens (a) AZ31/SS316L, (b) AA6103/SS202

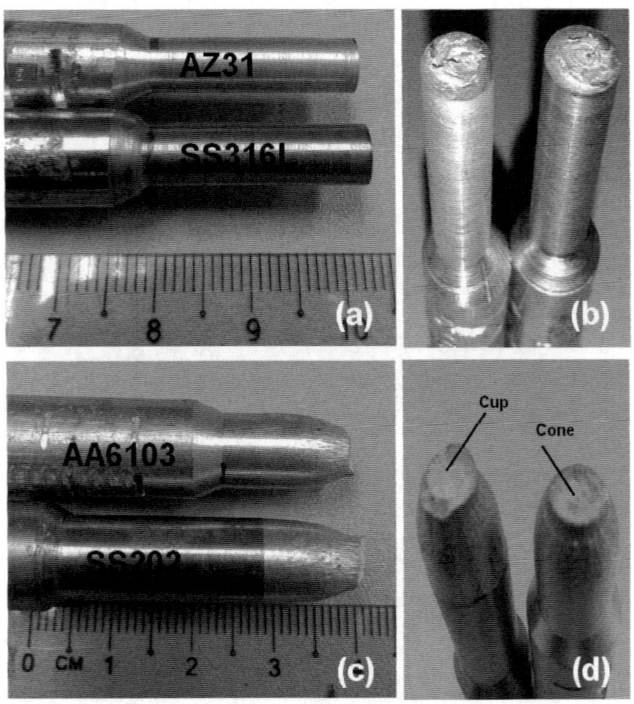

Figure 4. (a-b) The macro photographs of the tensile tested and the ductile fracture surface of tensile tested specimen AZ31/SS316L. (c-d) The macro photographs of the tensile tested and the brittle fracture surface of tensile tested specimen AA6103/SS202.

4.2. Hardness variations of welded parts

The hardness test of each sample was performed using the Vickers method under a 5 kg load. However, due to the narrow interface area, the hardness test near the area would be compared with the result of the microstructure. The hardness distribution in the weld zone of the friction welding joints is typically divided into five zones for different materials [18, 19]. First referred to as the undeformed zone (UZ), it is a zone that is not affected by heat (parent metal). The second zone is called the partially deformed zone (PDZ) or better known as the Heat Affected Zone (HAZ). The third zone is called a plasticized zone (PZ) where this zone is the second diffusion region of the material. The width of each of these zones depends on the thermal conductivity of the welded material [2]. The higher the thermal conductivity of the material the wider the zone is formed, and vice versa. In addition, welding parameters can also affect the width of the zone [20], as for the distribution of hardness in the connection zone can be seen in Fig. 5.

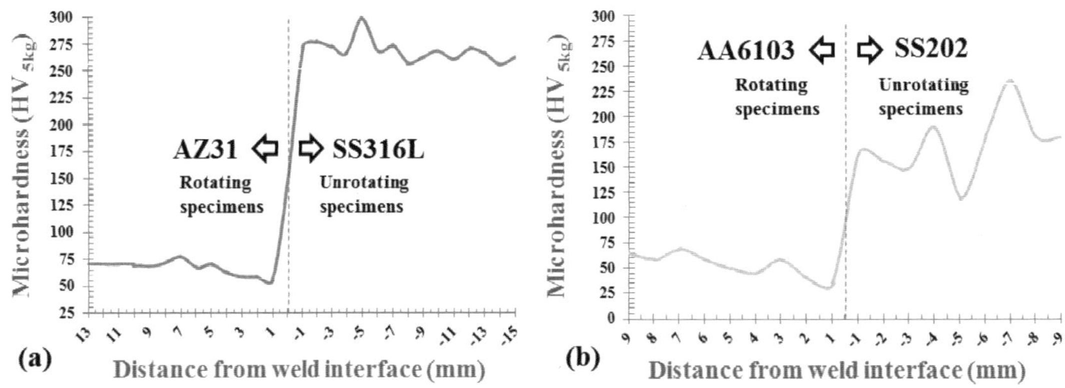

Figure 5. (a) Hardness distribution across the weld of AZ31/SS316L. (b) Hardness distribution across the weld of AA6103/SS202.

Appearance hardness distribution from joint AZ31/SS316L (Fig.5 (a)) does not fluctuate considerably when compared to joint A6103/SS202 (Fig.5 (b)). The results of microhardness testing seen in zone PZ (AA6103) shows recovery and recrystallization resulting from friction and deformation heat, so when approaching UZ, softening occurs. The softening process in the zone of PDZ (AA6103) is already visible in the turning process during flash discharges, as shown in Fig. 6.

Figure 6. Softening of the aluminium alloy after welding of AA6103/SS202.

IOP Conf. Series: Materials Science and Engineering **532** (2019) 012014 doi:10.1088/1757-899X/532/1/012014

4.3. Microstructure of welded parts

Based on the macrostructure the welded sample is shown in Fig. 7. It can be observed that for the joint AZ31/SS316L (Fig.7 (a)) there is no diffusion zone at the interface for all of the weld samples. As for the joint AA6103/SS202 (Fig 7. (b)) visible diffusion zone with an average width of 0.4 mm (sample trial 1) known as intermetallic zone (IMZ). Analysis performed for welding joint AZ31/SS316L using EDS, the diffusion process is hindered by the oxidation in the interface. According to Çelikyürek et al, (2011) the acceleration of diffusion depends on temperature rise, friction time and pressure and to minimize oxidation can be applied to preheat and spraying Argon gas [20]. This does not occur in the AZ31/SS316L welding joints because if preheat is done prior to welding, it will form a layer of oxide on the workpiece [10]. Post weld heat treatment (PWHT) process has also been performed on AZ31/SS316L welding results, but no diffusion zone is formed. So the steps to be taken for the joint AZ31/SS316L by conditioning the welding environment such as vacuum or spray an inert gas (Argon gas) [20] during welding. The absence of diffusion zone in the AZ31/SS316L welding joints becomes evidence or reason for the insignificant hardness distribution. While the weld joint AA6103/SS202 looks more fluctuating and this is the influence of the diffusion of atomic elements of both materials. This is also seen in Fig. 8 which is a the fractograph of welding joint AA6103/SS202.

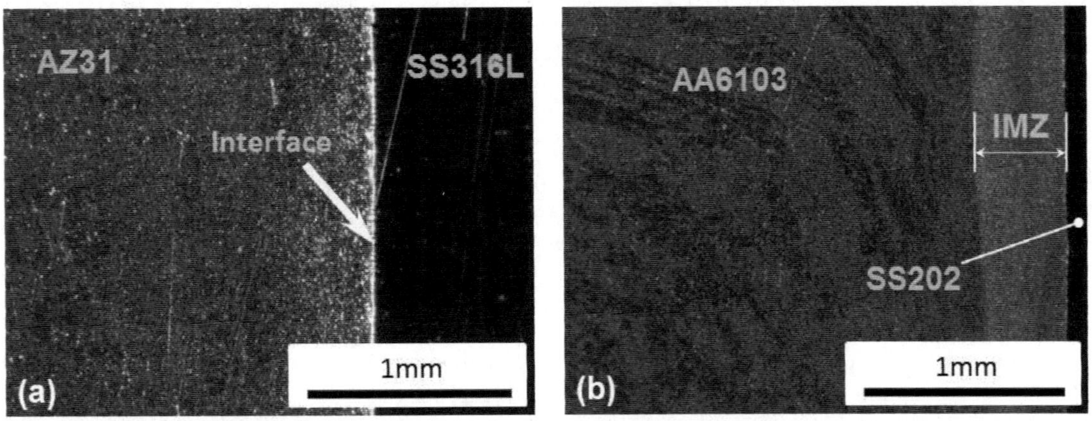

Figure 7. (a) Macrostructure of AZ31/SS316L, (b) Macrostructure of AA6103/SS202.

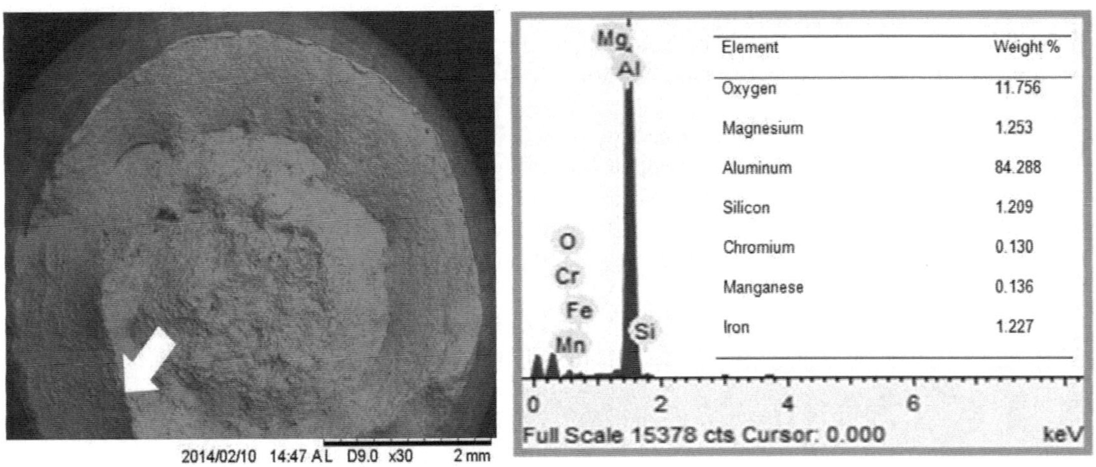

Figure 8. Fractograph of welding joint AA6103/SS202.

Analysis using EDS shows that this area has a large element of 1.25% Mg, 1.21% Si, 1.23% Fe and 11.76% O. If in the welding joint zone has a large number of foreign elements it will tend to form spiral defects and is the beginning the formation of the crack [21]. In contrasting dissimilar materials for medical implants such as the manufacture of orthopedic implants (pins) the distances of very much diffusion are undesirable since the alloy compositions of non-degradable metals contain many toxic elements such as Ni and Cr [22]. Therefore, this is the reason why different metal connections for medical implant applications use low heat input to anticipate large diffusion and away from the interface. According to Çelikyürek et al, (2011), the elements of Cr and Ni atoms can diffuse into the diffusion zone with a short friction time [20].

5. Conclusion

Friction welding can be used successfully to connect AZ31/SS316L. The welding joint shows bad characteristics. The effect of welding parameters on dissimilar materials greatly affects the magnitude and distance of atomic diffusion from the interface. This paper shows that the friction welding between AZ31/SS316L goes perfect but there is no intermetallic bond at the welded joint. Where the oxide layer blocks the diffusion process at the welded joint. Further research can be carried out using a vacuum process or providing a protective inert gas for welding AZ31 / SS316L.

Acknowledgements

The authors would like to thank the financial support from the Ministry of Research, Technology, and Higher Education of the Republic of Indonesia through Director of Research and Community Services with contract number: 017/KONTRAK-PENELITIAN/K10/KM/2018. Also, special thanks to The Institute for Research and Community Services Universitas Muhammadiyah Riau.

References

[1] V. V. Satyanarayana, G. M. Reddy, and T. Mohandas, "Dissimilar metal friction welding of austenitic–ferritic stainless steels," *Journal of Materials Processing Technology,* vol. 160, pp. 128-137, 2005.

[2] Kurt A., Uygur I., and U. Paylasan, "Effect of Friction Welding Parameters on Mechanical and Microstructural Properties of Dissimilar AISI 1010-ASTM B22 Joints," *WJ Supplements,* vol. 90, pp. 102-s-106-s, 2011.

[3] S. Celik and I. Ersozlu, "Investigation of the mechanical properties and microstructure of friction welded joints between AISI 4140 and AISI 1050 steels," *Materials & Design,* vol. 30, pp. 970–976, 2009.

[4] H. C. Dey, M. Ashfaq, A. K. Bhaduri, and K. P. Rao, "Joining of titanium to 304L stainless steel by friction welding," *Journal of Materials Processing Technology,* vol. 209, pp. 5862-5870, 9/19/ 2009.

[5] N. Özdemir, "Investigation of the mechanical properties of friction-welded joints between AISI 304L and AISI 4340 steel as a function rotational speed," *Materials Letters,* vol. 59, pp. 2504-2509, 2005.

[6] N. Özdemir, F. Sarsılmaz, and A. Hasçalık, "Effect of rotational speed on the interface properties of friction-welded AISI 304L to 4340 steel," *Materials & Design,* vol. 28, pp. 301–307, 2007.

[7] J. W. Elmer and D. D. Kautz, "Fundamentals of Friction Welding," in *Welding Brazing and Soldering*. vol. 6, 10 ed Materials Park, Ohio, USA.: ASM International Handbook, ASM Handbook Vol. 6 1993.

[8] R. Winiczenko and M. Kaczorowski, "Friction welding of ductile iron with stainless steel," *Journal of Materials Processing Technology,* vol. 213, pp. 453-462, 2013.

[9] H. Ochi, T. Sawai, Y. Yamamoto, K. Ogawa, and Y. Suga, "Evaluation of Strength of Friction Welded 6061 Alminum Alloy Pipe Joints by Heat Input and Burn-Off Length," *Proceedings of the Eleventh (2001) International Offshore and Polar Engineering Conference,* vol. IV, pp. 256-262, 2001.

[10] M. Sahin, "Joining of stainless-steel and aluminium materials by friction welding," *Int J Adv Manuf Technol* vol. 41, pp. 487–497, 2009.

[11] S. A. A. Akbarimousavi and M. GohariKia, "Investigations on the mechanical properties and microstructure of dissimilar cp-titanium and AISI 316L austenitic stainless steel continuous friction welds," *Materials & Design,* vol. 32, pp. 3066–3075, 2011.

[12] J. A. James and R. Sudhish, "Study on Effect of Interlayer in Friction Welding for Dissimilar Steels: SS 304 and AISI 1040," *Procedia Technology,* vol. 25, pp. 1191-1198, 2016.

[13] B. Darras and E. Kishta, "Submerged friction stir processing of AZ31 Magnesium alloy," *Materials & Design,* vol. 47, pp. 133-137, 2013.

[14] D. Wang, D. R. Ni, B. L. Xiao, Z. Y. Ma, W. Wang, and K. Yang, "Microstructural evolution and mechanical properties of friction stir welded joint of Fe–Cr–Mn–Mo–N austenite stainless steel," *Materials & Design,* vol. 64, pp. 355-359, 2014.

[15] F. Czerwinski, "Welding and Joining of Magnesium Alloys," in *MAGNESIUM ALLOYS-DESIGN, PROCESSING AND PROPERTIES*, T. Smiljanic, Ed., A free online ed. Croatia: InTech, 2011, pp. 469-490.

[16] A. M. M. García, "BLISK Fabrication by Linear Friction Welding," in *Advances in Gas Turbine Technology*, E. Benini, Ed., ed Croatia: InTech, 2011, pp. 411-434.

[17] ASTME8M, "Standard Test Methods for Tension Testing of Metallic Materials," ed. West Conshohocken-Philadelphia: ASTM International, 2009.

[18] B. S. Yilbaş, A. Z. Şahin, N. Kahraman, and A. Z. Al-Garni, "Friction welding of St-Al and Al-Cu materials," *Journal of Materials Processing Technology,* vol. 49, pp. 431-443, 1995.

[19] Sathiya P, Aravindan S, and A. N. Haq, "Effect of friction welding parameters on mechanical and metallurgical properties of ferritic stainless steel," *Int J Adv Manuf Technol,* vol. 31, pp. 1076-1082, 2007.

[20] İ. Çelikyürek, O. Torun, and B. Baksan, "Microstructure and strength of friction-welded Fe–28Al and 316 L stainless steel," *Materials Science and Engineering: A,* vol. 528, pp. 8530-8536, 2011.

[21] K. Uenishi, Y. Zhai, T. H. North, and G. J. Bendzsa, "Spiral Defect Formation in Friction Welded Aluminum " *WJ Supplements,* pp. 184-193, 2000.

[22] A. K. Nasution, N. S. Murni, N. B. Sing, M. H. Idris, and H. Hermawan, "Partially degradable friction-welded pure iron-stainless steel 316L bone pin," *J Biomed Mater Res B Appl Biomater,* vol. 103, pp. 31-8, Jan 2015.

IOP Conf. Series: Materials Science and Engineering **532** (2019) 012015 doi:10.1088/1757-899X/532/1/012015

Colour Behaviour and Stability of Anthocyanin in the Blue Flower of *Stachytarpheta jamaicensis* and *Hydrangea macrophylla*

AM Marpaung[1], I Zhang[1] and H Sutanto[2]

[1] Food Technology Department, Swiss German University, Tangerang, Indonesia

[2] Chemical Engineering Department, Swiss German University, Tangerang, Indonesia

Email: bangmuzi@yahoo.com

Abstract. The colour behaviour at pH 1-11 and the stability at pH 3 of anthocyanins extracted from the blue flower of *Stachytarpheta jamaicensis* (SJ) and *Hydrangea macrophylla* (HM) has been studied. The aqueous extract of both sources exhibited a wide range of colour from red to yellowish green. The colour intensity, total monomeric anthocyanin, and total phenolic content of SJ were higher than those of HM. Their colour was almost disappeared at pH 4-5 and rapidly degraded at pH 7-8, that made them a poor candidate for the colourant at the pH range. The stability of both extracts was tested at pH 3 in a 30°C incubator for 10 days. There were no significant differences in the degradation rate of colour intensity and the total phenolic compound of both extracts. The monomeric anthocyanin in HM degraded significantly faster than that in SJ. Overall, the SJ performed better than HM.

1. Introduction

Anthocyanins are the most interesting water-soluble plant-based natural colourant due to their wide range of colour including red, magenta, purple and blue [1]. The colour of an anthocyanin depends on many factors, especially pH. It is already well-known that, in general, anthocyanins are red at pH < 3 because of the presence of flavylium cation (**AH**$^+$) as predominant species. They are almost lost the colour at pH 4-5 due to the conversion of **AH**$^+$ to colourless hemiketal (**B**) through hydration. At the higher pH, they are purple or blue because of the existence of neutral and anionic quinonoidal base (A and A-, respectively) as the main species. Interestingly, certain anthocyanins are blue at pH 4, as found in butterfly pea flower (*Clitoria ternatea*) extract [1]. The anthocyanins in this flower are recognised as one of the most stable anthocyanins [2].

Other than C. *ternatea*, there are several plants grown in Indonesia that reproduce blue flowers, such as *Pecut Kuda* (*Stachytarpheta jamaicensis*) and *Kembang Bokor* (*Hydrangea macrophylla*), are naturally blue. S. *jamaicensis* is a weedy herbaceous plant that has a smooth and dark green stem, that turns woody towards the base of the stem [3]. S. *jamaicensis* is an underutilised plant that widely found in acclimatised tropics like in Indonesia and Malaysia. To our knowledge, there is no study related to the anthocyanin content of the blue flower of this plant. H. *macrophylla* is a landscape shrub that consists of many cultivars. The colour of the flower of this plant varies, depend on the pH. The main anthocyanin in the flower of H. macrophylla is delphinidin-3-glucoside [4], which classified as unacylated anthocyanin. The unacylated anthocyanins are very unstable in a low acidic and neutral condition. However, delphinidin-3-glucoside might configure a complex called metalloanthocyanin with Al^{3+} to provide stable blue colour [4, 5]. The configuration explained why the flower of H. macrophylla is blue in acidic soil. Only in the acidic soil the Al^{3+} presents and be transported through the roots to the flowers of the plant, as presented somewhere else [4].

Furthermore, to find the potential source of the anthocyanin-base blue colourant, it is important to study the colour characteristics and the stability of an anthocyanin extracted from the flower of S. *jamaicensis* and H. *macrophylla*.

2. Materials and Methods

2.1. Materials

The flowers of S. *jamicensis* (SJ) were obtained from plant wildly grow at Tabalong district, South Kalimantan, Indonesia. The flowers of H. *macrophylla* (HM) were obtained from Tangerang, Banten, Indonesia. The flowers were dried for 24 hrs at 45°C, pulverised and sieved through 250 µm screen, packed in a tight container and kept in a freezer until used. The deionised water was obtained from a local market (Amidis®). The Folin-Ciocalteu reagent, hydrochloric acid (HCl), sodium hydroxide (NaOH), acetic acid, gallic acid, sodium carbonate, ethanol 96%, sodium acetate, monopotassium phosphate and dipotassium phosphate were purchased from Merck kGaA, Darmstadt, Germany. All chemical used were analytical grade and directly used without further purification.

2.2. Extract Preparation and Colour Quality Determination

The extraction of anthocyanins from the flowers was arranged by following the method described in the previous research [1]. The extracts were adjusted to pH 1 to 11 by 0.1 M HCl or 0.1 M NaOH depend on the targeted pH with different dilution factor depend on their initial colour intensity. The light spectra of SJ and HM extracts was recorded at a visible region (400-700 nm) by the UV-Vis Spectrophotometer (Genesys 10uv Thermo Electron Corporation, USA). The spectroscopic data were used to determine the wavelength giving maximum absorption (λ_{max}), colour intensity (CI), and browning index (BI).

The CI was determined as ($A_{\lambda max}$ - A_{700}) x DF, while the BI as (A_{430}-A_{700})/ ($A_{\lambda max}$ - A700) [1]. A $_{\lambda max}$ was an absorbance at a wavelength with maximum absorbance, and DF was a dilution factor. A_{430} was an absorbance at 430 nm and A_{700} was an absorbance at 700 nm.

2.3. Stability Analysis

The stability test was exclusively studied at pH 3. Both extracts were packed in a 50-ml dark glass vial and stored in an incubator at 30°C without light exposure for 10 days. Each extract consisted of three replications. The CI, total monomeric anthocyanin (TA) and total phenolic content (TP) of the samples were determined every day.

The TA was determined as cyanidin-3-glycoside using the pH differential method [6]. A 0.2 ml sample was diluted with 1.8 ml buffer solution pH 1.0 and 4.5. The absorbance of both samples measured at 510 and 700 nm. The difference between absorbancies were calculated as the following:

$$A = (A_{510} - A_{700})_{pH\ 1.0} - (A_{510} - A_{700})_{pH\ 4.5} \quad (1)$$

The TA (cyanidin-3-glucoside equivalents in mg/L) was obtained by:

$$(A \times MW \times DF \times 1000)/(\varepsilon \times l) \quad (2)$$

MW is molecular weight of cyanidin-3-glucoside (449.2 g/mol), DF is a dilution factor, ε is molar absorptivity of cyanidin-3-glucoside which is equal to 26 900, and l is the cuvette widthThe TP was measured by the Folin-Ciocalteu (FC) method with gallic acid as the standard [6]. A 0.2 ml sample was reacted with 0.8 ml 20% sodium carbonate and 1 ml FC reagent. After an hour, the absorbance was observed at 765 nm. The calibration curve was constructed by the reaction of 0.2 ml gallic acid at several levels of concentration with 0.8 ml of 20% sodium carbonate and 1 ml FC reagent. The total phenolic was determined as gallic acid equivalent (GAE) through the following formula:

$$\text{TP (mg/L GAE)} = \text{Abs/m} \quad (3)$$

Abs was the absorbance (A) and m was the slope of the gallic acid standard curve.

2.4. Degradation Kinetics
The degradation kinetics of CI, TA, and TP were evaluated using the first order reaction (Eq. 4), then the $t_{1/2}$ was determined.

$$C = C_0 . e^{-kt} \quad (4)$$

$$t_{1/2} = \frac{\ln 2}{k} \quad (5)$$

C was the final concentration and C_o was the initial concentration, *k* was the constant of reaction rate (day^{-1})

2.5. Statistical Analyses
The statistical analyses involved were Pearson's correlation test and regression slope test. The analyses were conducted with the help of Microsoft Excel software.

3. Results and discussion

3.1. Colour Behaviour and Quality
Both SJ and HM extracts exhibited a wide range of hue at pH 1 to 11 as depicted at figure 1. This behaviour promotes them as a potential pH indicator. Two important colour quality of an anthocyanin-source extract that commonly evaluated are colour intensity (CI) and browning index (BI) [7]. The CI determines the intensity of coloured species which are red flavylium cation $\mathbf{AH^+}$, purple quinonoidal base **A** and blue anionic quinonoidal base $\mathbf{A^-}$. The increase of browning index indicates a decrease in the intensity of coloured species and the increase of pale-yellow species chalcone **C** (A_{420}). The measurement of BI is commonly used to monitor the colour degradation [1,7].

As seen in figure 2, SJ and HM extracts performed relatively similar colour quality. The highest CI appeared at pH 1 when the only species existed was $\mathbf{AH^+}$. At this pH, the CI of SJ was about two times of that of HM. Thus, the total anthocyanin content of SJ extract was approximately twice as high as that of HM extract. Both extracts were almost colourless at pH 4-5 due to the presence of hemiketal **B** as the predominant species. This was typical for the non-polyacylated anthocyanin as also observed in the extract *Melastoma malabathricum*, *Rhodomyrtus tomentosa*, *Vitex pinata*, and *Clidemia hirta* fruits and *Tibouchina semidecandra* flower [7]. This result was in accordance with the prior work that the main anthocyanin in HM extract was an unacylated anthocyanin (delphinidin 3-glucoside) [4]. The very low intensity at pH 4-5 also gave an early sign that the major anthocyanin in SJ extract was a non-polyacylated anthocyanin.

The relatively high CI of both extracts appeared at pH $\leq$ 3 and pH 7-8. Meanwhile, The BI was relatively low at pH $\leq$ 6 and dramatically increased as the pH rise to 7. Accordingly, both extracts are exclusively potential as a food colourant at a high acidic condition. Based on the CI, SJ extract was better than HM extract.

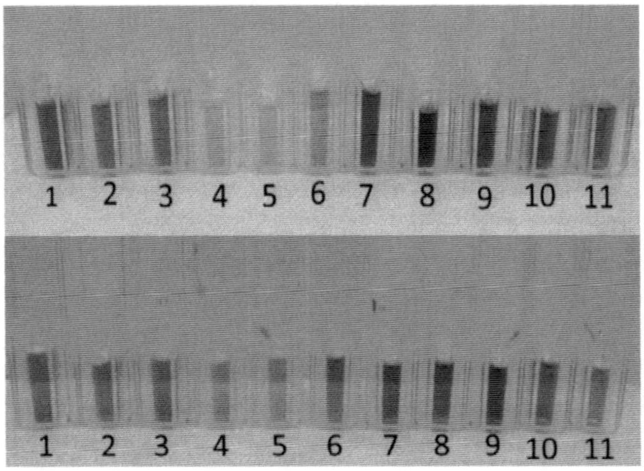

Figure 1. The colour of SJ extract (top) and HM extract (bottom) at pH 1 to 11

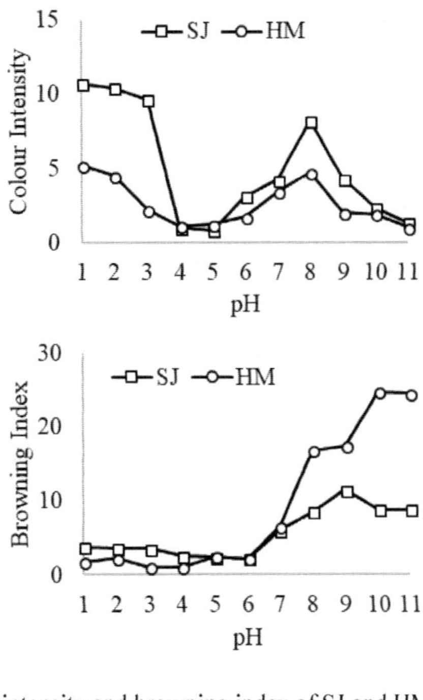

Figure 2. Colour intensity and browning index of SJ and HM extracts at pH 1 to 11

3.2. Stability Study

The stability of the extracts was studied by monitoring the decrease of their CI, total monomeric anthocyanin content (TA), and total phenolic content (TP) during storage at 30°C. Compare to HM extract, SJ extract was higher both in TP and TA. Total phenolic of SJ and HM extracts were 801.23 and 276.32 mg/l, respectively. Total monomeric anthocyanin in SJ was about 2.3 times than that in

HM extracts (100.32 mg/l vs 44.21 mg/l). Initially, based on the CI, there was three pH level chosen for the stability study: 3, 7, and 8. Unexpectedly, the colour of both extracts at pH 7 and 8 was very unstable and rapidly fade within hours. Prior to the research, we thought that the anthocyanin in HM extract would be in a complex with Al^{3+} to form a metalloanthocyanin. Hence, its stability at a neutral and low basic condition was expected to relatively high as performed by the other metalloanthocyanin like commelinin, protocyanin, protodelphin, and cyanosalvanin [8]. We also expected that as its colour was naturally blue, the SJ extract would relatively stable at a relatively high pH condition. However, the finding indicated that the anthocyanin in both extracts was in a monomeric state. In this state, the anthocyanin rapidly loses its colour at low acidic and neutral conditions [1]. Therefore, the probability of both extracts to be a blue food colourant in such condition was less likely.

As the consequence of the result above, the stability of both extracts was exclusively studied at pH 3. As a note, at pH 3, the CI of HM was only 42.3% of that at pH 1. It implied that more than 50% of $\mathbf{AH^+}$ in the extract was already converted to **B**. At the same pH, the SJ extract kept almost 90% of the $\mathbf{AH^+}$. Summarily, the SJ extract performed much better initial colour intensity than the HM extract at pH 3.

Table 1 summarises the first order degradation rate (k) and $t_{1/2}$ for the CI, TA, and PC of both extracts. The correlation test showed that there was a strong positive correlation among three parameters with the Pearson correlation coefficient ≥ 0.86. In another word, the total monomeric anthocyanin and phenolic compound decreased as the decrease of the colour intensity.

Table 1. Colour intensity and browning index of SJ and HM extracts at pH 1 to 11

Stability Parameter	k (day^{-1})		$t_{1/2}$ (day)		p-value of slope test
	SJ	HM	SJ	HM	
CI	0.021	0.020	32.76	34.35	0.76
TA	0.017	0.028	41.85	24.86	0.01
TP	0.018	0.018	39.58	37.68	0.78

Anthocyanin degradation consists of reversible colour degradation and irreversible chemical degradation [9]. An anthocyanin could undergo the colour fading without the chemical degradation. In this stage, the colour was dramatically decrease and the anthocyanin remains stable. As the result, the k of CI is much higher than the k of TA. In contrary, both extracts showed a small gap in the k of CI and TA. Rationally, the extracts lose their colour as they lose the anthocyanin structure. This was in alignment with the degradation mechanism of anthocyanin at a low pH that initiated by the chemical degradation of anthocyanin to its aglycon and followed by the colour degradation [10]. The aglycon was unstable and degraded further to 4-hydroxybenzoic acid and benzaldehyde derivative [10].

The regression slope test was applied to compare the k of all stability parameter between SJ and HM extracts. As seen in table 1, the colour and total phenolic stability of both extracts were statistically similar. Inversely, the monomeric anthocyanin in HM extract degraded significantly faster than in SJ extract. Based on the anthocyanin stability, the SJ was considered as better anthocyanin-source plant than HM.

4. Conclusion

The colour behaviour of the extract of *Stachytarpheta jamaicensis* and *Hydrangea macrophylla* flower was relatively similar. Their best performance as a colourant appeared at the high acidic condition when the predominant species was the red flavylium cation. The half-life of their colour at pH 3 was more than 30 days. Based on the initial colour intensity, total phenolic content and anthocyanin stability the S. *jamaicensis* was better than H. *macrophylla* extract. Both sources performed poorly at pH > 3. Their colour almost disappeared at pH 4-5 and degraded rapidly at pH 7-8. The improvement through a metal complexation and or intermolecular copigmentation might be considered to increase their chance to be a blue colourant that is highly stable at the low acidic and neutral condition.

References

[1] Marpaung AM, Andarwulan N, Hariyad P and Faridah DN 2015., "Spectral characteristics and color stability of Melastomataceae and *Clitoria ternatea* L.", in *Proc.* 17th *Food Innovation Asia Conference*, Bangkok, 2015.
[2] Marpaung AM, Andarwulan N, Hariyadi P and Faridah DN 2018 *Curr. Nutr. Food Sci.* **14** 1
[3] Liew PM and Yong YK 2016 *Evid. Based Complement Alternat. Med.* Article ID 7842340
[4] Schreiber HD, Wade SE, Mayhew KM and Cobb JA 2011 *J. Environ. Hortic.* **29** 131
[5] Goto T and Kondo T 1991 *Angew. Chem. Chem. Int.* **30** 17
[6] Marpaung AM, Andarwulan N and Prangdimurti E 2013 *Acta Hortic.* **1011** 205
[7] Marpaung AM, Andarwulan N, Hariyadi P and Faridah DN, "Study on spectral characteristics of selected unacylated, monoacylated and polyacylated anthocyanin source plant extract at various pH", in *Proc.* 15th *ASEAN on Food Science and Technology*, Ho Chi Minh City, Nov. 2017.
[8] Yoshida K, Mori M and Kondo T 2009 *Nat. Prod. Rep.* **26** 884
[9] Marpaung AM, Andarwulan N, Hariyadi P and Faridah DN 2017 *Nat. Prod. Res.* **31** 1
[10] Sun J, Bai W, Zhang Y, Liao X and Hu X 2011 *Food Chem.* **126** 1203

IOP Conf. Series: Materials Science and Engineering **532** (2019) 012016 doi:10.1088/1757-899X/532/1/012016

A mathematical model of the intravenous glucose tolerance test illustrating an *n*-order decay rate of plasma insulin in healthy and type 2 diabetic subjects

A Kartono[1], Nurullaeli[2], H Syafutra[1], S T Wahyudi[1] and T Sumaryada[1]

[1]Department of Physics, Faculty of Mathematical and Natural Sciences, Bogor Agricultural University (IPB), Bogor 16680, Indonesia
[2]Program Studi Teknik Informatika, Fakultas Teknik, Matematika dan Ilmu Pengetahuan Alam, Universitas Indraprasta PGRI, Jakarta Timur, Indonesia

E-mail: akartono@apps.ipb.ac.id

Abstract. The most accurate considered method to determine the insulin sensitivity (S_I) and glucose effectiveness (S_G), that is the intravenous glucose tolerance test (IVGTT). In this study, the modified minimal model with additions of an *n*-order decay rate of plasma insulin function and an insulin infusion rate function is introduced. This modified model enables estimation of two main feature of type 2 diabetes (T2D) diagnostic, that is the index of S_I and S_G. In this paper, the modified model has been was used to study four cases published IVGTT data, including normal glucose tolerance (NGT) or healthy subjects and T2D subjects with different types of insulin infusion rates. This modified model study has been shown that in NGT subjects had higher S_I and S_G index than T2D subjects. The R^2 value, the coefficient of determination of between measured and calculated plasma concentrations of these four cases, is above 0.95, which indicates good agreement.

1. Introduction

At this time, many thousands of individuals worldwide have been affected the T2D disease. The population of T2D sufferers is increasing and will keep on to increase rapidly in the future. Much medical research has been performed to prevent this disease and to slow down the number of affected individuals by the disease. Today, mathematical modeling has been applied widely for finding new ideas and for developing incrementally new methods to improve the management and control of T2D sufferers effectively. Someday, this models can become an inspiration of medical and clinical to prevent the number of T2D sufferers worldwide [1, 2].

The model basic idea is how to transform a complex physiological process into simple mathematical equations set and then this model can be appropriately validated to describe a real physiological system and can be used the process of a medical and clinical treatment. The development of a robust, valid and verified model allows its users to further observe, understand and explain a complex physiological system [2].

---

[1]Agus Kartono, Department of Physics, Faculty of Mathematical and Natural Sciences, Bogor Agricultural University (IPB), Jalan Meranti, Building Wing S, 2nd Floor, Kampus IPB Dramaga, Bogor 16680, Indonesia.

Glucose-insulin dynamics models have been widely applied to studying a better understanding of the T2D disease. A valid and verified model may describe some useful information on individuals or groups of subjects through appropriate clinical experiments, such as the rate of glucose disappearance, S_I and S_G index. All parameters of these models that unknown are commonly estimated from a subject experiment data [3, 4].

A simple blood test that usually is used to determine the blood glucose level of an individual cannot be used to diagnose whether they are normal, pre-diabetic or diabetic. Because the test does not directly provide S_I and S_G indexes information for subjects. Therefore, a further IVGTT test is often used to obtain more comprehensive information. An IVGTT procedure is a bolus injection of glucose to increase the glycemic level of an individual followed by frequent blood sampling to obtain some main information when the glucose disappearance in the blood. This test will provide the ability for observing the glucose decrease in the human body system [2].

An IVGTT test is one of the important experimental procedure for estimating the S_I and S_G indexes, since it is relatively easy to measure, and its analysis generates a lot of information. The procedure of IVGTT test is subject will be injected a glucose bolus afterward collecting a set of plasma glucose and insulin samples over a period of 2-3 h. For the T2D subject, it is sometimes necessary to be injected an exogenous insulin infusion due to insufficient endogenous insulin secretion from beta-cell in pancreas although it has been stimulated a glucose injection. The S_I index is one of the main parameters to diagnose T2D subjects. This index will measure how well glucose disappears after insulin is released. The opposite of S_I index is called the insulin resistance (IR). The S_G index measures the capability of glucose for suppressing endogenous glucose production. These indexes can be applied for studying and comparing outcomes among groups of different individuals.

For interpreting an IVGTT test, several physiological models have been developed. Mari proposed a circular model and combined it with an IVGTT experimental to describe plasma glucose and insulin profiles for healthy and non-insulin-dependent diabetic (NIDD) subjects [5]. Gaetano and Arino proposed a model of the global physiologic dynamical system, this model showed some numerical results obtained from fitting the dynamical model to plasma glucose and insulin concentrations measured in healthy subjects while undergoing the IVGTT [6].

The purpose of this paper, to develop the minimal model Bergman [7] and also to modify the modified minimal model in previous work [8] for describing how glucose and insulin together control the production and disposal of glucose in the NGT and T2D subjects while undergoing the IVGTT. In the new modified minimal model, the assumption is based on that the insulin decay rate is not always a first-order process and also add a function of the insulin infusion rate in previous literature [9]. The new modified model can be used to estimate two main indices of S_I and S_G indexes.

2. Materials and methods

2.1. The intravenous glucose tolerance test without an insulin infusion

For studying case 1 and 2, a case 1 data (Subject 6) and a case 2 data (Subject 7) which will be compared by the present results were obtained from the published literature [6]. A standard IVGTT was applied without an insulin infusion, so a function of insulin infusion rate is presented as follows: $u(t) = 0$, $t \geq 0$ min. When time 0 min, a glucose bolus of 0.33 g/kg body weight was rapidly infused for 3 min. Later blood samples were measured at time intervals: 0, 2, 4, 6, 8, 10, 12, 15, 20, 25, 30, 35, 40, 50, 60, 80, 100, 120, 140, 160 and 180 min.

2.2. The intravenous glucose tolerance test with an insulin infusion

For studying case 3, the averaged IVGTT test data from seven T2D subjects were obtained from the previous literature [5]. In the standard IVGTT test procedure, after the subjects fasting, they will be measured their basal glucose. Later at 0 min, a 0.3 g/kg body weight glucose bolus was infused immediately. The blood samples will be measured again at time intervals: 2, 3, 4, 5, 6, 8, 10, 15, 20, 25, 30, 40, 60, 80, 100, 120, 140, 160, 180, 210, and 240 min. When time at 20 min in this test, the insulin was infused at a constant rate of 50 mU/kg body weight for 5 min. The insulin infusion rates function is presented as follows: $u(t)$ = 50 (mU/kg body weight) at $20 \leq t \leq 25$ min.

For studying case 4, an averaged IVGTT test data from eight T2D patients are obtained from the previous literature [10]. In this an IVGTT procedure, a glucose bolus of 200 mg/kg was administrated over 60 seconds at time 0, and then blood samples were measured at time intervals: 2, 3, 4, 5, 6, 8, 10, 12, 14, 16, 19, 22, 25, 30, 40, 50, 60, 70, 80, 90, 100, 110, 120, 140, 160, 180, 200, 220, 240, 260, 280, and 300 min. The modified IVGTT procedure was applied to show the response of endogenous insulin secretion in theT2D patients and for reducing the insulin infusion rate to 50% which used usually in the T2D patients. After an injection of the glucose bolus, insulin infusion rates were given: phase 1 (2 to 4 min) 3.5 mU/(kg min); phase 2 (7 to 17 min) 0.5 mU/(kg min); phase 3 (17 to 50 min) 0.25 mU/(kg min). The insulin infusion rate functions are presented as follows:

$$u(t) = \begin{cases} 3.50\left(\frac{\text{mU}}{\text{kg}}\text{body weight}\right), 2 \leq t \leq 4 \text{ min} \\ 0.50\left(\frac{\text{mU}}{\text{kg}}\text{body weight}\right), 7 \leq t \leq 17 \text{ min} \\ 0.25\left(\frac{\text{mU}}{\text{kg}}\text{body weight}\right), 17 < t \leq 50 \text{ min} \\ 0.10\left(\frac{\text{mU}}{\text{kg}}\text{body weight}\right), 50 < t \leq 300 \text{ min} \end{cases}.$$

2.3. The modified minimal model with the n-order decay rate of plasma insulin rate

In physiological research, the minimal model has been used widely on the metabolism of glucose in the human body and also used in medical and clinical studies for estimating the index of S_I and S_G. In this study, the minimal model will be modified by an assumption of a decay rate of plasma insulin because of a stimulation by glucose is not always a first-order function. However, this decay rate is an *n*-order function. A mathematical model for describing the insulin infusion rate was introduced by Zheng and Zhao [9]. The new modified minimal model in this study will be proposed as follows:

$$\frac{dG(t)}{dt} = p_1(G_b - G(t)) - X(t)G(t), \qquad G(t_0) = G_0, \tag{1}$$

$$\frac{dX(t)}{dt} = -p_2X(t) + p_3(I(t) - I_b)^n, \qquad X(t_0) = 0, \tag{2}$$

$$\frac{dI(t)}{dt} = \gamma(G(t) - G_b)t - k(I(t) - I_b)^n + \frac{u(t)}{V}, \qquad \text{if, } G(t) > G_b \text{ and } I(t_0) = I_0, \tag{3}$$

$$\frac{dI(t)}{dt} = -k(I(t) - I_b)^n + \frac{u(t)}{V}, \qquad \text{if, } G(t) < G_b \text{ and } I(t_0) = I_0, \tag{4}$$

Parameter, unit, and definition for the new modified minimal model in above are shown in Table 1. The S_I index was calculated as:

$$S_I = \frac{p_3}{p_2}. \tag{5}$$

The new modified minimal model becomes identical to the modified minimal model [8] when the exogenous insulin infusion rate is set to zero so that the decay rate of the plasma insulin is considered as a first-order process.

Table 1. Parameter, Unit, and Definition for the New Modified Minimal Model

Name	Unit	Definition
$G(t)$	mg/dl	the glucose concentration in the plasma compartment at time t
$I(t)$	μU/ml	the insulin concentration in the plasma compartment at time t
$X(t)$	min^{-1}	the removal rate of insulin from the interstitial compartment at time t
Gb	mg/dl	the basal concentration of glucose
Ib	μU/ml	the basal concentration of insulin
G_0	mg/dl	the initial plasma glucose concentration after injection of the glucose bolus
I_0	μU/ml	the initial plasma insulin concentration after injection of the glucose bolus
p_1	min^{-1}	the glucose decay rate constant or called the S_G index
p_2	min^{-1}	the disappearance rate of remote insulin
p_3	(μU/ml) min^{-2}	the rate at which remote insulin declare increased by plasma insulin
k	min^{-1}	the insulin clearance fraction
γ	min^{-1}	a measure of the secondary pancreatic response to glucose
$u(t)$	(mU/kg) min^{-1}	the exogenous insulin infusion rate
V	L/kg	the distribution volume
n	no unit	the decay rate of the plasma insulin and the increase rate of the remote insulin

2.4. Particle Swarm Optimization parameter estimation techniques

Particle Swarm Optimization (PSO) is a new method of parameters estimation that is inspired by the behavior of animals, such as birds flock movements. Each animal is considered a particle. A particle in dimensional space has a position and velocity that is encoded as vector coordinates. This position and velocity vector is considered as a condition of the particle in the search space. Each position and velocity in the search space is an alternative estimation that can be evaluated using the new modified minimal model function.

In this research, the parameters of p_1, p_2, p_3, k, n, and γ in the new modified minimal model are estimated with IVGTT data using the PSO algorithm. These parameters to determine an index of the S_I and S_G. Because of the concept effectiveness and simplicity of PSO algorithm cause this algorithm widely be applied in practical problem solving of parameters estimation [11, 12]. The coefficient of determination, R^2, is calculated from parameters estimation. The coefficient of determination between the best-fit curve and the data, $y_i - \hat{y}_i$, is calculated by the equation as follows:

$$R^2 = 1 - \frac{\sum_{i=1}^{n}(y_i - \hat{y}_i)^2}{\sum_{i=1}^{n}(y_i - \bar{y}_i)^2}, \tag{6}$$

where y is experimental data, $\hat{y}$ is the prediction of the non-linear least-squares fit and $\bar{y}$ is the averaged experimental data. The experimental data of the glucose-insulin system of these four cases that obtained from the published literature, including NGT and T2D subjects during the IVGTT test will be estimated their parameters

3. Results and Discussion

The calculation results of the R^2 value between experimental data and simulation results of this new model to these four cases is above 0.95. It means that the new modified minimal model was good agreement with the experimental data. The better model could be explained due to occur the increased flexibility of this new model, particularly the assumption of insulin function part which describes the insulin decay rate is not always a first-order process, usually called nonlinearly. Another reason is the insulin injection function can stimulate the insulin secretion from beta-cells in a pancreas which exactly reflects the actual IVGTT test situations.

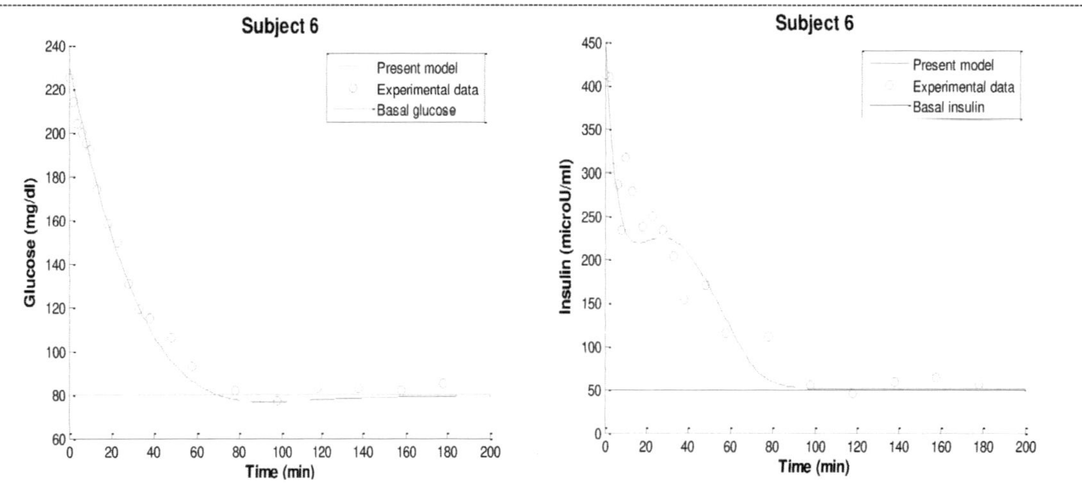

Figure 1. Profiles of plasma glucose and insulin level are produced by the new modified minimal model for NGT human in the case 1, with parameters: $k = 0.120$ [min^{-1}], $\gamma = 0.0145$ [min^{-1}], $G_b = 80$ [mg/dl], $I_b = 50$ [μU/ml], $p_2 = 0.075$ [min^{-1}], $S_I = 5.50 \times 10^{-5}$ [μU/ml) min^{-1}], $S_G = 0.025$ [min^{-1}], $u(t) = 0$, $n = 0.998$, R^2 for glucose = 0.98 and R^2 for insulin = 0.96.

In case 1 and case 2, this curve profile of the experimental and calculation results of blood glucose concentration and insulin concentration is shown in Figure 1 and 2. In this standard IVGTT, after an injection of glucose bolus occurs the blood glucose rise, so higher glucose concentration than before. Afterward, blood glucose shows an apparent decay immediately to the basal line for 1 h. The insulin concentration also increases due to stimulated by the injected glucose, afterward occur an exponential decay of insulin, and finally a secondary peak appears in case 1 and case 2.

The parameters estimation of S_I and S_G index for NGT subjects without an additional insulin infusion shown in Table 2. The precision of the S_I and S_G index were compared to the normal range of the S_I and S_G index from a previous literature [13]. The parameters estimation was not statistically different from normal ranges of the S_I and S_G index. The new modified minimal model provides an estimation of the S_I and S_G index accurately because these values are in good agreement with previously published values in healthy humans [13].

Table 2. Parameters Estimation of The New Modified Minimal Model for Healthy (NGT) Subject without an Insulin Infusion.

Parameters	Subject 6	Subject 7
G_b (mg dl^{-1})	80	90
I_b (μU ml^{-1})	50	30
S_I((μU/ml) $^{-1}$ min^{-1})	5.50×10^{-5}	1.75×10^{-4}
S_G (min^{-1})	0.025	0.035
n	0.998	0.925
R^2 for glucose	0.98	0.98
R^2 for insulin	0.96	0.96

The approximate normal range pancreatic responsivity of the S_I (5.0×10^{-5} - 2.2×10^{-3} (μU/ml) min^{-1}) and S_G (0.0012-0.045 min^{-1}) index [13].

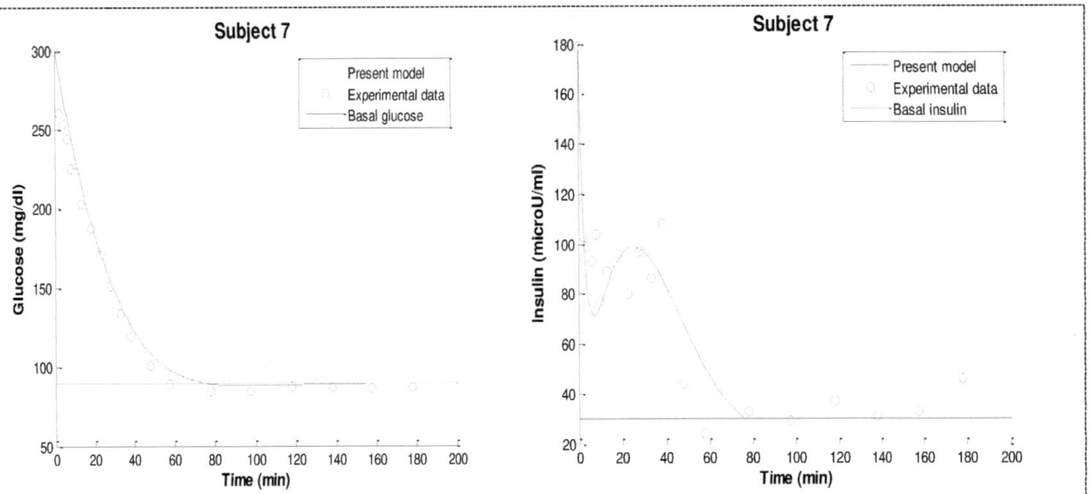

Figure 2. Profiles of plasma glucose and insulin level are produced by the modified minimal model for NGT human in the case 2, with parameters: $k = 0.46$ [min^{-1}], $\gamma = 0.0125$ [min^{-1}], $G_b = 90$ [mg/dl], $I_b = 30$ [μU/ml], $p_2 = 0.06$ [min^{-1}], $S_I = 1.75 \times 10^{-4}$ [μU/ml) min^{-1}], $S_G = 0.035$ [min^{-1}], $u(t) = 0$, $n = 0.925$, R^2 for glucose = 0.98 and R^2 for insulin = 0.96.

In case 3, the study of T2D patients using the new modified minimal model is the same as in the study of case 1 and case 2. A curve profile is shown in Figure 3. The blood glucose takes more than 1.5 h to return to the basal line even though given an insulin infusion for 5 min starting at 20 min. There is a great insulin peak from 20 to 30 min. The present model describes well the actual IVGTT test behavior and achieves the R^2 value of plasma glucose is 0.98 and the R^2 value of insulin concentration is 0.96.

In case 4, this study about the release rate of insulin from beta-cells in the pancreas by an administration of the exogenous insulin with gradually reduced infusion rates during the IVGTT operation in the T2D patients. Although the exogenous insulin was infused all the time during the IVGTT test, the plasma glucose levels decrease more than 2 h to return to baseline as shown in Figure

4. The calculation results of glucose and insulin concentration are in good agreement. The R^2 value for blood glucose is 0.98 and the R^2 value of insulin is 0.96.

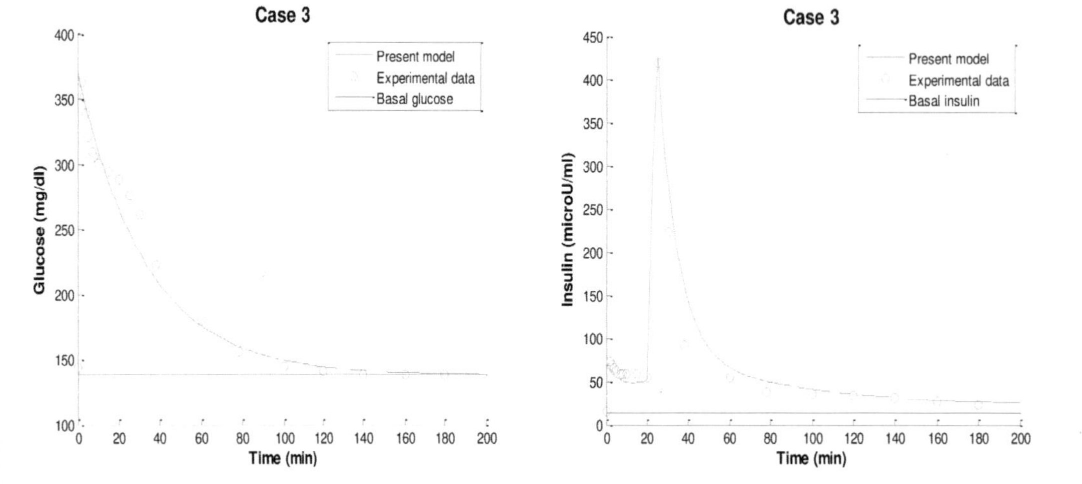

Figure 3. Profiles of plasma glucose and insulin level are produced by modified minimal model with insulin infusion for T2D subject in the case 3, with parameters: k = 0.109 [min^{-1}], γ = 0.0011 [min^{-1}], G_b = 138 [mg/dl], I_b = 24 [μU/ml], p_2 = 0.088 [min^{-1}], S_I = 8.5 × 10^{-7} [μU/ml) min^{-1}], S_G = 0.03 [min^{-1}], $u(t)$ = 50 (mU/kg body weight) at $20 \leq t \leq 25$ min, n = 0.980, R^2 for glucose = 0.98 and R^2 for insulin = 0.96.

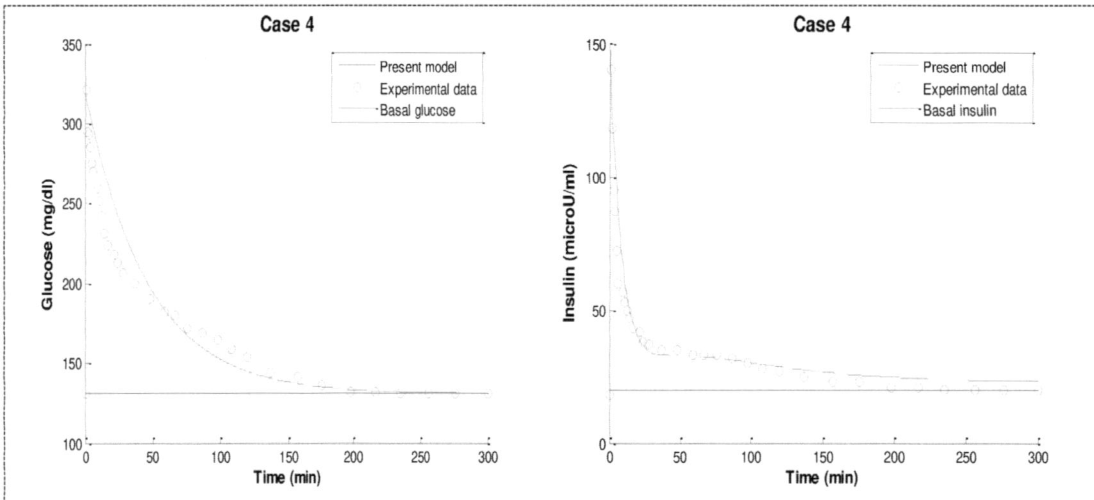

Figure 4. Profiles of plasma glucose and insulin level are produced by modified minimal model with insulin infusion for T2D subject in the case 4, with parameters: k = 0.13 [min^{-1}], γ = 0.0004 [min^{-1}], G_b = 131 [mg/dl], I_b = 20 [μU/ml], p_2 = 0.01 [min^{-1}], S_I = 4.5 × 10^{-7} [μU/ml) min^{-1}], S_G = 0.0225 [min^{-1}], $u(t)$ = 3.5 mU/(kg min) (at 2-4 min); 0.5 mU/(kg min) (at 7-17 min); 0.25 mU/(kg min) (at 17-50 min); 0.1 mU/(kg min) (at 50-300 min), n = 0.999, R^2 for glucose = 0.98 and R^2 for insulin = 0.96

The parameters estimation of S_I and S_G index for healthy and diabetic subjects with additional insulin infusion shown in Table 3. This study showed that healthy subjects had higher S_I and S_G index than diabetic subjects during an IVGTT, nevertheless, basal glucose was lower in healthy subjects than diabetic subjects. The parameters estimation was not in the interval from normal ranges of the S_I and S_G index.

Table 3. Parameters Estimation of The New Modified Minimal Model for Diabetic (T2D) Subject with an Insulin Infusion.

Parameters	Case 3	Case 4
G_b (mg dl^{-1})	138	131
I_b (μU ml^{-1})	24	20
S_I ((μU/ml) min^{-1})	8.50×10^{-7}	4.5×10^{-7}
S_G (min^{-1})	0.03	0.0225
n	0.980	0.999
R^2 for glucose	0.98	0.98
R^2 for insulin	0.96	0.96

The approximate normal range pancreatic responsivity of the S_I (5.0×10^{-5} - 2.2×10^{-3} (μU/ml) min^{-1}) and S_G (0.0012-0.045 min^{-1}) index [13].

4. Conclusions

The human body is much too complex to model in its entirety so that every model will necessarily require some simplifications and assumptions. The challenges of creating a model of this complex disease are equal by the challenges in designing an appropriate and predictive physiological of the human body. In the disease model, the researcher must make assumptions of physiological of the human body just to analysis which of these variations of tissues are likely to affect the hypothesis of the disease being examined.

The new modified minimal model is used to extend the capabilities of the minimal model to the diagnostic of patients with very low beta-cell sensitivity, including NGT and T2D subjects, by giving an exogenous insulin infusion. This is aiming for improved parameters estimation of S_I and S_G, especially for the T2D subject with very low beta-cell sensitivity. An exogenous insulin infusion is applied to extend the range of subjects in which the minimal model can commonly be used, this model is still limited to using data from an IVGTT.

During the IVGTT test, the interaction of glucose-insulin shows a great difference between NGT and T2D patients. Hence, the new modified minimal model with an n-order decay rate of plasma insulin function process is proposed for simulating the plasma glucose and insulin profiles, not only for healthy humans but also for type 2 diabetic subjects, especially with the addition of a different insulin infusion rates. The R^2 value of between experimental data and calculation results are above average 0.95, this indicates the simulation results are the good agreement. It also indicates that modify the classic minimal model improves the model more flexible.

Acknowledgments
This research was funded by Direktorat Riset dan Pengabdian Masyarakat Direktorat Jenderal Penguatan Riset dan Pengembangan Kementerian Riset, Teknologi dan Pendidikan Tinggi sesuai dengan Perjanjian Pendanaan Penelitian dan Pengabdian kepada Masyarakat Tahun Anggaran 2018 Nomor: 1624/IT3.11/PN/2018 tanggal 21 Februari 2018.

References

[1] Boutayeb A and Chetouani A 2006 Biomed. Eng. Online **5**(43) 1

IOP Conf. Series: Materials Science and Engineering **532** (2019) 012016 doi:10.1088/1757-899X/532/1/012016

[2] Caumo A, Simeoni M and Cobelli C 2001 Modelling Methodology for Physiology and Medicine (Academic Press, London) p 337

[3] Bergman RN, Ider YZ, Bowden CR, and Cobelli C 1979 J. Am. Physiol. **23**(6) E667

[4] Bergman RN and Bowden CR 1981 Carbohydrate Metabolism (John Wiley and Sons Ltd) p 269

[5] Mari A 1998 Diabetologia **41** 1029.

[6] Gaetano AD and Arino O 2000 J. Math. Biol. **40** (2) 136

[7] Pacini G and Bergman RN 1986 Comput. Methods Programs Biomed. **23** 113

[8] Kartono A 2013 Theory Biosci. **132** 195

[9] Zheng Y and Zhao M 2005 Comput. Methods Programs Biomed. **79**(1) 73

[10] Ward GM, Walters JM, Barton J, Alford FP and Boston RC 2001 Metabolism **50** 515

[11] Eberhart RC and Kennedy J 1995 A new optimizer using particle swarm theory: Proceedings of the Sixth International Symposium on Micro-machine and Human Science, Nagoya, Japan, pp 39-43

[12] Kennedy J and Eberhart RC 1995 Particle swarm optimization: Proceedings of IEEE International Conference on Neural Networks, Piscataway, NJ, pp 1942-1948

[13] Boston RC, Stefanovski D, Moate PJ, Sumner AE, Watanabe RM and Bergman RN 2003 Diabetes Technol. Ther. **5**(6) 1003

IOP Conf. Series: Materials Science and Engineering **532** (2019) 012017 doi:10.1088/1757-899X/532/1/012017

Functional cake characteristics of modified arrowroot starch (MAS) with the gelatinization-retrograde method

D Damat[1], A Tain[2], H Handjani[3], U Chasanah[4] and D D Siskawardani[1]

[1] Department of Food Science and Technology, University of Muhammadiyah Malang, Malang 65144 East Java Province, Indonesia
[2] Department of Agribusiness, University of Muhammadiyah Malang, Malang 65144 East Java Province, Indonesia
[3] Department of Fisheries, University of Muhammadiyah Malang, Malang 65144 East Java Province, Indonesia
[4] Department of Pharmacy, University of Muhammadiyah Malang, Malang 65144 East Java Province, Indonesia

E-mail: damatumm@gmail.com

Abstract. Healthy food consumption is a critical issue which enhances the functional food production demand. This research aimed to identify the optimum modified arrowroot starch for the best functional cake characteristics. Completely Randomized Design with one factor (arrowroot starch: wheat flour proportion) which consisted five levels (0:100; 25:75; 50:50; 75:25 and 100:0) was applied. The first step was MAS production (gelatinization-retrograde method) then applied to functional cake followed by proximate, resistant starch, expansion volume index and texture analysis. The result showed that 100% MAS was the best treatment with the highest carbohydrate and resistant starch (25.85% and 59.28%), and the lowest ash content, hardness, and expansion volume index (1.8%, 104.95N/mm2 and 45.32%). It can be concluded that functional cake characteristic becomes better and resistant starch (RS) level become higher with MAS application.

1. Introduction

Nowadays, people have a lot of activity but less healthy food consumption. Functional food isn't medicine but it's aimed to reduce diseases risk factors and enhance health performance. A functional food can be found as cake, bread, or beverages.

A cake is typically famous and acceptable dessert, because it's simple, consumable, and served with attractive physical form, color, and various flavor. A cake is typically baked and divided based on cooking technique and ingredients. Various research were done for improving cake characteristics alike texture and organoleptic improvement with potato flour [1], sorghum flour heat-moisture effect to functional cake [2], cake glycemic index [3], the effect of oil palm to the cake [4], dough formula effect to the cake microstructure [5], sponge cake shelf life extension by active packaging as an alternative chemical preservative to direct addition, and pastry glycemic index of arrowroot starch with red bean flour [6].

Arrowroot starch is alternative flour which can be used for producing functional food like cake. The arrowroot starch research has been done, such as butyrate synthesis [7], butyrate hypoglycemic, functional pastry microscopic and organoleptic of MAS [8], arrowroot starch physicochemical properties [9], functional properties of arrowroot starch in cassava and sweet potato composite starches [10], physical modification effect on granule morphology, pasting behavior, and arrowroot starch functional properties [11], heat-moisture treatment effect on the physicochemical, structural, and rheological arrowroot starch [12]. Meanwhile, the research of gelatinization-retrograde method which applied to modified arrowroot starch is less, even though it's possible to increase the resistant starch [13-14].

Resistant starch (RS) is starch molecules which resist digestion and has a lot of benefits for metabolic health. It can be fermented in the colon by the gut flora become short chain fatty acids (SCFA), especially acetate, propionate, and butyrate [15]. Resistant starch can be hypercholesterolemia and hypoglycemic, so it will be helpful for degenerative patients' alike diabetes or coronary heart [16] and it's well known that the food with high resistant starch means low glycemic index [17].

The functional cake perfection with high resistant starch alike arrowroot starch is an interesting topic to be analyzed and discussed. Therefore, the current study aimed to analyze the arrowroot starch effect to the functional cake characteristics.

2. Materials and Methods

Arrowroot tubers (10 months) were collected from Wagir-Subdistrict, Malang, East Java, Indonesia. Then the analysis specimens are a standard glucose solution, distilled water, alcohol, Nelson reagents, Arsenomolydate reagents, $Na_2S_2O_3$, HCl, H_2SO_4, NaOH, K_2SO_4, Na_2SO_4-HgO, boric acid saturated solution, ethanol, acetic acid, methyl red or methylene blue, and iodine. This research started with arrowroot starch extraction then followed with modification of arrowroot starch by using the gelatinization-retrograde method. The second stage was functional cake production from modified arrowroot starch and wheat flour (0: 100; 25:75; 50:50; 75:25; and 100: 0%).

2.1. Modified Arrowroot Starch Production

Arrowroot starch was suspended in distillate water (20% b/v), heated and mixed at 70oC for 10 minutes up to homogenous. After that, the suspended arrowroot starch gave high heat treatment 121oC for 60 minutes in an autoclave, and it was cooled for 1 hour in room temperature (24oC), then it was retrograded for 24 hours at 4oC in the freezer. The starch was dried in a cabinet dryer at 50oC for 28 hours, grounded used grinder and strained (80 mesh). Then MAS was packed used plastics polypropylene.

2.2. Functional Cake Production and Analysis Method

The functional cake production used several additional materials such as egg, margarine, sugar, and cake emulsifier. The cake microscopy was well documented by Scanning Electron Microscope (SEM) [18], while the chemical characteristics including RS content [19], water content, ash content, protein, fat, and carbohydrate were recorded by Different Method [19]. Texture analysis was analyzed by Texture Profile Analyzer type EZ-Test Model SM-SOON-168 [5].

2.2.1. Resistant starch analysis. The method which applied for resistant starch analysis was AOAC: 991.43. Total dietary fiber kit (Product code: 112979, Merck Millipore, Darmstadt, Germany) and digest enzyme alike amylase, protease and amyloglucosidase were used. Then the digestive was added alcohol and filtered. Afterward, the resistant starch residues were washed, dried and weighted.

2.2.2. Cake microscopy analysis. Analysis of starch granules was carried out using the Hitachi type 3000 SEM (Scanning Electron Microscop) brand. Samples in the form of powder were sprinkled thinly and evenly in the conductive double side tape that had been affixed to the specimen holder. Handle where the sample is drawn, and the specimen holder is placed in the space provided, then the handle is closed again. Then the EVAC / AIR button is pressed to make the chamber vacuum, until the WATER LED lights are constantly blue. The SEM software icon on the computer is clicked, the start icon is clicked to start the sample observation. Filaments in SEM will emit electrons that will hit the sample. Then the electrons are reflected by the sample towards the detector, then processed and displayed on the monitor. The results of the analysis and observation are stored on the hard drive. When the observation on the sample has been completed, the stop button is clicked.

2.2.3. Texture. Texture analysis was closely related to the mechanical evaluation characteristics, which depended on the control forced. The deformation curves were the last result that explained the material response. The cake texture included hardness and broken potency was analyzed used Texture Profile Analyzer type EZ - Test Model SM-SOON-168.

2.2.4. Statistical analysis. The quantitative data were subjected to analysis of variance (ANOVA) and followed by Duncan's test ($\alpha = 0.05$) using the program SPSS 10.0 for Windows.

3. Results and Discussion

3.1. Cake Chemicals Properties

The result showed that arrowroot starch and wheat flour formula's gave significantly effect ($p< 0.05$) on water content, protein, fat, ash, carbohydrate (Table 1) and resistant starch (Figure 1). Based on ISO standards, the cake meets the standards of water content. Treatment A (0% MAS and 100 %WF) showed the highest level of water content (31.60 %) compared to others. This is probably caused by the cakes are produced by 100% wheat flour, and it has a larger cavity and more porous than others, therefore it affects the water stored enhancement. In addition, the water content of wheat flour is higher than arrowroot starch and water absorption ability of wheat flour is higher too (56-58%) [20]. Water content is important because it indicates total water that could be absorbed and storage time, afterward it is usually better to choose the flour which can absorb the largest water in cake production.

Table 1. Cake chemicals properties (%).

Treatment (MAS:WF)	Water	Protein	Fat	Ash	Carbohydrate
A (0: 100)	31.60 ^{b}	2.48 ^{c}	6.61 ^{d}	3.29 ^{b}	56.32 ^{a}
B (25: 75)	30.59 ^{b}	2.36 ^{b}	6.50 ^{c}	3.85 ^{d}	56.70 ^{a}
C (50: 50)	30,03 ^{a}	2.31 ^{b}	6.20 ^{a}	3.52 ^{c}	57.94 ^{b}
D (75: 25)	30.16 ^{a}	2.27 ^{b}	6.30 ^{b}	3.24 ^{b}	58.03 ^{b}
E (100: 0)	30.19 ^{b}	2.02 ^{a}	6.31 ^{b}	1.80 ^{a}	59.28 ^{c}

Remark: the letters (a-d) in each column shows significant difference at Duncan's test ($\alpha = 5\%$).

Treatment A is known to have the highest protein content, which is 2.48%. This result is caused by the protein content of wheat flour was higher than modified arrowroot starch. Moreover, it is consistent that wheat flour contains gluten which is a blob and elastic when wheat flour unmixed in water. The largest protein contribution was given by 16.5% egg yolk, 10.9% egg white, and 10-11% wheat flour. It also indicates that the process of arrowroot starch modification does not affect the cake protein content [21].

Based on ANOVA results, it was known the significant effect of arrowroot starch and wheat flour formula's which against cake fat contents. It was clearly showed that treatment A (6.61 %) had the highest fat content. This is caused by arrowroot starch fat content (0.80%) was lower than wheat flour (1.11%), and the addition of additives such as liquid margarine which can't be dissolved properly [6]. Fat content is a necessary factor that affects the cake quality, because of it able to reduce the bitter taste and mildew potency, and further is able to increase consumer acceptability [22].

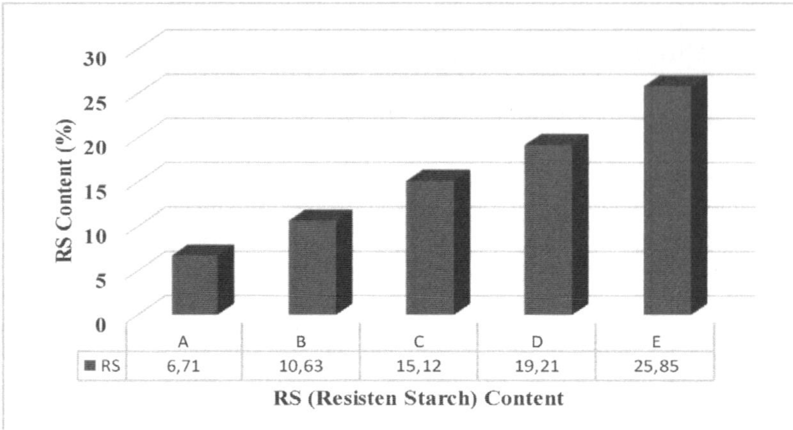

Figure1. Resistant starch content of functional cake.

The highest carbohydrate (59.28 %) and resistant starch (25.85%) were treatment E (100% MAS, 0% WF) and it was given significantly effects ($p<0.05$). While, the lowest carbohydrate and resistant starch were treatment A (0% MAS, 100% WF). It means, increasing modified arrowroot starch (MAS) application enhance the resistant starch and carbohydrate content. This caused by in gelatinization process, the starch granules absorbed water maximally and busted. Then amylose will burst bond that effect to the starch granules structure irregularity, increasing starch suspense viscosity, and birefringent disappearance. Further, the cooling process also makes the kinetic energy declines which less to combine the amylose molecules again. Therefore, the several amylose molecules will merge with amylopectin molecules in the outside of granules, thus swollen starch granules will be microcrystal and suspended. This modification structure (gelatinization-retrograde methods) is profitable because it produces higher resistant starch than original starch. So that, whenever produce the cake with this raw material (MAS) will enhance the resistant starch of cake too. The food product with high resistant starch content tends to be hypoglycemic and hypocholesterolemic [8].

These results were consistent with the research of heat-moisture treatment of rice will increase the resistant starch. In this case, it caused by heat-moisture treatment can produce starch fraction which resists to the enzyme hydrolysis as consequences of double helices formation and compartmentalization of amylose–amylose, amylopectin–amylopectin and amylose-amylopectin chains [13] [23].

3.2. Cake Expansion Volume Index and Texture

The arrowroot starch and wheat flour formula's gave significantly affect (Duncan's test $\alpha = 5\%$) to the expansion volume index (Figure 2) and texture of cake (Table 2). The highest expansion volume index was A treatment (0% MAS: 100% WF) and tend to decrease with enhancing modified arrowroot starch application. The protein content of wheat flour is higher (10-11%) than arrowroot starch, is the main reason. In other hands, the cake production process also affected by gluten composition in wheat flour. The arrowroot starch didn't support the increasing expansion volume index because it's only consisted 3.34% [9].

Wheat flour has the capability to create the great dough that able to detain the gas production in the form of air, vapor, and CO_2. As the result, the spoons structure will be created on the food product (cake). The high gluten level is the main component that responds to this case. Gluten has high elasticity, so it able to swell the dough and conduct the big product but floaty. Therefore, for producing the crumb, great structure and tight volume of cake, it is needed correct method that able to create stable dough with a small air bubble. This vesicle act as the center, that will swell when there is CO_2 which produced by baking powder during the cooking or baking process [24-25]. The gluten quality and quantity also gave a contribution to the expansion volume index. Gas retention, crumb structure of cake level, and cake volume were depended to the gluten in flour and gluten production mechanism during cooking process especially in mixing dough [25-26]. The aeration also gave the significantly affect to the dough viscosity [2]. In other hands, the expansion volume index also affected by heat treatment. According to the previous research, heat treatment could increase the dough viscosity, and it decreased the gravity specific level. As the result, there was a lot of vesicles that trapped inside the dough and automatically expansion volume increase significantly [5].

Table 2. Cake physicals properties.

Treatment	Texture (N/mm^2)
A (0% MAS:100% WF)	190,45 [c]
B (25% MAS:75% WF)	193,28 [c]
C (50% MAS:50% WF)	126,21 [b]
D (75% MAS:25% WF)	120,93 [b]
E (100% MAS:0% WF)	104,95 [a]

Remark: the letters (a-c) shows significant difference at Duncan's test ($\alpha = 5\%$).

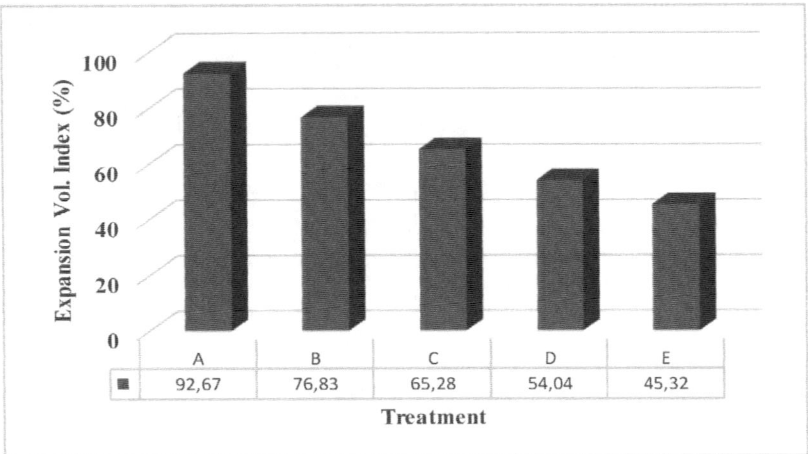

Figure 2. Expansion volume index (%).

The texture scale explains the hardness level of cake. The hardness scale is about 104.95-193.28 N/mm2 and it was a significant difference ($p<0.05$) (Table 2). The increasing modified arrowroot starch application decreased the hardness level. The lowest hardness level was cake with 100% arrowroot starch treatment (104.95 N/mm2) while the highest hardness level was cake with 100% wheat flour treatment (193.28 N/mm2). This result was consistent with the previous research, the addition of modified starch (5%) was able to decrease the hardness level and cake adhesion [27]. The decreasing of hardness level is caused by intramolecular interaction of starch molecules and supporting materials were decline. However, there was an analysis that explained cake hardness level had a negative correlation with expansion volume index, in other words, cake with small expansion volume had tight crumb structure [5]. A cake which created from pre gelatinization starch had a harder texture because it had lower porosity level [4].

4. Conclusion

The best result is cake with 100% MAS treatment with the highest resistant starch and carbohydrate level but has low fat, protein, ash, water, expansion volume index, and hardness. It means, modified arrowroot starch with a gelatinization-retrograde method is able to improve the cake chemical and physical properties by increasing resistant starch component. Therefore, modified arrowroot starch is useful for enhancing functional food production. Furthermore, it needs to analyze the effect to the glycemic index, digestion, hypoglycemic, and hypocholesterolemia.

References

[1] Jeddou B, Bouaziz F, Ellouzi S Z, Chaari F, Ellouz-Chaabouni S, Ellouz-Ghorbel R and Nouri-Ellouz O 2017 *Food Chem.* **15** 668-677
[2] Janjarasskul T, Tananuwong K, Kongpensook V, Tantratian S and Kokpol S 2016 *LWT-Food Sci. and Technol.* **72** 166-174
[3] Ferng L, Liou C, Yeh R and Chen S H 2015 *Food Hydrocolloids* **53** 01-08
[4] Mamat H, Hamid M A and Hill S E 2012 *Int. J. Advanced Sci. Eng. Information Technol.* **2(4)** 48-53
[5] Hesso N, Garnier C, Loisel C, Chevallier S, Bouchet B and Le-Bail A 2015 *Food Structure* **5** 31-41
[6] Istiqomah A and Rustanti N 2015 *J. Nutrition College* **4(2)** 620-627
[7] Damat D, Haryadi, Marsono and Cahyanto M N 2008 *Agritech. J.* **28(2)** 63-69
[8] Damat D, Haryadi, Marsono Y and Cahyanto M N 2008 *Majalah Farmasi Indonesia* (Yogyakarta: Fakultas Farmasi-UGM)
[9] Damat D, Tain A, Handjani H and Khasanah U 2017 *J. Apl.Teknologi Pangan* **6(4)** 161-166
[10] Charles A L, Cato K, Huang T and Chang Y 2015 *Food Hydrocolloids* **53** 187-191
[11] Astuti R M, Widaningrum, Asiah N, Setyowati A and Fitriawati R 2018 *Food Hydrocolloids* **81** 23-30
[12] Pepe L S, Moraes J, Albano K M, Telis V R and Franco C M 2015 *Food Sci. and Technol. Int.* **12** 505-513
[13] Hung P V, Vien N L and Phi N T L 2015 *Food Chem.* **191** 67-73
[14] Chen X, He X, Fu X and Huang Q 2015 *J. Cereal Sci.* **63** 109-115
[15] Topping D L and Clifton P M 2001 *Physiological Reviews* **81** 1031–64
[16] Elena S, Viuda-Martos, Manuel, Fernández-LÓpez, Juana, Pérez-Alvarez and Jose A 2015 *Polysaccharides Bioactivity and Biotechnology* (Switzerland-Springer International Publishing) pp 1911-1931
[17] Lockyer S and Nugent A P 2017 *Review: Health effects of resistant starch* (British Nutrition Foundation) pp 10-41
[18] Soral-Smietana M and Krupa U 2005 *Czech J. Food Sci.* **23** 74–83
[19] AOAC 1999 *Official Methods of Analysis of the Association of Official Analytical Chemist* (Washington-USA)
[20] Edmund J T, and Perry K W 2008 *Soft Wheat Quality. In: Food Engineering Aspects of Baking Sweet Goods* ed G S Servet and S Serpil (CRC Press) pp 1–30
[21] Murthado T 2002 *Bolu Gulung* (Jakarta: Gramedia Pustaka Utama)
[22] Breshears and Crowe 2013 *J. Food Res.* **2(4)** 1-10
[23] Chung H J, Liu Q and Hoover R 2009 *Carbohydrate Polymers* **75** 436–47
[24] Sahi S S and Alava J M 2003 *J. Sci. Food Agr.* **83** 1419-29
[25] Sakiyan O, Sumnu G, Sahin S and Bayram G 2004 *Eur. Food Res. Technol. J.* **219** 635-38
[26] Al-Dmoor H 2013 *Eur. Sci. J.* **9(3)** 166-80
[27] Sahnoun M, Ismail N and Kammoun R 2016 *J. Food Sci.* **53(1)** 481-90

Acknowledgments

We would like to thank the Ministry of Research and Higher Education, the Republic of Indonesia for providing funding support through the Higher Education Research Excellence SKIM No. 017/SP2H/P /K7/KM /2017.

IOP Conf. Series: Materials Science and Engineering **532** (2019) 012018 doi:10.1088/1757-899X/532/1/012018

Bioactive Compounds of Lemongrass (*Cymbopogon citratus*) essential oil from different parts of the plant and distillation methods as natural antioxidant in broiler meat

E S Hartatie, I Prihartini, W Widodo, A Wahyudi
Department of Animal Science, University of Muhammadiyah Malang Indonesia

E-mail: endang_s@umm.ac.id

Abstract. The purpose of this study is to investigate bioactive compounds of lemongrass (Cymbopogon citratus) essential oil from different parts of the plant and distillation methods as natural antioxidant in broiler meat. Distillation of lemongrass leaves, stalk and whole plant was done by using steam water distillation and water distillation. Physical analysis conducted to lemongrass EO. Bioactive compounds identification by GC-MS and antioxidant activity was analyzed by DPPH radical scavenging test. The EO composition of lemongrass leaves, stalks and whole plant which been distilled using steam-water distillation and water distillation are very varied and have different inhibition. On steam water distillation, the EO was distilled from lemongrass stalk have the highest antioxidant activity up to 72.724% inhibition. While in water distillation method, the EO was distilled from whole plant of lemongrass have the highest antioxidant activity up to 70.113 % inhibition. Lemongrass oil can use as natural antioxidant in broiler meat.

Keywords: bioactive compounds, distillation method, GS-MS, lemongrass oil, natural antioxidant

1. Introduction

Meat and meat products including broiler meat are susceptible to quality deterioration due to their rich nutritional composition [1]. The quality deterioration is caused by chemical and microbial changes. The oxidation of meat lipids is the most important form of chemical deterioration. Oxidation of lipid leads to the formation of several other compounds which have negative effects on the quality of fresh meat and meat products. It causes changes in color, texture and flavor (sensory) and nutritional quality decreases [2] [3]. The main factors that determine the loss of food quality and shelf-life reduction are lipid oxidation and bacterial contamination [4]. The short shelf-life of refrigerated packed meat cause these products commercialization more difficult [5]. Therefore, it happened postpone meat oxidation and preventing bacterial cross-contamination are highly to food processors. Lipid oxidation is an important determinant of shelf-life of meat and meat products, so that the control of lipid oxidation in fresh and meat products is the focus of meat scientists[6] .

Lipid oxidation can be reduced by the using antioxidants in meat and meat product, so that the quality of product and shelf life can be increase [3][5]. The antioxidants can make from Synthetic or natural material. Synthetic antioxidant such as butylated hydroxytoluene (BHT), butylated hydroxyanisole (BHA), propyl gallate (PG) and tert- butylhydroquinone (TBHQ) have been widely used in meat and poultry products [7] [8][9] [10]. Recently, consumers have refused synthetic antioxidants because of their carcinogenicity and this has resulted in increased demand for natural antioxidants, especially of plant origin. It has increased in the recent years due to growing worried of consumers about potential toxicological effect of synthetic antioxidants [11] [12] [13] [5]. The main advantages of natural antioxidant in foods including in meat are high consumers preference and other reason is their save use but disadvantages are their higher production costs and low effectiveness [10]. So that the most of the reason investigations have been directed towards the identification of natural antioxidants from various plant sources and plant materials. The plants that rich in phenolics are a good alternative to synthetic antioxidants [3]. The majority of natural antioxidants are phenolic compounds and the most important are the tocopherols, flavonoids and phenolic acids antioxidant [14].

Lemongrass is an aromatic plant and it belongs to the gramineae family [15]. Lemongrass (*Cymbopogon citratus*) is widely used herb in tropical countries especially in southeast Asia [16]. It has a citrus flavor and widely used as an herb in Asian cuisine, can be used fresh, dried or powdered. It is commonly used in soups, curries and teas, also is suitable for fish, beef, poultry and seafood. Moreover, it is used as a preservative. Lemongrass essential oil is applied for its medicinal value [17]. Lemongrass is so interesting due to its commercially valuable EOs and widely used in traditional medicine as well as in food technology [18]. The biologically active constituent of lemongrass is citral constituting more than 75% (w/w) of its essential oil [19]. The chemical composition of the essential oil *Cymbopogon citratus* varies according to the geographical origin [16]. Lemongrass grows in almost all tropical and subtropical countries [20]. In Indonesia, lemongrass (*Cymbopogon citratus*) is a plant commonly used as ingredient. Indonesians use lemongrass as spicy limited to the stalk only, while the leaves are just thrown away as garbage or waste and usually use in fresh form. The aim of this study was investigated the bioactive compounds of essential oil from lemongrass leaves, stalk and whole plant which extracted using steam-water distillation and water distillation as natural antioxidant in broiler meat.

2. Research Methods

2.1 Plant Materials and Preparation

Lemongrass (Cymbopogon citratus) plant used in the study were collected from Ngantang area, East Java, Indonesia. Research material is divided into 3 parts, namely leaves, stalks and whole plant (leaves + stalks). Lemongrass plant hrvested at 6 months of age.

2.2 Extraction of Essential oil

Extraction of lemongrass essential oil was done by two methode of hydrodistillation are using steam-water distilation (indirect distillation) and water distilation(direct distillation). Distillation was done (carried out) by using extraction tool with a capacity of 5 kg. The were extracted from leaves, stalk and whole plant and lemongrass was extracted in fresh condition after harvesting. The extraction was performed for 4 h with the temperature maintained at approximately 100°C. Lemongrass oils were kept at room temperature in glass bottle away from sunlight.

2.3. Identification of lemongrass oil

The resulting lemongrass oils were identified to find out its appearance, color, odor, rendement, specific gravity, optical rotation, refractive index and solubility. Analysis procedure identification of lemongrass oil refers to SNI No. 06-3953-1995.

2.4 Determination of bioactive components

Bioactive component in lemongrass EO was analyzed using Gas chromatograpy- mass specctrometry (GC-MS) was performed using "GC-2010" first step, with the column oven temperature and injection temperature was set at 80°C and 250°C, respectively. The flow control mode used pressure 24,9 kPa with the total flow and column flow at 61.0 mL/min and 0,57 ml/min, respectively. Purge flow 3 ml/min and split ratio 100 were used. The oven temperature was programmed at 80.0°C for 3 minutes, and at 250°C for 5 minutes. For the second step using "MS-QP2010 Ultra", ion source temperature was set at 200°C and interface temperature 250°C, solvent cut time: 0,50 min, detector gain mode: relative to the tuning result, detector gain: 0.95 kV + 0.00 kV, with a threshold: 0. Start time: 1.00 min, end time: 29.25 min. ACQ mode used for the scan, with event time at 0.3 seconds, scan speed 3333, start m / z: 20 and end m / z: 800.

2.5 Antioxidant activity (DPPH radical scavenging activity)

DPPH scavenging activity of lemongrass oil was determined according the procedure of [21]. For assay, 1gram sample was taken in test tube and 9 ml ethanol (96 %) was added to the sample tube, masses for 24 hours. Then 4 ml filtrate was taken in test tube and 1 ml of the fresh prepared solution of DPPH was added, let stand for 30 min. The absorbance of the sample was measured at 517 nm using spectrophotometer UV-VIS double bin (Shimadzu, Japan).

DPPH solution 1ml plus 4 ml ethanol was used as a control. All analyses were run in duplicates. The DPPH radical scavenging effect was calculated as inhibition of percentage, according to the following formula:

$$\text{Inhibition of percentage } (\%) = [\, Ac(o) - Aa(t) \,/\, Ac(o)] \times 100 \quad (1)$$

where: Ac(o) is an absorbance of control DPPH solution at 0 min and Aa(t) is absorbance of test sample after 20 min.

The data of this research were analyzed using qualitative and quantitative description.

3. Results and Discussion

3.1 Characteristics of lemongrass oil

The results of identification to lemongrass oil are shown in Table 1.

Table 1. Characteristics of lemongrass oil

Method of hydro distillation	Characteristics of lemongrass oil							
	appearance	Color	Odor	Relative Density	Refractive index	Optical rotation	Solubility	tenement
Steam-water distillation								
Leaves	Liquid	Yellow	Citrus	0,8933	1.4773	-0,06	1:2	0,05%
Stalk	Liquid	Yellow	citrus	0.8712	1.4775	+0.38	1:2	0.04%
Whole plant	liquid	yellow	citrus		1.4780	+0.05	1:2	0.04%
Water distillation								
Leaves	Liquid	Yellow	Citrus	0.8705	1.4725	-0.01	1:2	0.23%
Stalk	Liquid	Yellow	Citrus	0.8719	1.4745	-0.03	1:2	0.07%
Whole plant	Liquid	Yellow	Citrus	0.8876	1.4775		1:2	0.04%

The results of the study showed the essential lemongrass oil distilled from leaves, stalk and whole plant liquid, yellow color, citrus odor, average relative density 0.8705 – 0.8933 and relative index 1.4725 -1.4780. This physical characteristic of lemongrass oil is in accordance with the requirement of lemongrass oil by SNI lemongrass oil no. 06-3953-1995.

3.2 Bioactive Compound of lemongrass oil by GCMS

The test results of GCMS method in leaves, stalk and whole plant lemongrass oil at steam- water distillation and water distillation are displayed in TABLE II. The compounds shown are compounds having concentrations greater than 1 percent.

The results of the study generally show the composition of EO of lemongrass leaves , stalks and whole plant which been extracted using steam-water distillation and water distillation are very varied. A total of compounds in oil from lemongrass'leaves at steam-water distillation are more than lemongrass' stalk and whole plant, but a total compounds lemongrass oil at water distillation are relatively the same. The compounds of lemongrass leave EO are different from lemongrass stalk EO.

Lemongrass is a rich source of citral. Citral is the main chemical component of lemongrass EO. Citral or 3,7-dimethyl-2,6-octadienal is the name given to a natural mixture of two isomeric acyclic

IOP Conf. Series: Materials Science and Engineering **532** (2019) 012018 doi:10.1088/1757-899X/532/1/012018

monoterpene aldehydes are geranial (transsitral, citral A) and neral (cis-citral, citral B). Chemically, citral is a mixture of two aldehydes that have the same molecular formula, $C_{10}H_{16}O_2$ but different structures. In this study, citral in the leaves lemongrass oil was exist in either cis form (Z-citral) and trans form (E-citral), but in the stalk lemongrass oil was exist in cis form (Z-citral) and citral at both distillation methods. Besides that the leaves lemongrass oil contains trans- geraniol, 6-methyl-3 hepten-2-one, beta-myrcene and longipinenepoxide in different amounts between steam-water distillation and water distillation methods [22] as reported the citral content of essential lemongrass oil made from leaves was higher than made from stalk at steam distillation method [18], which indicated that phenolic content of the essential oil extracted using hydro-distillation method from lemongrass stalk higher amount than extracted from lemongrass leaves.

It is noteworthy from several studies have shown that the composition of EOs are influence by climatic, seasonal, local and experimental conditions, etc. The maturity of lemongrass when harvesting influence, the quality essential oil and citral contents [23}.

Table 2. Bioactive compounds of lemongrass oil with GC-MS

Part of plant and Distillation method	Compounds	Retention Times	Con. Unit (%)
Leaves- Steam water distillation	1. 6-Methyl-5-hepten-2-one	5.745	14.137
	2. beta-Myrcene	6.118	18.771
	3. Linalool	8.182	3.351
	4. Citronella	9.151	1.302
	5. Longipinenepoxide	9.704	1.759
	6. Z-citral	10.965	19.225
	7. Trans-geraniol	11.323	10.686
	8. E-Citral	11.555	26.522
	9. Geranyl acetat	13.577	1.359
	10. Patchouli alcohol	18.507	1.060
Stalk: Steam water distillation	1. Z-Citral	10.939	36.501
	2. Trans-Geraniol	11.287	1.832
	3. Z-Citral	11.518	55.172
	4. 1,2-Benzenedicarboxylic acid, diotyl ester	21.989	8.949
Whole plant Steam water distillation	1. 6-Methyl-5-hepten-2-one	5.739	1.252
	2. beta-Myrcene	6.112	1.641
	3. Z-Citral	10.932	36.043
	4. Trans-Geraniol	11.276	2.272
	5. Z-Citral	11.506	52.846
	6. 1,2-Benzenedicarboxylic acid,diotyl ester	21.979	3.756
Leaves- water distillation	1. 6-Methyl-5-hepten-2-one	5.740	1.614
	2. beta-Myrcene	6.115	3.000
	3. Longipinenepoxide	9.705	1.071
	4. Z-citral	10.961	39.397
	5. Trans-geraniol	11.324	4.064
	6. E-Citral	11.531	48.730
Stalk: water distillation	1. 6-Methyl-5-hepten-2-one	12.063	2.986
	2. L-LINALOOL	16.150	1.217
	3. TRANS-CARAN,4.5-EPOXI	18.543	1.365
	4. Z-Citral		
	5. Citral	20.402	9.902
	6. Juniper camphor	21.318	72.896
	7. 1,2-Benzenedicarboxylic acid, diotyl	30.797	2.170
	ester (CAS) Dioctyl phthalate	49.369	2.560
Whole plant water distillation	1. 6-Methyl-5-hepten-2-one	12.057	2.447
	2. Z-Citral	20.368	24.042
	3. Trans-geraniol	21.056	11.670
	4. E-Citral	21.292	45.123
	5. Geranyl acetat	24.505	1.251
	6. Juniper camphor	30.793	1.714
	7. 1,2-Benzenedicarboxylic acid,bis(2-ethylhexyl) ester (CAS) Bis(2-ethylhexyl) phthalate	49.401	5.622

3.2 Antioxidant Activity

The result of DPPH scavenging activity of lemongrass oils are shown in Table 3.

Table 3. Antioxidant activity (DPPH Radical Scavenging activity) of Lemongrass oil

Hydrodistillation method	Part of plant	Average of antioxidant activity (%)
Steam-water distillation	leaves	60.808
	Stalk	72.274
	Whole plant	48.026
Water distillation	leaves	57.331
	Stalk	51.034
	Whole plant	70.113

The results of this study of antioxidant activity with DPPH free radical scavenging test show that the EOs were distilled from lemongrass leaves, stalk and whole plant have different percentage inhibition, both on steam-water distillation and water distillation. On steam water distillation, the essential oil was distilled from lemongrass stalk have the highest antioxidant activity up to 72.724% inhibition. While in water distillation method, the essential oil was distilled from whole plant of lemongrass have highest antioxidant activity up to 70.113 % inhibition. Antioxidant activity of essential oil made from leaves was higher than made from stalk by using steam distillation method [22]. Other study shown that The highest percentage inhibition was obtained by EOs is distilled from lemongrass stalk [18] The application of the method is based on several considerations such as the type of plant raw material, oil characteristics, oil diffusion process with hot water, oil decomposition due to heat effects, production efficiency and reasons for economic value and production effectiveness.

4. Conclusion

The conclusion that the bioactive compounds of lemongrass (*Cymbopogon citratus*) essential Oil from different parts of the plant and different distillation method are very varied and have different percentage inhibition. On steam water distillation, the the highest value of antioxidant activity come from lemongrass stalk. While in water distillation method, the highest value of antioxidant activity come from whole plant of lemongrass. Based on average of antioxidant activity, Lemongrass oil can use as natural antioxidant in broiler meat.

References

[1] Devatkal, S.K., Thorat,P.,and Manjunatha,M. 2012 Effect of vacuum packaging and pomegranate peel extract on quality aspects of ground goat meat and nuggets. Journal of Food Science and Technology. Doi:10.1007/s13197-012-0753-5

[2] Karakaya,M., Bayrak,E., Ulusoy,K. 2011. Use of natural antioxidants in meat and meat products. Journal of food and Engineering. (1): 1-10

[3] Shah,M.A., Bosco, S.J.D. and Mir, S.A. 2014 Plant extracts as natural antioxidants in meat and meat products. Meat Science. Doi:10.1016/j.meatsci.2014.03.020

[4] Lopez,J.F.,Zhi.N.,Carbonell, L.A.Alvarez,J.A.P. and Kuri,P. 2005 Antioxidant and antibacterial activities of natural extracts : application in beef meatballs. Meat Science 69 (2005) pp371-380

[5] Velasco,V. and Williams,P. Improving meat quality through natural antioxidants. 2011. Chilean Journal of Agricultural Research 71 (2) pp313-322

[6] Olorunsanya, A.O., Olorunsanya,E.O., Bolu,S.A.O., Adejumobi,C.T. and Kayode,R.M.O. 2010 Effect of Granded Level of Lemongrass (Cymbopogon citratuss) on oxidative Stability of Raw or Cooked Pork Patties. Pakistan Journal of Nutrition. 9(5) pp467-470
[7] Biswas,A.K., Keshri, R.C., and Bisht,G.S 2004 Effect of enrobing and antioxidant on quality characteristics of precooked pork patties under chilled and frozen storage condition. Meat Science,66 pp733-741.
[8] Formanek,Z., Kerry,J.P.,Higgins,F.M.,Buckley,D.J.,Morrissey,P.A., Farkas,J. 2001 Addition of synthetic and natural antioxidants to alpha-topheryl acetat supplemented beef patties: Effects of antioxidant and packaging on lipid oxidation. Meat Science. 58 pp337-341
[9] Jayathilakan,K.,Sharma,G.K.,Radhakrishna,K., and Bawa,A.S. 2007 Antioxidant potential of synthetic and natural antioxidants and its effect on warmed-over-flavour in different species of meat. Food Chemistry, 105 pp908-916.
[10] Fasseas,M.K., Mountzouris,K.C., Tarantilis, P.A., Polissiou,M., and Zervas,G. 2007 Antioxidant activity in meat treated with oregano and sage essential oils. Food Chemistry 100 pp732-735
[11] Juntachote, T., Berghofer,E., Siebenhandl,S., and Bauer,F. 2006. The oxidative properties of Holy basil and Galangal in cooked ground pork. Meat Science 72 pp446-456
[12] Naveena,B.M., Sen, A.R., Vaithiyanathan,S., Babji,Y., and Kondaiah,N. 2008 Comparative efficacy of pomegranate juice, pomegranate rind powder and BHT as antioxidants in cooked chicken patties. Meat Science. 95 pp195-202
[13] Nunes de Gonzalez,M.T., Hafley,B.S.,Boleman,R.M., Miller,R.K., Rhee,K.S., and Keeton,J.T. 2008 Antioxidants properties of plum concentrates and powdwe in precooked roast beef to reduce lipid oxidation. Meat Science, 80 pp 997-1004
[14] Kumar,Y., Yadav,D.N., Ahmad, T. and Narsaiah, K. 2015 Recent Trends in the use of Natural Antioxidant for Meat and Meat Products. Comprehensive Reviews in Food Science and Food Safety,14 pp 796-812
[15] Akhila,A. 2010 Essential Oil-bearing Grasses: the genus Cymbopogon. Medical and aromatic plants-industrial profile. Taylor and Francis Group,LLC.
[16] Shah, G., Shri, R., Panchal, V., Sharma,N., Singh,B., and Mann,A.S. 2011 Scientific basis for the therapeutic use of cymbopogon citratus, Staps (lemongrass). Journal of advanced Pharmaceutical Technology & Research, 3 pp 3-8
[17] Pearson,O., 2010 The antibacterial properties of essential oil http:www.livesstrong.com/article/168697-the-antibacterialproperties-of-essential-oil/#ixzz1W7u9hn6K.2010.
[18] Mirghani, M.E.S., Liyana, Y,Y. and Parveen, J. 2012. Bioactivity analysis of lemongrass (Cymbopogon citratus) essential oil. International Food Research Journal 19(2) pp569-575
[19] Huynh, K.P., Maridable, J., Gaspillo, P., Hasika, M., Malaluan, R., Kawasaki, J. 2008 Essential oil from lemongrass extracted by supercritical carbon dioxide and steam distillatin. The Philippine Agric. Sci.19 pp36-41
[20] Cheel, J., Theoduloz, C., Schmeda-Hirsechmann. 2005 Free radical scavengers and antioxidants from lemongrass (Cymbopogon citratus). J.Agric.Food Chem.53 pp 2511-2517
[21] Huang,D., Ou, B. and Priop, R.L., 2005 The chemistry behind antioxidant capacity assays. J.Agric. Food Chem 53 pp1841-1856
[22] Putra, I.N.K., antara, N.S. and Wartini, N.M. 2013 Bioactive components of leaf and stalk of lemongrass (Cymbopogon citratus) essentials oil and its antioxidant activity. Udayana University Denpasar Bali.
[23] Tajidin, N.E., Ahmad, S.H., Rosenani,A.B., Azimah.H. and Munirah,M. 2012. Chemical composition and citral content in lemongrass (Cymbopogon citratus) essential oil at three maturity stages. African Journal of Biotechnology. 011 Vol.11(11):2685-2693.

IOP Conf. Series: Materials Science and Engineering **532** (2019) 012019 doi:10.1088/1757-899X/532/1/012019

Integration Distance Similarity with Keyword Algorithm for Improving Cohesion between Sentences in Text Summarization

Rizki Darmawan[1] and Adi Wijaya[2]

[1] Computer Engineering Department, Sekolah Tinggi Teknik Muhammadiyah Cileungsi, Indonesia
[2] Computer Engineering Department, Mohammad Husni Thamrin University, Indonesia

Abstract. In recent time the exponential growth of textual information available on the Web, end user need to be able to access information in summary form. Commonly the method to get the summary is extraction method. One of extraction method that easier and commonly used is Keyword Algorithm, but this algorithm has a weakness in the cohesion between the sentences. Distance similarity method is one method used for solving the cohesion problem. The idea of this paper is to improve cohesion between sentences based on extraction of keyword algorithm. The hybrid keyword algorithm and the distance similarity method is proposed. The proposed method was compared three distance similarity such as Cosine, Dice and Jaccard that looking for the cohesiveness between sentences according to keyword algorithm extraction and performance as standard of evaluation. The result showed that Dice has the highest cohesion degree is 45.87 %. Although the best performance is Cosine that performance is influenced with gold standard of abstractive human summary.

1. IntrIntroduction

Recent research on who investigates a summary, how to make it and how to evaluate it is being intensively conducted. This can be happened because of an increase in it [1]. The summary becomes a very important thing because end users get the meaning of a reading that can be obtained quickly and does not destroy the existing meaning. The summary that get by extracting is obtained from a document after searching through the search engines [2]. The number of text-summarization research conducted by researchers proves that research in this field becomes a trend worthy of research.

The summary has the main purpose of presenting the main ideas in a document that create the document less. If a sentence has value as important as the other sentence will result in ineffective summaries in the absence of a reduction in the size of information. On the other hand the reduction of information will result in a proportional decrease in a document [3]. The information content in document appears in bursts, and one can therefore distinguish between more and less informative segments.

There are two ways for creating the summary: manually or automatically. Summaries that are done automatically are called auto summarization. For Creating auto summarization there are two methods that are extraction and abstraction. [4].

The method of making a summary extraction is a method that takes an important sentence from a reading while the abstraction method takes the essence of a reading text, information fusion, sentence compression and reformulation [5].

The method of extraction is easier to make than the abstraction method. The method of extraction is done by three steps that is making a representation between the original text, scoring sentence and choosing the highest sentence scoring. Some methods are used on extractives such as Naïve-Bayes keyword extraction, Hidden Markov Model, Graph Method, Latent Sematic Indexing [4].

The important thing of Keyword extraction is mining the text of the document [6]. By extracting the appropriate keywords, we can easily select sentences in documents that have relationships with each other. The key extract calculation uses TF / IDF calculation, where this calculation indexes words in a sentence in a paragraph. The calculation is within the keyword extraction algorithm [7]. Research similar with these topics is called automatic term recognition in the context of computational linguistics, automatic indexing or automatic keyword extraction in research in information retrieval [8]. If someone wants to know the main statement of a paper, someone wants to have some keywords. The number of words is sometimes sufficient for the document overview; However, more powerful tools are desirable [9].

To represent a document, there are steps that are divided into paragraphs, sentences and tokens. Sometimes some preprocessing, such as stop word removal is also performed. The second step is to try to determine which sentence is important on a document [10]. Conversely abstractive summaries require machines to generate language and are difficult to replicate or extend to a wider domain.

Keyword Algorithm is easier and common in extracting summary. However, the algorithm has a problem in the lack of cohesion or correlation between sentences [11-12]. The correlation between sentences can be seen from how similar a sentence with other sentences in a text and to what extent the ideas in the text are expressed clearly by avoiding confusing information [11].

One way to resolve the problem of cohesion between sentences in extract summary is with determine the optimal combination between sentences [13]. The determination and cohesion optimization can be applied by using similarity method [14]. The function for similarity measure should be easy to compute, it should implicitly capture the relatedness of the documents, and it should also be explainable [15]. The similarity between two sentences, according to the vector representation described is calculated as the distance similarity [16].

The purpose of this research is to improve cohesion in the summary results based on keyword algorithm extraction combine with distance similarity method. Finally, our work of this paper is summarized in the last section.

2. Material and Method

The keyword algorithm uses the calculation of TF / IDF, which calculates the weight of a term that will determine performance a term in a document is in the corpus. Weighting is done by calculating a positive value for several times that the term occurs in a particular document, as well as a negative term to the number of documents containing the term. Consider term t and document d, where t appears in n of N documents in D. The TF-IDF function is of the form as follows:

$$TFIDF\ (t,d,n,N) = TF\ (t,d)\ x\ IDF\ (n,N)$$

When the TF-IDF function is run against all terms in all documents in the document corpus, the words can be ranked by their scores. A higher TF-IDF score indicates that a word is both important to the document, as well as relatively uncommon across the document corpus. This is often interpreted to mean that the word is significant to the document, and could be used to accurately summarize the document. TF-IDF provides a good heuristic for determining likely candidate keywords, and it has been shown to be effective after several decades of research.

Similarity measured such as Jaccard (sim (a,b) = $|A \cap B| / |A \cup B|$), Dice (sim (a,b) = $2.|A \cap B| / |A|+|B|$) and cosine coefficient (sim (a,b) = $|A \cap B| / \sqrt{|A|.|B|}$) are defining as arithmetic expression of cardinality of the set $|A|$, $|B|$, $|A \cup B|$ and intersection $|A \cap B|$ [17]. This coefficient among the most used similarity measures in science and are subject active research in many field such as approximate reasoning, decision making and fuzzy sets.

The summarization stage consists of three components i.e. the keyword extraction algorithm, the compression ratio selector and the cosine equation method. These three components will summarize the text being fed as a result of the final text summarized.

The first pre-processed document is tokenized by keyword extraction algorithm and then calculates TF/IDF for each term. Then sum all of TF/IDF term for each sentence and get sum of each sentence the next process is rank all of sentence based on sum of TF/IDF. After sentence is selected then perform calculation of their similarity with cosine, jaccard and dice method. After the calculation of the similarity method, the next process is re-arranging all of sentence based on the similarity from the highest to the lowest similarity. The compression ratio determines the position of sentence rank. In this study using a compression of 50% that means the sentence summary shrinkage 50% from the original text [18]. The new text with new sentence arrangement will be the final summarized text.

The resulting of extractive summary will be evaluated using two points of view, performance and degree of cohesion. F-Measure represents an evaluation for the performance of a summary while the degree of cohesion evaluates how closely the relationship of a sentence to another sentence. F-Measure is measuring how far the technique is capable of predicting of correct sentence. Evaluation can be classified into intrinsic and extrinsic evaluation [19]. Intrinsic evaluation judges the summary quality by its coverage between machine-generated summary and human generated summary. Extrinsic evaluation focuses mainly on the quality by its effect on other tasks. In intrinsic evaluation, Precision (P), recall (R), and F-measure (F) are used to judge the coverage between the manual and the machine generated summary:

$$P = \frac{|S \cap T|}{|S|} \tag{1}$$

$$R = \frac{|S \cap T|}{|T|} \tag{2}$$

$$\mathrm{F} = \frac{|2 * \mathrm{P} * \mathrm{R}|}{|\mathrm{R} + \mathrm{P}|} \tag{3}$$

Where S is the machine generated summary and T is the manual summary [19]. For the cohesion evaluation, we can measure with the formula as follows:

$$CoH = \frac{Log(C_s * 9 + 1\ )}{Log\ (M * 9 + 1)} \qquad Ns = \frac{(o) * (o - 1)}{2} \tag{4}$$

$$C_s \ = \frac{\sum_{\forall S_i, S_j \in Summary} Sim_{cos}\left(S_i, S_j\right)}{N_s} \tag{5}$$

$$M = \max Sim_{cos}\ (i, j)\ , i, j\ \leq N$$

$$Ns = \frac{(o) * (o - 1)}{2} \tag{6}$$

Where CoH corresponds to the cohesion of a summary, Cs is the average similarity of all sentences in the summary S, $Sim_{cos}(Si,Sj)$ is the cosine similarity between sentences Si and Sj, Ns is the number of nonzero similarity relationships in the summary, O is the number of sentences in the summary, M corresponds to the maximum similarity of the sentences in the document and N is the number of sentences in the document. In this way, CoH tends to zero when the summary sentences are too different among them, while that CoH tends to one when these sentences are too similar among them. Thus, this feature tends to favor the summaries that contain sentences about the same topic [12].

The dataset used in this research is collected from UCI Dataset containing documents of Reuters-21578 that has collection appeared on the Reuters newswire in 1987. The documents were assembled and indexed with categories by personnel from Reuters Ltd. (Sam Dobbins, Mike Topliss, and Steve Weinstein) and Carnegie Group, Inc. (Peggy Andersen, Monica Cellio, Phil Hayes, Laura Knecht, Irene Nirenburg) in 1987, https://archive.ics.uci.edu/ml/datasets/Reuters21578.
In Figure 1 can be explained that after the data from UCI Reuters- 21578 completed prepared then the data will be tested into summarization stage.

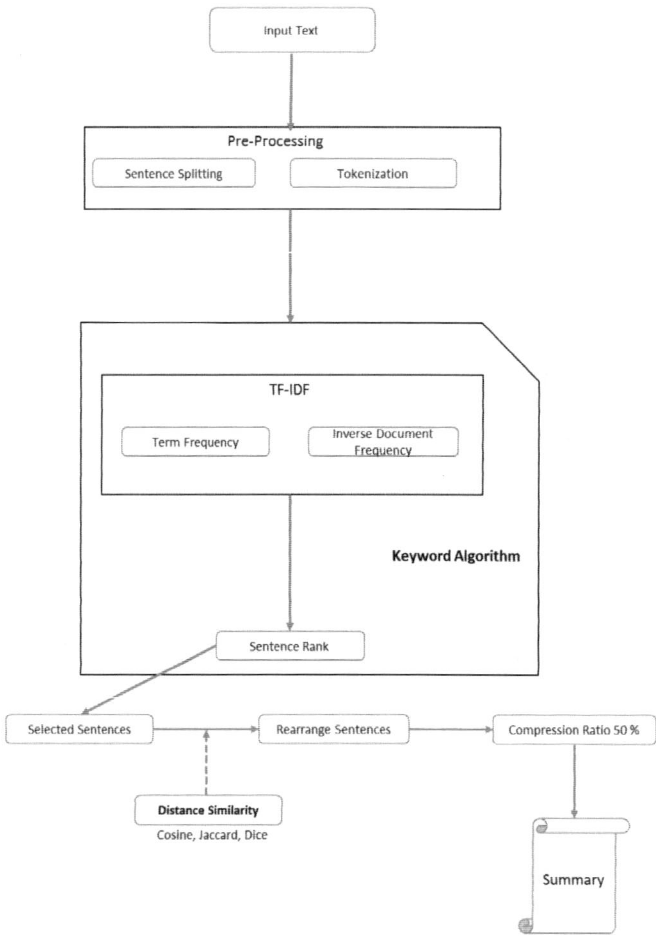

Fig. 1. Block Diagram Proposed Model

3. Result and Discussion

The research using computer platform with specification based on Inter Core I5 2.70 Ghz 8 Gb RAM Microsoft Windows 7 Ultimate 64 Byte. The Software is using Java with Netbeans IDE 8.2.

The evaluation of the result is using cohesion degree [12] that to know about how high related the sentences using combination of keyword extraction and distance similarity and using F-Measure to know performance of the summary as condition of intrinsic evaluation. In this experiment we use three distance similarity such as Cosine, Dice and Jaccard. This Similarity is very common used by researcher in many aspect [17].

After summarized using keyword algorithm with, the summary then re-calculated using the distance similarity for finding the best cohesion between sentences then compress half of the passage. The best cohesion then become candidate of sentence summary. Then we calculated the cohesion degree of the summary than can be seen in equation (4). Table 1 show the result of cohesion degree of summarization with distance similarity and without them

Table 1. Cohesion Degree from Distance Similarity (in percent)

	Jaccard	Cosine	Dice
cohesion degree	41.50	44.30	45.87

From Table 1 can be seen that Dice similarity is the best cohesion than another distance similarity such as Jaccard and Cosine that has value 45.87%. The cohesion before re-calculated using distance similarity is 37.47%. The improve cohesion using Dice is 8.40 %, Cosine is 6.83% and Jaccard is 4.03%. Although in another experiment conclude that Cosine has best performance when be compared with Dice and Jaccard [20] but another reason perhaps Dice similarity measure cannot be induced when one vector is zero which overcomes the disadvantage of the cosine measure [21] and also its depend with document categorization from datasets [15] .

Evaluation the performance of summarization using recall, precision and F-measure that the formula can be seen in equation 1 until 3 for three distance similarity show in Table 2 below.

Table 2. Recall, Precison and F-Measure

	Jaccard	Cosine	Dice	keyword
Recall	39.67%	52.91%	39.81%	40.56%
Precision	35.42%	46.50%	39.13%	37.46%
F-Measure	37.12%	48.54%	38.82%	37.41%

From Table 2 can be seen that Jaccard similarity such has Recall is 39.67 %, Precision 35.42 % and F-Measure 37.12 %. Cosine similarity has Recall 52.91 %, Precision 46.50 % and F-Measure is 48.54 %. Dice similarity has Recall 39.81 %, Precision 39.13 % and F-Measure 38.82 %. Based on this data, the best performance of the Distance similarity is Cosine similarity that has F-Measure is 48.54 % since the keyword itself has F-measure 37.40 %. So that the best improve performance from Cosine is 14.14 %. Although the best cohesion degree is Dice similarity but the best performance is Cosine. One reason to explain about this phenomenon is intersection human summary and machine summary is high. Intersection means how many words in machine summary have same similarity with number of word in human summary. If intersection is high, automatically make the high result although length of word in machine and human has big influence contribution to the result. This study has confirm that intersection between human summary and machine play big influence for evaluation measurement such as recall, precision and F-measure[11].

4. Conclusion

The main goal of summary is to present the main ideas in document in less space. If all sentences in a text document were equal importance, producing summary would not very effective, as any reduction in the size of a document would carry a proportional decrease in its informative.

In this research is used keyword algorithm model with three distance similarity. In this experiment, the best distance similarity that improved the cohesion degree is Dice similarity although the best performance is Cosine. Summarization with Cohesion degree not influence the performance because performance sometimes is measured compare with gold standard that be abstractive.

Acknowledgment

Thanks to Mr. Romi Satria Wahono that has guide us to make good research and encourage to present this research. Thanks to Sekolah Tinggi Teknik Muhammadiyah Cileungsi that support us to finish this paper.

Reference

[1] E. Hovy and C. Lin, “Automated and Text Summarization and The Summarist System,” *Assoc. Comput. Linguist. (Proceeding TIPSTER ’98 Proc. a Work. held Balt. Maryland)*, pp. 197–214, 1998.

[2] A. Porselvi & S.Gunasundari, “Survey on web page visual summarization,” *Int. J. Emerg. Technol. Adv. Eng.*, vol. 3, pp. 26–32, 2013.

[3] D. Radev, E. Hovy, and K. McKeown, “Introduction to the special issue on summarization,” *Comput. Linguist.*, vol. 28, no. 4, pp. 399–408, 2002.

[4] D. Das and A. F. T. Martins, “A Survey on Automatic Text Summarization,” *Eighth ACIS Int. Conf. Softw. Eng. Artif. Intell. Netw. ParallelDistributed Comput. SNPD 2007*, vol. 4, pp. 574–578, 2007.

[5] R. M. Aliguliyev, “A new sentence similarity measure and sentence based extractive technique for automatic text summarization,” *Expert Syst. Appl.*, vol. 36, no. 4, pp. 7764–7772, 2009.

[6] W. M. Wang, Z. Li, J. W. Wang, and Z. H. Zheng, “How far we can go with extractive text summarization? Heuristic methods to obtain near upper bounds,” *Expert Syst. Appl.*, vol. 90, pp. 439–463, 2017.

[7] M. Rajman and R. Besan, “Text Mining - Knowledge extraction from unstructured textual data,” *Proc. 6th Conf. Int. Fed. Classif. Soc.*, 1997.

[8] Y. MATSUO and M. ISHIZUKA, “Keyword Extraction From a Single Document Using Word Co-Occurrence Statistical Information,” *Int. J. Artif. Intell. Tools*, vol. 13, no. 01, pp. 157–169, 2004.

[9] H. P. Luhn, “A Statistical Approach to Mechanized Encoding and Searching of Literary Information,” *IBM J. Res. Dev.*, vol. 1, no. 4, pp. 309–317, 1957.

[10] R. Ferreira, L. De Souza Cabral, R. D. Lins, G. Pereira E Silva, F. Freitas, G. D. C. Cavalcanti, R. Lima, S. J. Simske, and L. Favaro, “Assessing sentence scoring techniques for extractive text summarization,” *Expert Syst. Appl.*, vol. 40, no. 14, pp. 5755–5764, 2013.

[11] K. Nandhini and S. R. Balasundaram, “Improving readability through extractive summarization for learners with reading difficulties,” *Egypt. Informatics J.*, vol. 14, no. 3, pp. 195–204, 2013.

[12] M. Mendoza, S. Bonilla, C. Noguera, C. Cobos, and E. León, “Extractive single-document summarization based on genetic operators and guided local search,” *Expert Syst. Appl.*, vol. 41, no. 9, pp. 4158–4169, 2014.

[13] M. A. Fattah and F. Ren, “GA, MR, FFNN, PNN and GMM based models for automatic text summarization,” *Comput. Speech Lang.*, vol. 23, no. 1, pp. 126–144, 2009.

[14] Y. Bestgen, “Improving Text Segmentation Using Latent Semantic Analysis: A Reanalysis of Choi, Wiemer-Hastings, and Moore,” *Comput. Linguist.*, vol. 32, no. 3, pp. 455–455, 2006.

[15] M. Rafi and M. S. Shaikh, “An improved semantic similarity measure for document clustering based on topic maps,” *Comput. Sci. Dep. Karachi Pakistan*, no. March, 2010.

[16] I. C. Mogotsi, "Christopher D. Manning, Prabhakar Raghavan, and Hinrich Schütze: Introduction to information retrieval," *Inf. Retr. Boston.*, vol. 13, no. 2, pp. 192–195, 2010.

[17] S. Jimenez, F. A. Gonzalez, and A. Gelbukh, "Mathematical properties of soft cardinality: Enhancing Jaccard, Dice and cosine similarity measures with element-wise distance," *Inf. Sci. (Ny).*, vol. 367–368, pp. 373–389, 2016.

[18] R. Darmawan and R. Wahono, "Hybrid Keyword Extraction Algorithm and Cosine Similarity for Improving Sentences Cohesion in Text Summarization," *J. Intell. Syst.*, vol. 1, no. 2, pp. 109–114, 2015.

[19] K. Nandhini and S. R. Balasundaram, "Extracting easy to understand summary using differential evolution algorithm," *Swarm Evol. Comput.*, vol. 16, pp. 19–27, 2014.

[20] V. Thada and V. Jaglan, "Comparison of Jaccard, Dice, Cosine Similarity Coefficient To Find Best Fitness Value for Web Retrieved Documents Using Genetic Algorithm," *Int. J. Innov. Eng. Technol.*, vol. 2, no. 4, pp. 202–205, 2013.

[21] J. Ye, "Multicriteria decision-making method using the Dice similarity measure based on the reduct intuitionistic fuzzy sets of interval-valued intuitionistic fuzzy sets," *Appl. Math. Model.*, vol. 36, no. 9, pp. 4466–4472, 2012.

A New Geological Map of Formation Distribution on Southern Part of South China Sea:; Natuna Island, Indonesia

H Kausarian[1,*], S Lei[2], G T Lai[3], Y Cui[2], A Suryadi[1]
[1]Geological Engineering Program, Faculty of Engineering, Universitas Islam Riau, Jalan Kaharuddin Nasution no. 113 Pekanbaru, Riau 28284, Indonesia
[2]Laboratory of Marine Geology, Tongji University, 1239 Siping Road, Shanghai 200092, China
[3]School of Environment and Natural Resource Sciences, Faculty of Science and Technology, Universiti Kebangsaan Malaysia, 43600 UKM Bangi, Selangor, Malaysia

[*]E-mail: husnulkausarian@eng.uir.ac.id

Abstract. Natuna Island is located at the outer part of the Indonesian territory and also as the end part of South China Sea at the south. Natuna consists of the islands that are surrounded by coral reefs and composed by sedimentary rock and sediment surface (alluvium). Natuna is a part of the Sunda group (Sundaland) which evolved through tectonic events (on the east) and volcanic activity, the formation of this island occurred in late Triassic time. Survey (field observation) applied for geological mapping by sample analyzing and tracking using handheld GPS. This research uses Natuna geological map from geological research and development center of Indonesia in 1994 with the scale of 1: 250,000 and interpretation of satellite imagery from the year 2015 with scale 1: 50,000 as reference material. The new spreading of formation distribution is seen in the Raharjapura Formation which has the lithology of alternating sandstone and repeating limestone spread across the north to the southwest and dominates the arrangement of formations in this island. Another new spreading occurs in the Eastern for Pengadah Formation, this formation has conglomerate lithology, sandstone aligned with Raharjapura's formation. Hydrocarbon outcrop also found in the northeastern region of Natuna, this is a common thing because the Natuna islands are famous for its natural wealth of oil and gas. The final result is an updated geological map showing significant changes to the distribution of rock formations and alluvium and also propose the idea for developing this island as the vital area for Indonesian territory border.

1. Introduction

Study area located in Natuna Island, Riau Island Province [1] is the outer part of Indonesian territory and also as the end part of South China at the south and connected to the Melaka Strait at the south [2]. Natuna island is configured by two tertiary basins, namely West Natuna Basin and East Natuna Basin, Separated by Natuna Arch with trend north-south (Figure 1). Natuna Area is part of the Sundaland, consequently the evolution of this area can be traced back to the forming of Sundaland. Sundaland was formed and cratonised in Late Triassic time that caused Sundaland to become a stable continent of South East Asia [3]. Natuna Island surrounded by an active subduction zone, its actions resulted in the forming mélange complex, as is found in parts of Natuna Island in the Natuna Arch [4]. West Natuna Basin is related with spreading of South China Sea. The collision of Pacific plate with Eurasia resulting in the

stop of subduction activity during Eocene. Then that plate collapse because of the gravity effect. The final result of those events is spreading which form South China Sea [5]. Meanwhile, East Natuna Basin in past is forearc basin encounter shallow sediment deposit and deep marine facies during Oligocene subduction activity desist [6]. The oldest sediment at East Natuna Basin is Late Oligocene which are deposits from alluvium and delta plain.
Geological information of Natuna Island is mostly from exploration in oil and gas industry (offshore). The Geological information for Natuna Island is very least. The study area is focused onshore on Natuna Island which has a very complex geological condition. An updated research is needed to improve the information on Natuna Island to increase the accuracy of geological condition at Natuna Island. The objective of this study is trying to provide the information for the border of Indonesian territory in term of the location of Natuna Island as the outer part of Indonesian territory. another objective is for proposing as the Geopark by its resource of geo-diversity.

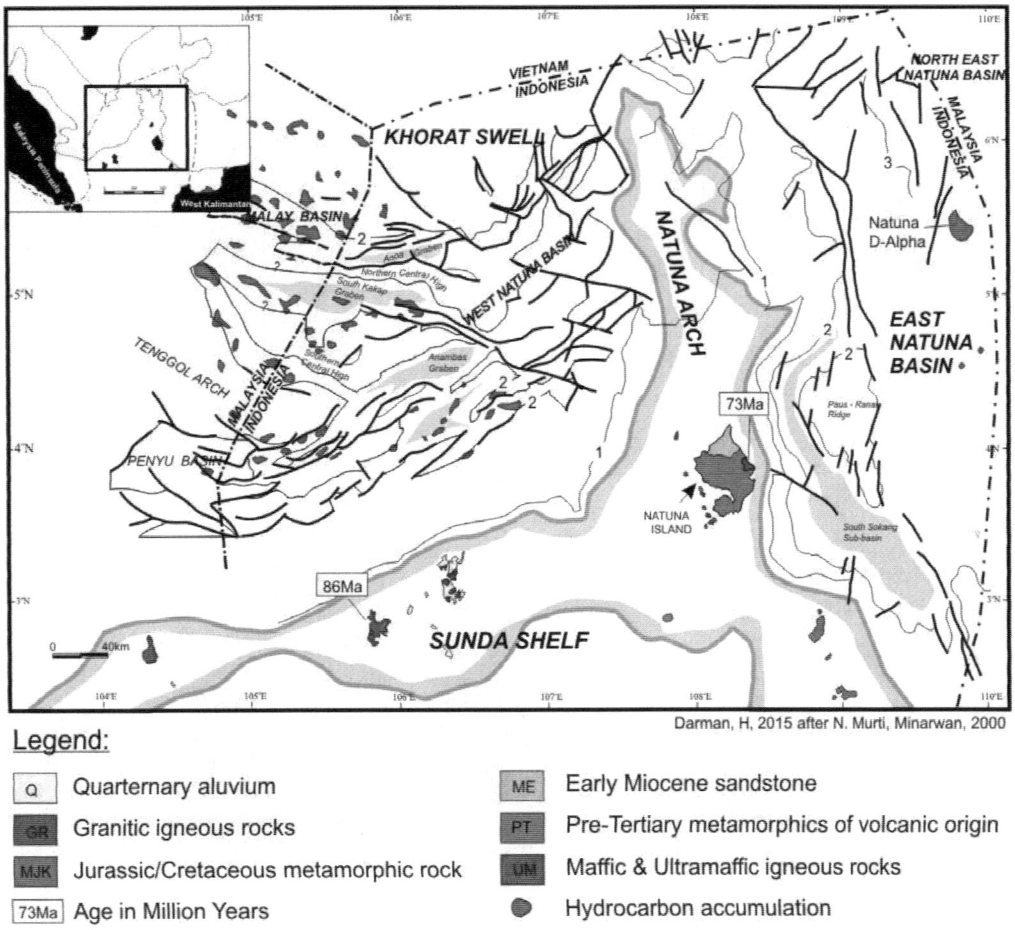

Figure 1. Structural elements of West Natuna Basin, East Natuna Basin and Natuna Arch (Natuna Island).

2. Material and methodology

This research conducted in 3 steps, literature study, field observation and laboratory analysis. The first step is literature study to get information of regional geology at study area which consists of some formations namely Alluvium, Raharjapura Formation, Pengadah Formation, Ranai Granite, Bunguran Formation and Mafic/Ultramafic Rock.

The next step is fieldwork or field observation to get geological authentic data of study area. Field observation performs with tracking the road using handheld GPS [7] – [11]. Geological data collected focused on identification of lithology at the outcrop to determine the rock formation of it.
The final step is laboratory analysis done by comparing the result of field observation with geological maps of Natuna Island which has existed. In this case the map that made as references are Geological Map of The Teluk Butun & Ranai Sheet, Sumatera [12] with scale 1:250.000 and Geological Map from interpretation of Satellite Imagery [13], [14] X Sheet Riau Island Province [15] with scale 1:50.000. From the comparing field data and a geological map will result in a new updated geological map of Natuna Island.

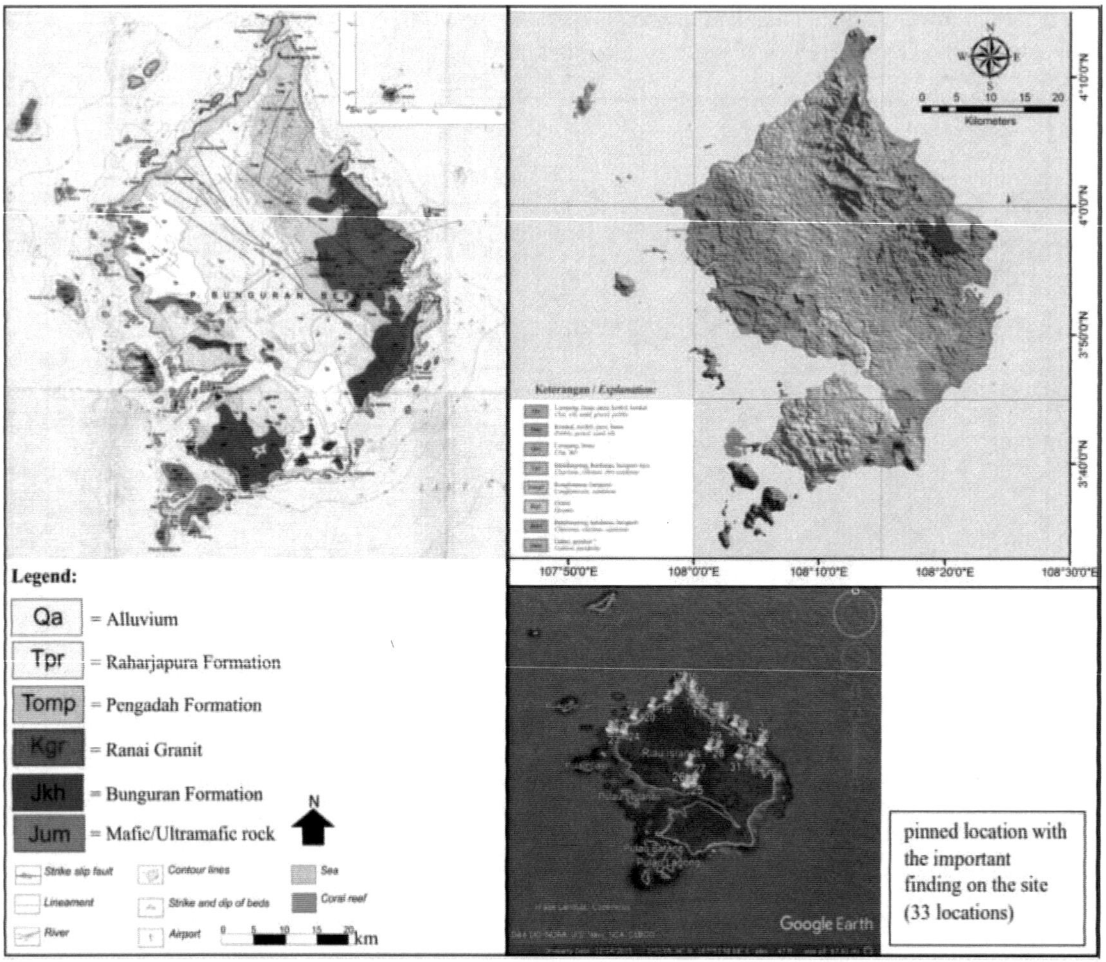

Figure 2. Comparing references maps with field observation, clockwise; Geological Map of The Teluk Butun & Ranai Sheet, Sumatera (scale 1:250.000); Geological Map from interpretation of Satellite Imagery X Sheet Riau Island Province scale 1:50.000 (combined); Map of field observation at Natuna Island modified from Google Earth.

3. Result and discussion

The result of 36 field station and compartment with reference maps show some changes to the boundary between some formations of the study area. An updated geological map of formation distribution on Natuna Island can be seen in Figure 3.

3.1. The update distribution of Alluvium (Qa)

Alluvium deposit in the study area consists of gravel, sand, silt and peat deposits. Alluvium spreads on west and south part of Natuna Island. Silica sand is easy to find along the coastline on the east part of Natuna Island which closed with the downtown of Ranai until the Alif Stone tourism area. The age of alluvium is Quaternary Holocene.

3.2. The update distribution of Raharjapura formation (Tpr)

Raharjapura formation (Figure 4) widespread from northern to southern part of Natuna Island. Lithology that dominated in this formation consists of the alternation of sandstone, siltstone, and mudstone. The various sandstone is yellowish-white sandstone, well sorted with few of organic material (STN11 and STN14), sandstone with rock fragment and traces fossil (STN12), and sandstone with pure white color (ST15). Besides that, an outcrop of black sandstone also found with high TOC (Total Organic Carbon) that allegedly as seeping of hydrocarbon outcrop (STN10). In others hand, mudstone and siltstone mostly blackish gray with the high organic material.

Stratigraphically, Raharjapura Formation and Pengadah Formation have unconformable contact. It was shown at STN14 and STN15 outcrops where sandstone of Raharjapura Formation overlies unconformable sandstone of Pengadah Formation with the angle (angular unconformity). The age of Raharjapura Formation is correlated with the Udang Formation (Pliocene) exposed in the West Natuna Basin [16].

3.3. The update distribution of Pengadah formation (Tomp)

Distribution of Pengadah Formation (Figure 5) is from the northeast until the east part of Natuna Island. Pengadah Formation dominated by coarse grain lithology such as conglomerate (STN23), coarse grain sandstone (STN6) with some siltstones. Clast of conglomerate has various clasts (polymictic) with different size ranging from 2 cm to 20 cm. furthermore, interbedded of red sandstone, conglomerate and siltstone found at STN21. Trace fossil easily found in mudstone of this formation (STN32) with some lenses of sandstone. In others hand, unconformity contact between Pengadah Formation and Bunguran Formation can be seen at STN23 where the conglomerate of Pengadah Formation overlying the chert layer of Bunguran Formation. Unconformity contact between both marked by an erosional surface. The age of Pengadah Formation estimated from Oligocene to Miocene in the West Natuna Basin [16].

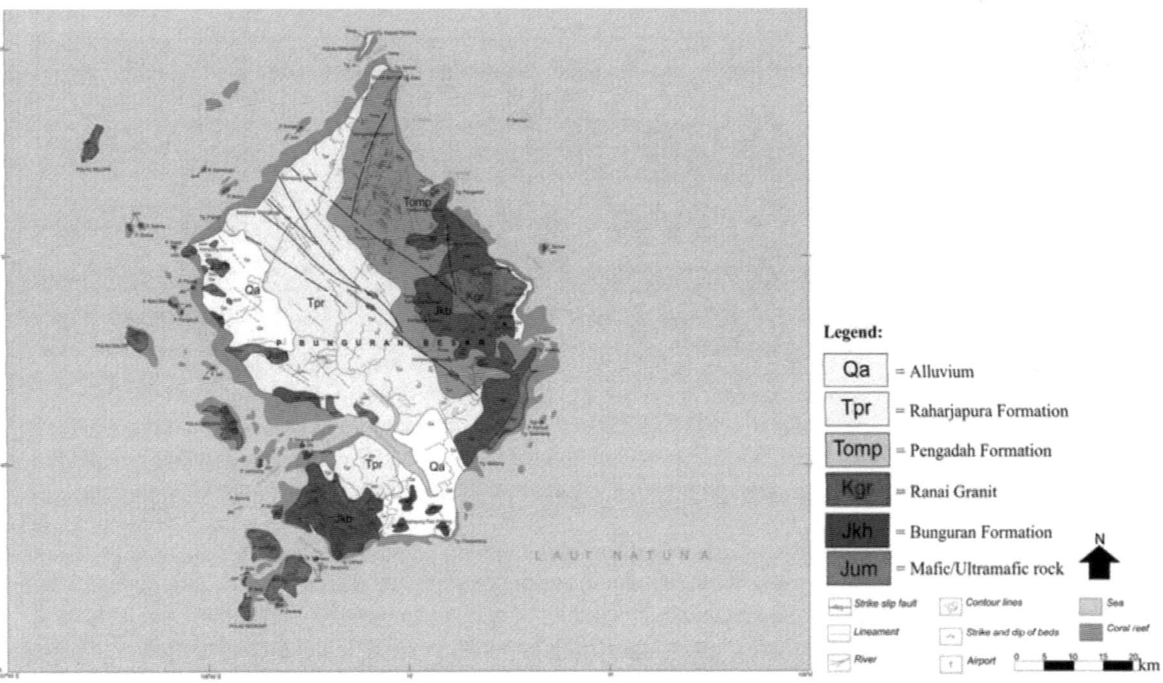

Figure 3. Geological map of formation distribution on Natuna Island resulting from this study.

Figure 4. Raharjapura Formation outcrop; A. yellowish sandstone STN11; B. Sandstone with few organic materials STN14; C. Yellowish sandstone with rock fragment STN12; D. sandstone with trace fossil STN12; D. High TOC sandstone as seeping layer of hydrocarbon outcrop STN10; E. Angular unconformity between Raharja Formation and Pengadah Formation.

Figure 5. Pengadah Formation outcrop; A. coarse grain sandstone with gravels; B. unconformity contact between conglomerate of Pengadah Formation and chert of Bunguran formation; C. Interbedded of conglomerate, mudstone, and sandstone STN21; D. mudstone with trace fossil and sandstone lenses; E. coarse grain sandstone with few organic materials STN25.

3.4. The update distribution of Ranai granite (Kgr)

Ranai Granite (Figure 6) exposed at Ranai mountain and tourism area of Alif Stone located at the western eastern part of Natuna Island. Characteristic of Ranai Granite is white with porphyritic and aphanitic texture. From field observation, minerals contain that identified are quartz, feldspar, plagioclase, biotite, and muscovite. There are 3 periods of granite at STN33 which shown in Figure 6 (C). The relative ages of those granite determined by stratigraphy law (Inclusion and cross-cutting). Granite 1 is the oldest granite because it trapped in granite 2. Meanwhile, granite 2 intruded by aplite 3 (fine grain of granite) that indicate granite 2 is older than aplite. The age of Ranai Granite around Cretaceous [17].

3.5. The update distribution of Bunguran formation (Jkb)

Bunguran Formation (Figure 7) mostly exposed at the eastern part of Natuna Island. Lithology that found is chert interbedded with siltstone and tuff. Some kinds of chert were found at the field like massive chert (STN3) with white color, reddish chert interbedded with graded bedding sandstone and mudstone (STN24), and black chert with pyrite contain. Tuff deposit characterized by loose material which interbedded with chert shown at an outcrop of STN23. Chert and graded bedding sandstone indicate that environment of depositional this formation is deep sea deposit.

Contact between Bunguran Formation and Ultramafic Rock found at STN17. Chert of Bunguran Formation overlies unconformably on weathered and eroded basalt with reddish brown. The age of Bunguran formation is late Jurassic up to mid-Cretaceous [18]-[19].

3.6. The update distribution of Mafic / Ultramafic rock (Jum)

The Mafic/Ultramafic rock (Figure 8) exposed randomly on the western and southern part of Natuna Island. Some field stations that expose mafic/ultramafic rock namely STN5, STN17, STN22, and STN30. From reference map, mafic/ultramafic rock did not expose at the eastern part of Natuna Island but from field observation basalt outcrop (fresh and weathered) found at STN5, STN17, and STN24 which located at eastern part of Natuna Island. STN5 near Kelanga Village basalt outcrop found with aphanitic texture and blackish green in color. At Akar Island (STN22), basalt found in pillow basalt structure that only exposed when low tide. At Mekar Jaya Village, the complex outcrop of mafic/ultramafic rock exposed where in that outcrop consist of some kinds of mafic/ultramafic rock. Based on field observation, mafic/ultramafic rock that identified were Peridotite, Gabbro, and Basalt.

Figure 6. Ranai Granite outcrop; A. white granite STN1; B. Granite outcrop at tourism area Alif Stone; C. three periods of granite intrusion (STN33).

Figure 7. Bunguran Formation outcrop; A. Massive white chert STN3; B. reddish chert interbedded with graded bedding sandstone and mudstone STN24; C. Graded bedding sandstone; D. layered chert STN27; E. Contact between chert of Bunguran Formation and mafic rock (weathered basalt) STN17; F. Black Chert STN20.

Figure 8. Mafic/ultramafic rock outcrop; A. basalt at Kelanga Village STN5; B. Contact between weathered basalt and chert (Bunguran Formation) STN17; C. Basalt at Akar Island STN22; D. complex mafic/ultramafic rock at Mekar Jaya Village STN30.

4. Conclusion

Generally, this study shows significant changes in the boundary for some formation. On the western and southern part of Natuna Island covered by Alluvium. Raharjapura Formation exposed from northern until southern part of Natuna Island. Pengadah Formation commonly found on the northeast to the center part of Natuna Island. The angular unconformity contact between Raharjapura formation and Pengadah Formation clearly shown on the field. Ranai granite boundary extends to southwest from the previous map produced. There are three periods of intrusion Ranai Granite in the study area. Bunguran Formation (late Jurassic up to mid-Cretaceous) consists of chert, graded bedding sandstone, siltstone, and tuff is well distributed mostly at the eastern part of Natuna Island. Meanwhile, mafic/ultramafic rock from references map used only exposed randomly at south and west of the study area. However, the outcrop of mafic/ultramafic rock (Basalt) also found at the eastern part of Natuna Island from this study. This study also propose a clear information for the border concern of Indonesian territory, because Natuna Island is the outer part of Indonesian territory, beside this Island is very suitable for proposing as the Geopark by its resource of geo-diversity.

References

[1] Lubis M Z, Pujiyati S, Pamungkas D S, Tauhid M, Anurogo W, Kausarian H 2018 *Biodiversitas* **19**(4) 1451
[2] Kausarian H, Sumantyo J Tetuko S, Kuze H, Karya D, Panggabean G Fried 2016 *Int. J. on Adv. Sci. Eng. and IT* **6**(5) 568
[3] Hutchison C S 1989 *Geological evolution of South-east Asia* Oxford: Clarendon Press.
[4] Hamilton W B1979 *Tectonics of the Indonesian region* US Govt. Print. Off.
[5] Doust H, Sumner H S 2007 *Petroleum Geoscience* **13**(2) 127
[6] White Jr J M, Wing R S 1978 *Structural development of the South China Sea with particular reference to Indonesia* 159
[7] Kausarian H, Sumantyo J Tetuku S, Kuze H, Karya D, Wiyono S 2017 *ARPN J. of Eng. and Apl. Sci.* **12**(4) 980
[8] Kausarian H, Sumantyo J Tetuko S, Kuze H, Aminuddin J, Waqar M M 2017 *Can. J. of R. S.* **43**(2) 95
[9] Kausarian H, Batara B, Putra D B, Suryadi A, Lubis M Z 2018 *Int. J. on Adv. Sci. Eng. and IT* **8**(3) 856
[10] Suryadi A 2016 *J. of Geos. Eng. Env. and Tech.* **1**(1) 1
[11] O'Rourke T D, Palmer M C 1996 *Earthquake Spectra* **12**(3) 493
[12] Hakim A S, Suryono N 1994 *Peta Geologi Lembar Teluk Butun dan Ranai. Sumatera* Bandung: Pusat Penelitian Pengembangan Geologi
[13] Lubis M Z, Anurogo W E, Hanafi A D, Kausarian H, Taki H M, Antoni S A 2018 *Biodiversitas* **19**(3) 1117
[14] Kausarian H, Batara B, Putra D B 2018 *J. of Geos. Eng. Env. and Tech.***3**(1) 39
[15] Suwijanto 2015 *Geological Map Interpretation of Remote Sensing Image Natuna Island Riau Island Province* Badan Geologi Indonesia
[16] Daines S R 1985 *Structural history of the W Natuna basin and the tectonic evolution of the Sunda Region*
[17] Tjia H D 1998 *Origin and tectonic development of Malay-Penyu-West Natuna basins*
[18] Haile N S 1970 *ECAFE, Tech. Bull.* **3** 55
[19] Lubis M Z, Anurogo W, Kausarian H, Choanji T, Antoni S, Pujiyati S 2018 *IOP Conf. Ser.: Ea. and Env. Sci.***176**(1) 012025

Molecular Reidentification of Endophytic Fungus LBKURCC40 Isolated from Dahlia's Tuber (*Dahlia Variabilis*) Using the Sequence of ITS rDNA

Nabella Suraya[1], Finna Piska[1], Nova Wahyu Pratiwi[2], Aulia Ardhi[3], Saryono[1]
[1]Department of Chemistry Faculty of Mathematics and Natural Sciences, Universitas Riau
[2]Department of Biology Faculty of Mathematics and Natural Sciences, Universitas Riau
[3]Department of Food and Agricultural Product Technology Faculty of Agricultural Technology, Universitas Gadjah Mada

Corresponding author's email: saryono@lecturer.unri.ac.id

Abstract: Endophytic fungus LBKURCC40 isolated from dahlia tubers (*Dahlia variabilis*) was morphologically and molecularly identified. Molecular identification was carried out using ITS4 and ITS5 primers, furthermore, the homology level of LBKURCC40 isolates was checked using BLAST program and the suitability of BLAST results was tested with the TrichOKEY program. The results of showed that LBKURCC40 isolates were identified as *Trichoderma asperellum* with a homology level of 99%. Phylogenetic analysis was carried out with N-J tree bootstrap test with 10,000 repetitions, and LBKURCC40 isolate was obtained to have the closest kinship to *Trichoderma asperellum* strain Tr266B.

Keywords: dahlia, endophytic fungus, ITS, Trichoderma

1. Introduction

Dahlias are tubular horticultural plants that contain compounds with high bioactivity. Endophytic microbes in dahlia plants are known to produce compounds with the same bioactivity as dahlia plants. Continuous interaction of dahlia tuber with microbes such as symbiotic mutual fungi produces endophytic fungi with the same ability as the host. The ability of these endophytic fungi provides an alternative to overcome drug resistance, and obtain the most desirable plant properties [11, 12, 14].

Endophytic fungi are known to correlate with their hosts in the ability to produce secondary metabolites. The type and amount of secondary metabolites produced by endophytic fungi is influenced by ecological or environmental conditions, such as temperature, humidity, and soil nutrients [8]. Some compounds produced by endophytic fungi have biological activities, such as alkaloids, terpenoids, and phenolics [16]. In agriculture, endophytic fungi can be used as biocontrol agents for various pathogenic microbes, insects, nematodes and pests [17]. In industry, endophytic fungi can be used to produce enzymes such as amylase, lipase, protease, and inulinase [2, 15].

In this study the identification of endophytic fungus in orange-flowered dahlia was carried out morphologically and molecularly. Morphological identification is done macroscopically and microscopically. Morphological identification can only identify to the genus level, and is not a reliable indicator for genetic identification between species, and is not sufficient to accurately identify species [7]. For that use advanced identification using molecular identification at the ITS region using primers ITS4 (forward) and ITS5 (revers). The ITS area is one of the fastest evolving regions and varies between species in the genus [6]. This area has several features so that it is easy to identify fungal molecules,

IOP Conf. Series: Materials Science and Engineering **532** (2019) 012021 doi:10.1088/1757-899X/532/1/012021

namely: (i) the length of the ITS area between 600 and 800 bp and can be easily strengthened using universal primers specific to the rRNA gene; (ii) has multicopy properties from rDNA repetition so that the ITS area is relatively easy to strengthen small, dilute or highly degraded DNA samples, and (iii) the ITS area varies greatly between fungal species with different taxonomies or even isolates from the same species [9].

The identification of endophytic fungus LBKURCC40 from orange flowering dahlia tubers has previously been carried out by previous researchers, but further identification is needed to ascertain the species of these isolates. Previous identification morphologically by Lorenita (2013) identified LBKURCC40 was *Monilia* sp. Morphological identification can only identify to the genus level, and has a low level of accuracy. This is because some species have almost the same morphological characteristics, so it's difficult to distinguish. To ensure the species from the isolates obtained, then the reidentification of LBKURCC40 isolates was morphologically and molecularly.

2. Methods

2.1. Morphology

Culture was grown on Potato Dextrose Agar (PDA) media at room temperature for 7 days. Morphological identification is done macroscopically and microscopically. Macroscopic morphology observation is the characterization of colonies and colony colors, while the microscopic identification of morphology is performed on conidia, hyphae, conidial heads, conidiophores and spores.

2.2. DNA extraction

LBKURCC40 isolate's DNA used is a collection of Enzyme, Fermentation, and Molecular Research Laboratory Universitas Riau which has been isolated Lorenita, *et. al* (2013) isolated from orange flowering dahlia tubers. Isolates were grown on Potato Dextrose Agar (PDA) media on petri dishes using the spread plate method. Micellia of fungi are isolated using the Wizard Genomic DNA Purification Kit (Promega). The results of DNA extraction continued to the electrophoresis stage using 0.8% agarose gel.

2.3. PCR amplification and DNA sequencing

The DNA obtained was used as a template for PCR amplification. PCR amplification using the ITS4 universal primer (5'-TCC GCT TAT TCC TGA TAT GC-3 ') as forward and ITS5 (5'-AGT AAA AGT ACA AGG-3' CGT) as revers. PCR amplification was carried out with a total volume of 50 μL containing 2 μL of DNA isolate, 10 μL of each primer concentrating 2 olmol, 5 μL dNTP concentrating 2mM, 10 μL buffer + 5 μL MgCl2, 7.75 μL H2O and 0.25 μL Go Taq DNA Polymerase. The reaction begins with a hot star at 95°C for 5 minutes, denaturation at 94°C for 1.5 minutes, annealing process at a temperature of 53°C for 1 minute, and an extension process at 72°C for 3 minutes. This PCR reaction process lasts for 35 cycles.

Sequencing results were verified with each other with the help of the BioEdit Sequence Alignment Editor and Analysis program for Windows 95/98/NT (Hall, 1999) and the proximity between each sequence was performed using clustal X version 2.0.9 (http://www.clustal.org) (Thompson, *et. al.*, 1997). Phylogenetic analysis was carried out using the Neighbor-joining method using MEGA v software. 6. The confidence value of each branch is determined by 10,000 times the bootstap analysis.

3. Results and discussion

LBKURCC40 isolate was identified morphologically by looking at the growth and fungi colonies on PDA media. Morphological identification is done macroscopically and microscopically. Macroscopic morphological observations were performed by observing colony color and colony growth, while microscopically observing hyphae whether septic or not, transparent or cloudy, colored or not.

Morphological identification results showed that the LBKURCC40 isolate was *Tricoderma* sp, with macroscopic as follows, at the beginning of the incubation the isolate would be white, and then it would turn into dark green, with rapid growth. Whereas the results of identification microscopically indicate characteristics of mycelium having septa, branching conidiophores, and conidia in round shape. This characteristic is in accordance with that reported by Saryono, *et. al* (2018). Saryono, *et. al* (2018) reported *Trichoderma* had conidiophore branching down the recurrent lateral branch, conidia round to oval-shaped spring, early colonies white and greenish, and dark yellow or dark green mycelium color. The results of the identification of the morphology of isolates LBKURCC40 can be seen in **Figure 1.**

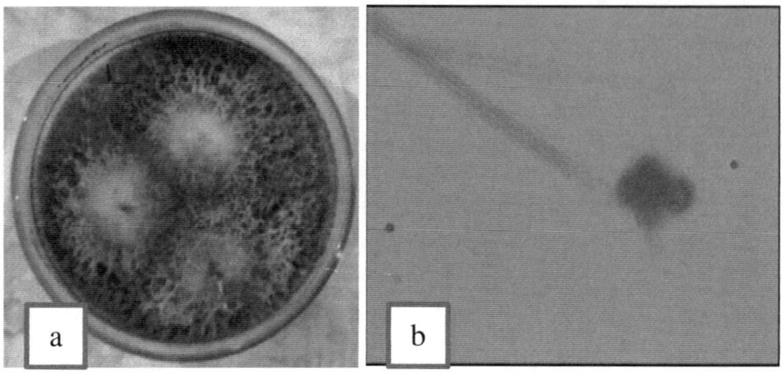

Figure 1. Morphological characteristics of LBKURCC40 fungi (a) Macroscopic (b) Microscopic with 40x magnification

Figure 1. Macroscopic and microscopic identification of morphological and LBKURCC40 isolates was *Trichoderma* sp. The results of this identification are different from the identification carried out by Lorenita, *et. al* (2013). The difference results with what has been done by Lorenita, *et. al* (2013) because morphological characteristics alone are not sufficient to identify the fungus to the species level appropriately, and molecular identification needs to be acknowledged because the *Trichoderma* species in addition have relatively few morphological characteristics and limited variations that cause species identification errors [6]. For this reason, further identification is needed using the TrichOKEY program.

Molecular identification of LBKURCC40 isolates was carried out in the ITS area using ITS4 and ITS5 primers, and the obtained DNA homology sequence compared to the data contained in GenBank using the BLAST program on NCBI. BLAST analysis was carried out to see the level of sequence homology by comparing the sample sequences of LBKURCC40 fungi isolates in the ITS rDNA area with sequences contained in the database by alignment. Alignment analysis can be done to compare two or more sequences. The results of the homology analysis of sequences from LBKURCC40 isolates are shown in Table 1.

Table 1. Homology identification of the LBKURCC40 isolate sequence

Species name	Strains	GenBank Access Code	Reference	% Identity of BLAST results
T. asperellum	Tr266B	KF737411.1	**GenBank**	99%
T. asperellum isolat 2730		EU272534.1	**GenBank**	99%
T. asperellum	T34	EU077227.1	**GenBank**	99%
T. pubescens	DAOM 166162	NR_077179.1	**GenBank**	99%
T. hamatum	CIB T144	EU280105.1	**GenBank**	99%
T. viride		X93981.1	**GenBank**	99%
T. velutinum	TUB F-801	AF149873	[3]	86%
T. fertile	DAOM 167161	AF400260	[3]	89%
T. oblongisporum	CBS 343.93	AF149854	[3]	95%
T. longipile	CBS 340.93	AY865630	[3]	92%
T. stromaticum	CBS 101875	AF098287	[3]	91%
T. spirale	CBS 346.93	AF400262	[3]	89%

T. ovalisporum	Dis70a	AY380897	[3]	98%
T. erinaceum	DAOM 230015	AY266658	[3]	98%
T. strigosum	CBS 348.93	AY387661	[3]	98%
T. hamatum	DAOM 167057	Z48816	[3]	99%
T. asperellum	CBS 433.97	AY380912	[3]	99%
T. longibrachiatum	CBS 816.68	Z31019	[3]	96%
T. ghanense	G.J.S. 95-137	Z69588	[3]	97%
T. saturnisporum	CBS 330.70	Z48726	[3]	96%
T. sinensis	TUB F-1043	AF486014	[3]	93%
T. effusum	TUB F-354	AF149858	[3]	95%
T. konilangbra	G.J.S. 96-146	AF400261	[3]	96%

Alignment of the BLAST program is carried out statistically by the system by shifting each sequence, and from the number of suitable base pairs. The alignment sequence obtained is used to determine the homology of the sequence being compared. BLAST alignment results in Table 1 show that LBKURCC40 isolates have more than 85% homology with some *Trichoderma* species. The percentage of homology is obtained by comparing the overall amino acid residue equal to the length of the DNA sequences of isolates LBKURCC40.

The results of BLAST identification showed that the isolates obtained were indeed *Trichoderma*. Further identification of advanced species was carried out using the TrichOKEY program. TrichOKEY is a Trichoderma molecular identification program at the species level based on DNA BarCode nucleotide DNA [6]. TrichOKEY uses an online method to identify molecules quickly from an isolate at the genus, clade and species level based on a diagnostic combination of several oligonucleotides (special marks) specifically found in ITS1 and ITS2 [4]. The results of the identification of LBKURCC40 isolates using the TrichOKEY program can be seen in Figure 2.

The results of species identification with the TrichOKEY program (Figure 2) show that the species obtained were *Trichoderma*, in accordance with morphological data both macroscopically and microscopically. These results are reinforced by the 5 anchors shown in **Figure 2 (a)** and the species with the closest homology match, namely *Trichoderma asperellum*. These nucleotides among *Trichoderma* species have a very high degree of similarity, making it difficult to distinguish species and frequent species identification errors.

The kinship relationship between fungi at all levels is determined by phylogenetic trees. This difference is due to significant nucleotide differences [5]. Alignment in the filogram tree shape is shown in Figure 2.

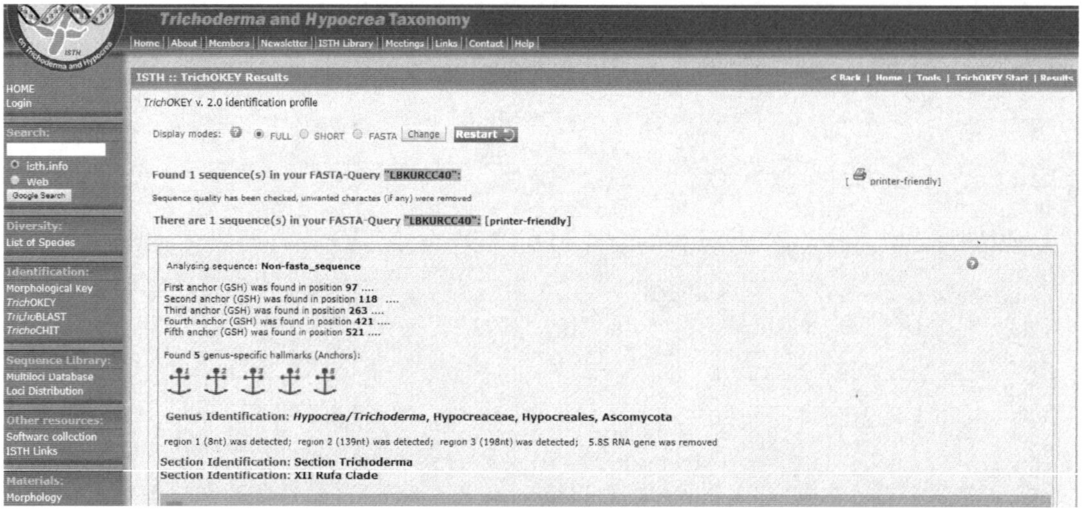

(a)

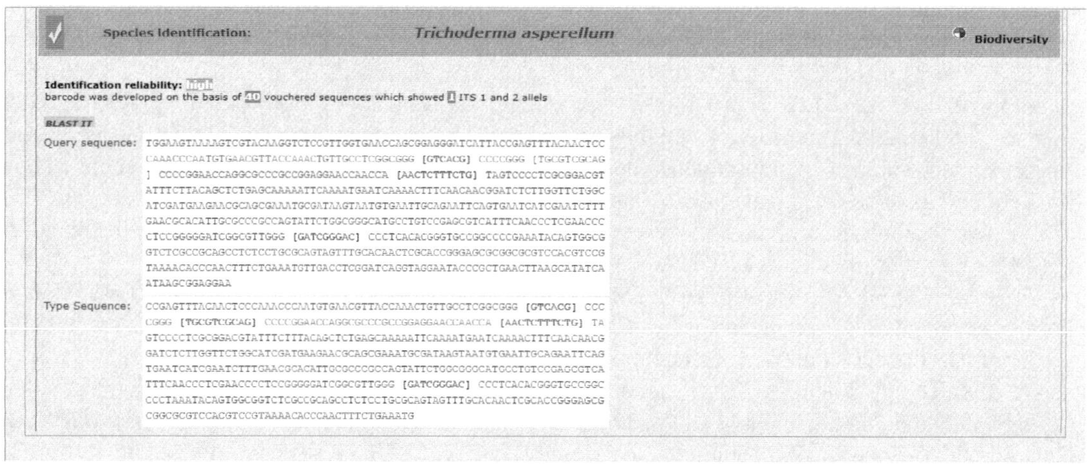

(b)

Figure 2. Identification of LBKURCC40 isolates using the TrichOKEY program

Figure 3 is a filogram tree of LBKURCC40 fungi that shows kinship based on nucleotide changes from evolutionary time. Based on Figure 3, *Trichoderma* LBKURCC40 fungi have the closest kinship to Trichoderma asperellum strain Tr266B fungi with branching number 8558. Bootstrap analysis is carried out with 10,000 repetitions, aimed at testing phylogenetic tree branching and validation. The bootstrap value obtained is more than 70%, which shows that the grouping done is accurate [1].

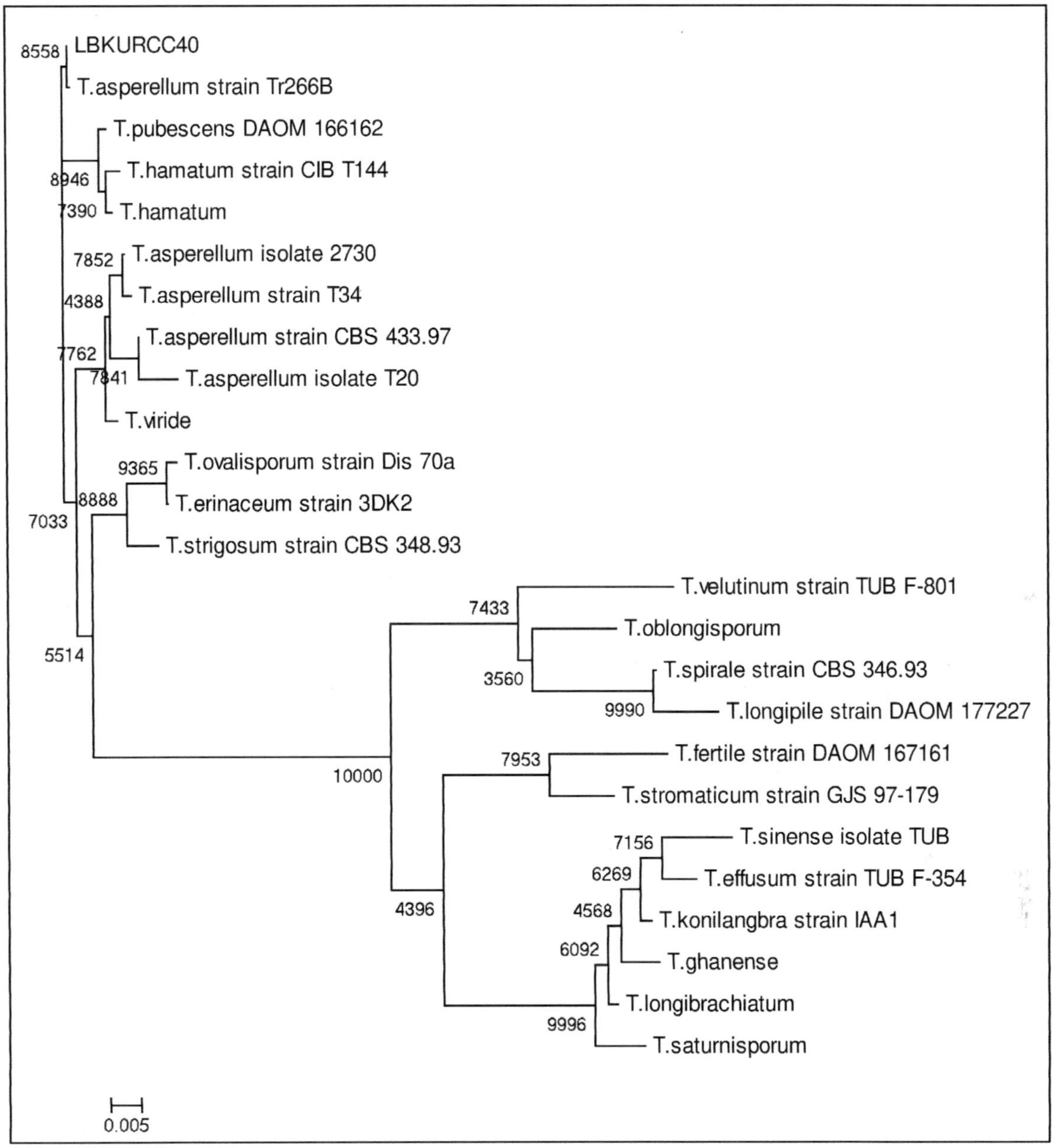

Figure 3. Neighbor-joinin tree (NJ) from the ITS sequence

4. Conclusions

The identification showed that LBKURCC40 isolate was *Trichoderma asperellum* based on morphological and molecular identification characteristics. Phylogenetic analysis with bootstrap 10,000 repetitions at each branch showed endophytic fungus LBKURCC40 had the closest kinship with *Trichoderma asperellum* strain Tr266B. This identification proved that morphological identification could not identify up to the species level, but it needed to be carried out with molecular continued identification.

Acknowledgements

This study was funded by the Postgraduate Research Grant from the Ministry of Research, Technology and Higher Education of the Republic of Indonesia on behalf of Prof. Dr. Saryono. M. S.

Reference

[1] Baldauf, S. L. 2003. Phylogeny for the faints of heart: a tutorial. *Elsevier.* 19 (6): 345-351.

[2] Correa, R. C. G., Rhoden, S. A., Mota, T. R., Azevedo, J. L., Pamphile, J. A., Souza, C. G. M., Polizeli, M. L. T. M., Bracht, A., and Peralta, R. M. 2014. Endophytic fungi: expanding the arsenal of industrial enzyme producers. *J Ind Microbiol Biotechnol.*

[3] Druzhinina, I. S., Kepchinsky, A. G., Komon, M., Bissett, J., Szakacs, G, dan Kubicek, C. P. 2005. An oligonucleotide barcode for species identification in *Tichoderma* and *Hypocrea*. *Fungal Genetics and Biology.* 42: 813-828.

[4] Druzhinina, I. S., Kepchinsky, A. G., dan Kubicek, C. P. 2006. The first 100 *Trichoderma* species characterized by molecular data. *Mycoscience.* 47: 55-64.

[5] Devi, T. P., Prabhakaran, N., Kamil, N., Borah, J. L., and Pandey, P. 2012. Development of species specific markers for detection of *Trichoderma* species. *Vegetos.* 25 (2): 207-217.

[6] Fahmi, A. I., Eissa, R. A., El-Halfawi, K. A., Hamza, H. A., and Helwa, M. S. 2016. Identification of *Trichoderma* spp. by DNA Barcode and Screening for Cellulolytic Activity. *Journal of Microbial & Biochemical Technology.* 8 (3):202-209.

[7] Hoyos-Carvajal, L., Orduz, S., and Bissett, J. 2009. Genetic and metabolic biodiversity of Trichoderma from Colombia and adjacent neotropic regions. *Fungal Genetics and Biology.* 46: 615–631.

[8] Jia, M., Chen, L., Xin, H. L., Zheng, C. J., Rahman, K., Han, T., and Qin, L. P. 2016. A friendly relationship between endophytic fungi and medicinal plants: A systematic review. *Frontiers in Microbiology.* 7: 1-14.

[9] Larena, I., Salazar, O, Gonza'lez, V., Julian, M. C., and Rubio, V. 1999. Design of a primer for ribosomal DNA internal transcribed spacer with enhanced specificity for ascomycetes. *Journal of Biotechnology.* 75: 187–194.

[10] Lorenita, M., Haryani, Y., Puspita, F., Trihartomo, D., and Sikumbang, S. 2013. Screening of endophytic fungi from tubers of *Dahlia variabilis. Journal of Agricultural Technology.* 9 (3): 565-570.

[11] Saryono[a]., A. M. Chainulfiffah., Ardhi Aulia., and Pratiwi, N. W. 2015. Production, purification, and characterization of inulinase from dahlia rizhosphere-isolated *Aspergillus clavatus.* Journal of Chemical and Pharmaceutical Research. 7 (9S): 165-176.

[12] Saryono[b]., Hendris, S., Fitriyah, D., Jose, C., Nugroho, T. T., and Ardhi, A. 2015. Antimicrobial activity and molecular characterization of endophytic fungi strain isolated from dahlia *(Dahlia variabilis).* Journal of Chemical and Pharmaceutical Research. 7 (9S): 201-208.

[13] Saryono., Finna, P., Nurmala, S., Wahyu, P. N., and Aulia, A., 2018. phological identification and hydrolytic enzymeproducing abilities of fungi associated with wilting banana plants (*Musa* sp.). Res. J. Chem. Environ. 79-86.

[14] Saryono., Rakhmana, S., Rahayu, F., Ardhi, A., Rusli, Pratiwi, N. W., and Nugroho,T. T. 2017. Molecular identification of endophytic fungi isolated from the tuber of Dahlia variabilis and exploration of their ability in producing ßgalactosidase. *Biodiversitas.* 18 (1): 145-152.

[15] Silvera, D., Luthfin, I., Aulia, A., Wahyu, P. N., and Saryono. 2018. Optimization of process parameters for inulinase production from endophytic fungi *Fusarium* solani LBKURCC67, *Humicola fuscoatra* LBKURCC68 and *Fusarium oxysporum* LBKURCC69. Res. J. Chem. Environ.71-78.

[16] Tan, R. X., and Zou, W. X. 2001. Endophytes: a rich source of functional metabolites. The Royal *Society of Chemistry 2001.* 18: 448-459.

[17] Potshangbam, M., Devi, S. I., Sahoo, D., and Strobel, G. A. 2017. Functional Characterization of Endophytic Fungal Community Associated with Oryza sativa L. and Zea mays L. *Frontiers in Microbiology.* 8: 1-15.

IOP Conf. Series: Materials Science and Engineering **532** (2019) 012022 doi:10.1088/1757-899X/532/1/012022

Developing organic feed stuff as the main component of broiler chickens' feed with canvas model in East Java: *A Case Study of Analysis on Organic Rice and Corn in Malang and Tulungagung*

SutantoA, WidodoW, and RahayuI D

E-mail: sutanto@umm.ac.idwahyuwidodohandayani@gmail.com
dwirahayuimbang@gmail.com

Abstract. In general, this research aimed to: [1] analyze the market profile and supporting institution regarding the obtainability of productions and demands; [2] investigate the agri-business institution and organization on empowering the organic feed availability especially for the broiler chickens.This research belonged to surveying research through accommodating surveys respectively piloted in Malang and Tulungagung. Therefore, the method for this research referred to survey recruiting 62 respondents in total that was assisted by interview to the main informants, group discussion, and document analysis (converted as secondary data). Moreover, for analyzing the data, the descriptively qualitative and interpretative design was accommodated by means of Business Canvas Model (BCM) approach. By and large, the conclusions of this research were formulated as follows: [1] in accordance with the market profile and supporting institution regarding the availability of productions and demands, the market reinforcement was of necessity in order to advance the productions' values as well as the farmers' incomes in supplying the organic feed; and [2] regarding the agri-business institution and organization on providing the broiler chickens with organic feed, the highly serious exertion from all of elements, exclusively all the stakeholders, was of compulsory in addition to empower the progress and success of organic farming. In addition, the management of farmers' institutions centralizing on production was required to be maintained. Therefore, the collaboration of various farmers' institutions was of urgency to be reinforced.

1. Introduction

The number of various problems that will be formulated as the main concern of analysis in developing organic rice and corn cover: (1) the lack of capital authorization, (2) the lack of technological mastery in production, (3) the bumpiness of supporting facilities, and (4) the lack of marketing or institutional supports. Those aforementioned cases are presumed: (1) to drive the productivities, both quantity and quality, low; (2) to affect the occurrence of price fluctuation; and (3) to lower the incomes and additional values gained by the farmers. Moreover, it has been clearly explicated that the development on agri-business of organic rice and corn, particularly on the upstream one, has not significantly met the expectation yet. This occurrence is affected by: (a) the dynamic supplementary performance on the productions' institutions that are supported by the capital institutions; (b) the mid-leveled agri-business institutions tending to be powerless due to the implementation of technological productions; (c) the lack of farmer groups' capabilities in functioning the organic rice and corn resources and farming production factors; and (d) the lack of distributional institutions in supporting the facilities of production referring to the lack of the facility availability in terms of quantity and farmers' accessibilities.

East Java constitutes one out of central organic feed providers that does need to be intervened by the government in search of significantly being developed. However, all this time, there have not been any development model that allows supporting organic feed as the main component of broiler chickens' feed in which the organic feed is supposed to be the mainstay for establishing the robust betterment.

Therefore, the research on developing the organic feed as the main component of broiler chickens' feed in East Java by means of canvas model is of urgency to pilot. There are 4 reasons we cut our teeth on the Business Model Canvas: (1) a design goal when creating the Lean Canvas was ensuring it was accessible to anyone in the company (not just the business folks) because good ideas can come from anywhere. (2) Lean Canvas prioritizes getting your customer-problem-solution foundation in order first -- making it ideal for early stage innovation projects and startups. It's easy to see how if your customer and problem assumptions aren't in order, everything else on your canvas falls apart. (3) The Business Model Canvas attempts to overcome it's lack of customer-centricity by pairing it with the Value Proposition Canvas. (4) Creating a better business modeling tool or a better business model canvas was never the point. Lean Canvas is part of a bigger Continuous Innovation framework that helps you systematically uncover what customers want and build products they cannot refuse.

2. Research Method

This research was originally piloted in Malang and Tulungagung. Those areas were selected purposively by considering the fact that both are the centrals of developing the organic rice and corn in East Java. Moreover, the sub-districts that were selected as the location for piloting the research consisted of Sumber Gempol, Kalidawir, and Rejotangan sub-districts by officially engaging 31 respondents in total covering 27 farmers and 4 collector sellers. Whereas, in Malang, the recruited sample comprised 31 respondents consisting 30 farmers and 1 single collector seller in which all of whom were originally from Lawang and Tajinan sub-districts. The qualitative research by means of multi-methodic case study was accommodated through various survey methods, they are: field observation, in-depth interview to the case respondents, interview to the main informants, group discussion, and data analysis (on the secondary data). In addition, the in-depth interview (including to the main informants) was conducted to dig out the information related to the availability of organic feed in both research locations. At last, the descriptively qualitative and interpretative analysis by means of BMC approach was accommodated to fulfill the need of the data analysis.

3. Results and Discussion

3.1 Customer Segments

This research exhibited that, instead of widely marketing their products, farmers preferred to stagnantly offer their rice and corn to only the wholesalers, livestock farmers, and the middlemen who had been their regular customers around the locations of the research due to the production which was relatively lower than the total needs and to the fact that their areas were identically renowned as the livestock producers. Consequently, offering the products to the groups of the target customers as well as the market segments seemed impossible since the production was moderately limited. Precisely, the markets' segmentation based on the customers' economic levels (standard, high, or even be based on the incomes or other aspects) was complicated to be mapped.

Moreover, the products were actually rice brans and corns. The farmers mostly sold the brans to their closest neighbors since there were some farmers who had their own livestock. In fact, the need of brans, particularly in one of the research locations (Tulungagung), was significantly high due to the fact that Tulungagung was familiarly known as the central producer of laying hens. As the consequence, the demands of brans were dramatically high. This explicated the positive values on making business of organic rice because brans were mostly expended by the farmers themselves.

3.1.1 Value Propositions The products that were offered to the consumers were relatively restricted on the kinds of products, such as: rice bran and peeled corns. In Tulungagung, one of the research locations, the corn produced by the farmers was unpeeled and peeled ones. The unpeeled corn priced 4,400 rupiahs per kilo while the peeled one was 4,700 rupiahs per kilo. Meanwhile, the brans that were produced by the farmers consisted of two kinds, namely: (1) soft brans which were resulted from the highly standardized grinding process for producing the livestock feed, particularly known as bekatul, and (2) hard brans which were resulted from the rice grinding process, regularly renowned as dedak, and were usually distributed to the surrounding livestock farmers.

3.2 Channels

The organic feed products, from both rice and corn, were allowed to be distributed through innumerable procedures depending on the customers' segmentations that had been successfully established by the farmers. Occasionally, the distribution was executed through directed and coordinated marketing engaging the middlemen (such as: wholesalers, non-distributors, marketing agents, or sometimes websites) of which marketing distribution was not via selling and buying forums anymore. In conclusion, the information that had been successfully established by the farmers in marketing the rice bran and corn would be based on their own spoken information.

3.3 Customer Relationships

Farmers were to make and maintain the intact relationship with the customers in search of continually increasing the number of customers and establishing the nice engagement with the customers by means of informational technology-based patterns, especially for informatively and communicatively approaching the customers. The mentioned procedures were supposed to help the farmers engage with the customers well. By having a fully intact relationship, farmers were allowed to easily express anything to the consumers when they both were in the transaction process, exclusively when talking about price of rice bran and corn. Besides, the farmers selling the rice brans were to only set the price based on the fee of grinding process. Therefore, in term of price, it was equitably cheap.

3.4 Revenue Streams

The income resource of the organic feed business was of considerable strategy in determining the success of organic feed business. The profit that had been acquired from this business was accumulated by the deviation between the income of whole turnovers and the whole budgets. Moreover, the result of this research explored that the positive value that had been successfully achieved in making business in organic rice and corn was relatively low because the production scale was also intended to other non-organic plants.

3.5Key Activities

The number of activities that were executed by the farmers to make profits within their businesses were still limited in terms of cultivation aspects covering planting, maintaining, and harvesting. Afterwards, the plan on the production or the harvesting session and the standardization on the volume of market products to keep the stability of the market price had not been realized by most of farmers. This explicated that the incompetency of farmers in production drove them problematic, particularly in keeping the stability of production as well as quality and the continuity of the particular quota of availability demanded by the markets.

3.6 Key Resources

The farmers in Tulungagung had the averagely excellent backgrounds of education, 45% of whom graduated from senior high school. Besides, there was a great potential of productive ages that allowed the farmers to upgrade the quality of their farming businesses. The farmers who developed the organic feed were more productive because 40.90% of whom aged 40-55 years old averagely.

The field areas that were planted by the farmers were fairly narrow accumulating 0.1 to 2.5 hectare (there were 81.82% of farmers). Consequently, it was of complication for farmers to develop their businesses' scales. In addition, 78.12% of those who developed the organic feed, consisting of rice bran and corn, did not join the recommended formal institutions, Farmers Association for instance, since they prescribed that most of formal institutions were intended for merely formality. This was also supported by the experience backgrounds of the farmers that were relatively high, 5 to 40 years.

3.7 Key Partnerships

The farmers' success in developing the organic feed was entailed by the pleasant teamwork having been engaged with various parties, institutional organizations and others. The farmers' organizations held the very crucial role in cultivating the organic farming. This alluded the socialization and certification matters. In addition, the agri-business products of low-leveled farmers would be of difficulty to be realized if there was no support from other farmers' organizations. In some areas, farmers' organizations had been well-established, but, it was different from other areas of which farmers were getting difficult in establishing the expected one. The research revealed that optimization on farmers' formal institutions related to the organic cultivation was still in shortage. These institutions tended to only act as the media for effortlessly coordinating and organizing the programs that were incorporated with the official agencies so that the establishment would be government-based.

3.8 Cost Structures

The various kinds of costs which had to be paid for by the farmers in developing the organic feed covered planting, maintaining, and harvesting costs. Practically, the costs of transportation were assured by the collector sellers or those farmers who voluntarily came to haul the feed. However, it had been clearly comprehended that almost all of activities that were executed by the farmers did need costs.

4. Conclusions

In general, the conclusions regarding this research were formulated as follows: In accordance with the market profile and the supporting institutions regarding the availability of productions and demands, it is of urgency to establish the market greater in search of increasing the productions and the farmers' incomes during the production of organic feed.

Alluding to the agri-business institutions and organizations of organic feed, the serious exertion and full attention from all the stakeholders were of inevitability to support the implementation and success of organic farming, especially in empowering the availability of broiler chickens' organic feed. In addition, the organization of farmers' institutions in production was required to be maintained while the farmers' associations as well as the other similar associations were obligatory to be strengthened.

5. References

[1] Ahmed M H, Atti K A, Malik H E, Elamin K M, and Dousa B M 2014Ginger (Zingiber officinale) Root Powder As Natural Feed Additive For Broiler Chicks *Global Journal of Animal Scientific* 383-389.

[2] Ameh G, Eze S, and Omeje F 2013Phytochemical Screening and Antimicrobial Studies on the Methanolic Bulb Extract of Allium sativum*L. Afr. J. Biotechnol.* 1665-1668.

[3] Kader G, Nikkon F, Rashid M A, and Yeasmin T 2011 Antimicrobial Activies of The Rhizome Extract of Zingiber zerumbet Linn *Asian Pacific Journal of Tropical Biomedicine*409-412.

[4] Kusnandar, Padmaningrum D, Rahayu W, and Wibowo A 2013 Rancang Bangun Model Kelembagaan Agribisnis padi Organik dalam Mendukung Ketahanan pangan*Jurnal Ekonomi Pembangunan* vol 14 chapter 1*(Indonesia,June 2013)*pp 92 – 101.

[5] Onyimonyi A E, Chukwuma P C, dan Igbokwe C 2012 Growth and hypocholesterolemic properties of dry garlic powder (Allium sativum) on broilers. *African Journal of Biotechnology*vol 11 chapter 11 pp 2666–2671.

[6] Karangiya V K, SaysaniH H, Patil S S, Patil D D, Garg K S, Murthy N K, Ribadiya andVekariyaS J 2016 Effect of Dietary Supplementation of Garlic, Ginger and Their Combination on Feed Intake, Growth Performance and Economic in Comercial Broilers*Vet Word* vol 9 chapter 3 pp 245-250.

[7] Rahayu I, Widodo W, and Sutanto A 2015 Produksi Pakan Organik Ayam Broiler Untuk Keamanan dan Ketahanan Pakan dan pangan Di Indonesia *Laporan Penelitian Uanggulan Perguruan Tinggi*(Malang: Universitas Muhammadiyah Malang).

[8] Rahayu I D, Widodo W, and Sutanto A 2018 The effect of fermented organic feed usage on the health status of kampong chicken *International Journal of Biosciences(IJB)* vol 12 chapter 4 p 35-42.

[9] Selim N, Youssef S, and AF. Abdel Salam A S 2013 Evaluation of Some Natural Antioxidant Sources in Broiler Diets : 1-Effect on Growth, Physiological, Microbiological and Immunological*International Journal of Poultry Science*vol 12 pp 561-571.

[10] Singh C B et al 2012 Biological and chemical properties of Zingiber zerumbet Smith: A review. *Phytochemistry Reviews*

[11] Stoilova I et al 2007 Antioxidant activity of a ginger extract (Zingiber officinale) *Food Chemistry* vol 102 (3) pp 764–770.

[12] Suriya R,ZulkifliI and AlimonA R 2012 The Effect of Dietary Inclusion of Herbs As Growth Promotore in Broiler Chicken *Journal of Animal and Veterinary Advances* vol 11 (3) pp 346-350.

[13] Takeli A, KutluH R, Celik L and DoranF 2010 Determination Of The Effects Of Z. Offinale And Propolis Extracts On Intestinal Microbiology And Histological Characteristics In Broilers*Int. J. Poult. Sci* vol 9 pp 898-906.

[14] Wahyuningsih S 2007 Pengembangan Agribisnis Ditinjau dari Kelembagaan vol 3 (1) pp 9-20.

[15] Widodo W 2008 Ketahanan Pakan Unggas di Tengah Krisis Pangan *Naskah Pengukuhan Guru Besar* (Malang: Universitas Muahmmadiyah Malang).

[16] Widodo W, Rahayu I D, Sutanto A 2018Nutritive Content Assessment of Organic Feedstuff from Various Regions in East Java Province: Indonesia asPotential Exploration Efforts of Local Poultry Feed. *International Journal of Biosciences (IJB)* vol 12 (3) pp 18-23.

IOP Conf. Series: Materials Science and Engineering **532** (2019) 012023 doi:10.1088/1757-899X/532/1/012023

Agronomic character of several local genotypes of sorghum (*Sorghum bicolor* l.) at East Java

Sulistyawati[1,5], Dyah Roeswitawati[2], Jabal Tarik Ibrahim[3], Maftuchah[4]

Abstract. Sorghum classified as an annual crop that is very easy to be cultivated but its production until now classified as lagging compared to rice, corn and other cereals. In an effort to increase the fulfillment of food needs, sorghum plants can be developed in Indonesia as an alternative to local food other than rice. The development of sorghum plants is considered capable of supporting government programs to prevent the policy of importing rice and wheat.

This study aims to determine the character of several genotypes of local sorghum East Java grown on paddy fields in Purutrejo Village, Purworejo Sub-District, Pasuruan City at an altitude of 5 m ASL with alluvial soil type. The average temperature is 29°C - 34°C with an average rainfall of 1430 mm / year. The study was conducted from June to August 2017. The experiments were arranged in a Randomized Block Design, using nine local sorghum genotypes replicated three times. Nine genotypes of local sorghum are Pasuruan, Lamongan 1, Lamongan 2, Tuban, Sampang 1, Sampang 2, Tulungagung 1, Tulungagung 2 and Jombang.

The results showed that nine characterized genotypes had various agronomic characteristics on plant height, leaf number, stem diameter, panicle length, number of seeds per panicle, seed weight per panicle and 100 seed weight.

Keywords: Characterization, Local Genotypes, Sorghum

1. Introduction

Efforts to increase food production in Indonesia to meet the increasing population needs are more difficult. Limitations of suitable land to produce food and unpredictable global climate change are among the obstacles to be faced. Marginal land becomes an alternative in an effort to increase food production with various limitations of soil physical and chemical properties. Under these conditions, sorghum can still grow and produce, so that it can be developed as an alternative to local food other than rice [1]. In areas that often experience drought or get inundated with floods, sorghum plants can still be cultivated. Therefore, there is considerable opportunity to increase sorghum production and obtain superior varieties of sorghum [2]. Sorghum is a multipurpose plant, both as food, feed, and industrial materials processing. As a food, sorghum nutrition is not much different from other cereals [3]. In general, the level of sorghum protein is higher than maize, brown rice and barley but lower than wheat. Sorghum fat content is higher than brown rice, wheat, millet but lower than corn [4].

[1] Faculty of Agriculture, Merdeka Pasuruan University, Pasuruan, East Java, Indonesia.(corresponding author to provide phone : +6281-334-216-786, e-mail : mommyandri@gmail.com).

[2] Departement of Animal Science, Faculty of Agriculture and Animal Science, University of Muhammadyah Malang, Malang, East Java, Indonesia.

[3] Departement of Animal Science, Faculty of Agriculture and Animal Science, University of Muhammadyah Malang, Malang, east Java, Indonesia.

[4] Departement of Animal Science, Faculty of Agriculture and Animal Science, University of Muhammadyah Malang, Malang, East Java, Indonesia

[5] Corresponding author to provide phone : +6281-334-216-786, e-mail : mommyandri@gmail.com

One of the problems faced in the development of sorghum commodities in Indonesia is the lack of development of superior varieties, especially the result of the development of local genotypes. In East Java, for example, there are still unidentified and characterized wild genotypes of sorghum [5]. The identification of wild genotypes and accessions is necessary in order to develop local sorghum cultivars. Identification and characterization are the first steps used to discover the genetic variation of plants in the development of a superior cultivar type through breeding. Without diversity, improvements in the properties of a plant are impossible [6].

Food Security Agency includes sorghum as one of the commodities supporting the diversification of national food. The results of the Cereals Plant Research Institute showed that sorghum can substitute rice up to 30% with a taste that is acceptable to consumers [7]. In food self-sufficiency program, the Agency for Agricultural Research and Development has made efforts to procure new varieties of sorghum, but because the priority of development is still on rice and corn commodities, starting from 1960 until 2001 only 15 varieties were released, then 2013-2016 increased 6 varieties.

Based on the development of superior varieties of sorghum data in Indonesia and the fact that the government still needs the effort to support the success of food security, it is necessary to conduct research on the characterization of several local sorghum genotypes found in East Java.

2. Materials and methods

This research was conducted in Purutrejo Village, Purworejo Sub-District, Pasuruan City at an altitude of 5 m ASL with alluvial soil type. The average temperature is 29°C - 34°C with an average rainfall of 1430 mm / year. Research has been carried out in June - August 2017. The tools used in the research include cultivation tools and measuring tools. The materials used are local genus of sorghum genotypes obtained from six districts in East Java, manure and NPK fertilizer (Urea, SP-36, and KCl).

The study was arranged in Randomized Block Design with three replications and used nine local genotypes of sorghum, namely Pasuruan, Lamongan 1, Lamongan 2, Tuban, Sampang 1, Sampang 2, Tulungagung 1, Tulungagung 2 and Jombang. Each unit (plot) of the experiment contained 15 plants with 3 sample plants, selected plants that are in the middle row.Characters observed included plant height, number of leaves, stem diameter, panicle length, number of seeds per panicle, seed weight per panicle and 100 seed weight.

All genotypes were planted simultaneously in Pasuruan district with consideration: a. All genotypes will get a uniform treatment, b. Facilitate in the observation and retrieval of data. The planting was done by cultivation with a spacing of 70 cm X 30 cm in the prepared plots. Treatments include watering, weeding, hatching, pest and disease control and fertilization.

3. Results and discussion

Based on the analysis of the variety of genotypes for the seven characters observed, i.e. plant height, number of leaves, stem diameter, panicle length, number of seeds panicle^{-1}, seed weight panicle^{-1} and 100 seed weight showed significantly different results in 5% level. Differences in plant growth are influenced by internal factors such as genes and hormones that affect growth through inherited traits. External factors such as nutrients, water, temperature, humidity, and light also give different effects on the characteristics of a plant [8].

3.1 Plant Height, Leaf Number, and Stem Diameter

Further Tukey Test assay results, several genotypes showed differences in plant height, leaf number and stem diameter. Table 1 shows that genotype Tulungagung 1 is higher than the others with a mean of 389.38 cm. The lowest plants were Lamongan 1 genotype (168,10 cm) and were not different from Sampang 1 and Sampang 2.

Similarly, the number of leaves and stem diameter, the highest mean was owned by Tulungagung 1 genotype (strands respectively 13.56 and 2.44 cm), although there are several genotypes that match. The lowest number of leaves is Tulungagung 2 genotype, which is 5.78 pieces, while the lowest diameter is found in Sampang 2 genotype (1.32 cm).

From the data mentioned above can be explained that the high plant does not necessarily have many leaves because the stem of sorghum consists of segments that are leaf seating. Plant height is affected by the length of the segment while the number of leaves depends on the number of segments [9].

The growth of stem diameter in addition to the genetic properties of each genotype is also influenced by environmental factors and the photosynthesis process in the leaves.

Table 1. Plant height, number of leaf and diameter of stem

Genotype	Plant Height (cm)	Number of Leave	Stem Diameter (cm)
Pasuruan	225.48 b	8.78 abc	1.53 a
Lamongan 1	168.10 a	12.78 cd	2.32 bcd
Lamongan 2	218.70 b	11.11 bcd	2.42 cd
Tuban	319.28 c	10.56 bcd	1.87 abc
Sampang 1	200.69 ab	7.00 ab	1.32 a
Sampang 2	195.21 ab	8.78 abc	1.51 a
Tulungagung 1	389.38 d	13.56 d	2.44 d
Tulungagung 2	232.17 b	5.78 a	1.48 a
Jombang	304.78 c	10.78 bcd	1.80 ab
Tukey Test 5%	45.64	4.47	0.57

Note : The numbers followed by the same letter in the same column show no significant difference in the 5% Tukey Test

3.2 Length of Malai, Number of Seeds per Malai, Seeds Weight per Malai, Weight of 100 Seeds
Based on the observation, genotype diversity can be recognized more clearly in the generative phase. Quantitatively, there are differences in morphology of nine panicles studied (Figure 1). Table 2 shows that Tulungagung 2 genotype has a longer panicle compared to other genotypes, which is 50.92 cm but not followed by the number of seeds per panicle. The lowest average length of the panicle is in the Jombang genotype (20.51 cm) and is not different from the Tuban genotype.
The highest number of seeds per panicle is owned by Sampang 2 genotype, which is 3,002.00 and not different with Lamongan 1 genotype while the lowest is Sampang 1 genotype, as much as 471,78.
The Tuban genotype had a higher seed weight per panicle than the other genotype, which was 96.74 g while the Sampang 1 genotype had lower seed weight than the other genotypes.
The highest 100 seed weight was also achieved by Tuban genotype, of 3.20 g, although not unlike the Tulungagung 2 genotype. Sampang 1 genotype was lower in weight than the other genotype, which is 1.89 g.

IOP Conf. Series: Materials Science and Engineering **532** (2019) 012023 doi:10.1088/1757-899X/532/1/012023

Figure 1. Morphology of Nine Panicles Sorghum Local at East Java

Table 2. The size of panicles, number of seeds panicle-1, seed weight panicle-1 and 100 seeds weight

Genotype	Panicles Length (cm)	Number of Seeds Panicle^{-1}	Seed Weight Panicle^{-1} (g)	100 Seed Weight
Pasuruan	30.48 c	2005.11 b	51.61 b	2.81 e
Lamongan 1	26.89 b	2871.66 cd	61.21 cd	2.36 c
Lamongan 2	39.47 de	2606.78 c	58.38 cd	2.24 b
Tuban	22.00 a	2197.44 b	96.74 f	3.20 f
Sampang 1	41.09 e	471.78 a	23.57 a	1.89 a
Sampang 2	37.03 d	3002.00 d	56.48 bc	2.28 bc
Tulungagung 1	41.61 e	2726.22 c	65.57 e	2.61 d
Tulungagung 2	50.92 f	2257.78 b	64.34 de	3.13 f
Jombang	20.51 a	2096.56 b	55.39 bc	2.61 d
Tukey Test 5%	3.40	277.19	6.48	0.12

Note : The numbers followed by the same letter in the same column show no significant difference in the 5% Tukey Test

4. Conclusion

Based on the results of the characterization, there are eight genotypes that have the potential to be developed as a parent in the development of superior cultivars, because they have good agronomic character, namely Pasuruan, Lamongan 1, Lamongan 2, Tuban, Sampang 2, Tulungagung 1, Tulungagung 2 and Jombang genotype.

5. References

[1] Subagio, H., dan Suryawati. 2013. Wilayah Penghasil dan Ragam Penggunaan Sorgum di Indonesia. *Sorgum: Inovasi Teknologi dan Pengembangan, oleh Kementerian Pertanian Badan Penelitian dan Pengembangan Pertanian*, disunting oleh Sumarno, D. S. Damardjati, M. Syam dan Hermanto. Jakarta. IAARD Press. 24-37.

[2] Subagio, H, dan M Aqil 2014. Perakitan dan Pengembangan Varietas Unggul Sorgum untuk Pangan, Pakan, dan Bioenergi. *J. Iptek Tanaman Pangan* **9** (1): 39-50.

[3] ICRISAT 2004. *Sorghum, a Crops of Substance*. International Crops Research Institute for the Semi-Arid Tropics. Andhra Pradesh. India.

[4] Mejia, Danilo and Lewis, B., 1999. *Sorghum: Post-harvest Operations.* Food and Culture Organization of The United Nations.

[5] Susilowati, S. H., dan H. P. Saliem 2013. Perdagangan Sorgum di Pasar Dunia dan Asia serta Prospek Pengembangannya di Indonesia. *Sorgum: Inovasi Teknologi dan Pengembangan, oleh Kementerian Pertanian Badan Penelitian dan Pengembangan Pertanian*, disunting oleh Sumarno, D. S. Damardjati, M. Syam dan Hermanto, 7-23. Jakarta: IAARD Press.

[6] Mofokeng, M. A., G. Watson, H. Shimelis, and P. Tongoona 2012. Comparison between random amplified polymorphic DNA (RAPD) and simple sequence repeat (SSR) markers with high-resolution melt analyses in genetic variation analysis among selected sorghum genotypes. *African Journal of Biotechnology* **11** (102): 16697-16707.

[7] Suarni dan Firmansyah I. U 2005. *Struktur, Komposisi Nutrisi dan Teknologi Pengolahan Sorgum.* Balai Penelitian Tanaman Serelia.

[8] Elvira S. D, Muhamad Yusuf, dan Maiyuslina 2015. Karakter Agronomi Beberapa Varietas Sorgum pada Lahan Marginal di Aceh Utara. *J. Agrium* **12**(1):1-4.

[9] House, L.R. 1985. *A guide to Sorghum Breeding*. International Crops Research Institute for the Semi-Arid Tropics. ICRISAT Patancheru, Andhra Pradesh, India.

Chemistry and Structure Characterization of Bamboo Pulp with Formacell Pulping

S Hidayati[1], E Suroso[1], W Satyajaya[1], D A Iryani[2,3]

[1]Department of Agroindustrial Technology, Agriculture Faculty, University of Lampung
[2]Department of Chemical Engineering, Engineering Faculty, University of Lampung
[3]Research and Development Centre for Tropical Biomass, University of Lampung

*Corresponding Author: srihidayati.unila@gmail.com

Abstract. Bamboo is one of the non-wood materials which can be utilized as raw materials of pulp. The purpose of this work was to study the effect of the ratio of acetic acid: formic acid on the chemical properties of pulp from bamboo. The results of the study showed that increases of the ratio of formic acid able to degraded of cellulose, hemicellulose, lignin content and decreases the pulp yield. The best condition of this research showed on the ratio of acetic acid: formic acid = 90:10. This condition produced the pulp with composition consist are cellulose 66,98%, hemicelluloses 16,83%, and lignin 3,59%, with the yield of pulp is 42,279%

Keywords : acetic acid, bamboo, formacell, formic acid, pulping.

1. Introduction

Recently, there are growing interest in the utilization of alternative fibrous material derived from lignocellulosic materials such as agricultural waste, and bamboo to pulp production. The advantages of using non-timber raw materials are ease to pulping, excellent fiber for the special type of papers and high quality bleached pulp [2]. Non wood material can also utilized as an effective substitute for the over-exploited forest wood resources [3]–[6]. Bamboo is one of the non-wood materials which can be utilized as raw material for making pulp. There are several advantages of bamboo compared to other woody plants, such as grow rapidly (up to 120 cm per day), easily to spread, highly productivity and easily to harvest. [7]. The productivity of bamboo is reach into 2-6 times higher than pine plants [8].

Moreover, the characteristic of bamboo almost similar with woody plant. It has high cellulose content which about 40% -60.and fibre lenght is about 2-3 mm. The fibre lenght is equivalent to red spruce wood, but longer than pine, with average of fibre length is 1980 μm. Bamboo chip consist of 22,4% lignins, 19,5% xylans, 49,3% cellulose, 16,8% extractive and 1,5% - 3% ash [10], [11]. This properties suggests that bamboo is the best material for papermaking compared to other lignocellulosic materials such as bagasse and rice straw[9], [14].

The common process for pulp-making process in Indonesia using chemical pulping such as sulfate process or is known as Kraft pulping. This process is known as not an environmentally friendly cause its produce several waste problem [15], [16] mainly due to the presence of sulfur compounds [17]. In order to minimize the waste from the pulping process, it is necessary to find the alternative environmentally friendly pulping process. In all parts of the world many alternative pulping processes have been introduced. One of alternative enviromentally friendly paper making process is organosolv pulping method. Organosolv pulping is a process for separation of macromolecular components, comprises the delignification which is the solubilization of the lignin fragments resulted from the breakage of chemical bonds of protolignin [18]. An organic solvents which is usually associated with water in the 10% to 50% ratio (by volume), such as acetone, ethanol, metahol, formic acid, acetic acid etc [5]. Most of organosolv process by using volatile solvent carried out at high temperatures (185 – 210 °C). However, organosolv by using an organic acids such as acetic acid and formic acid require lower temperatures and pressure which is closer to atmospheric pressure [19 –21].

The formacell process is one of the pulping methods using acetic acid and formic acid as a cooking chemical [2]. The formacell process has many advantages, ie: produce high pulp yield, low lignin content, high brightness and good strength[22] – [29]. The pulping process use an organic solvent which is easy recycled. This process produces a low lignin content; thus, it can be bleached by using ozone or peroxy acetic acid. In addition, the lignin removed from the pulping liquor can be recovered in the recyling process of organic solvent and used for raw material for adhesives production or other products. Sundquist (2000) reported that the formacell pulping process for the herbaceous plant produces the pulp stronger than soda processes [19]. The formacell pulping process is applied to herbacious material such as bagasse [20], bamboo [32], and wheat straw [31].

There are several factors have to be considered on paper-making by using acetosolv pulping process, such as ratio of solvent to water, ratio of solvent to raw materials, concentration of catalyst, temperature and reaction time. [32]–[35]. The catalyst concentration is the important factor for degragading lignin. Mormanne (2009) reported that formacell pulping of *Cephalostachyum virgatum* kurz bamboo produced pulp product with the yield of pulp, kappa number, freeness, breaking lenght, bursting strength, tearing strength, and ISO brighteness are 42.88%, 22.6%, 40 °SR, 5,702.23 m 431.43 kpa, 88.8 cN and 20.7%, respectively[32]. However, this reseach was not reported the effect of acetic acid ratio: formic acid on the chemical properties and the morphology of pulp product. Thus, the purposes of the present work was to evaluate, the effect of acetic acid: formic acid ratio on chemical characteristic and pulp yield. The scanning electron microscopy (sem) was also conducted in order to study the physical features of fibers as well as structures and pulp bamboo morphology.

2. Materials and Methods

2.1 Materials

This experiment was used bamboo Betung sp as raw material which colected from Bandar lampung area. All of chemicals used in this study such as formic acid, glacial acetic acid, HCl, H_2SO_4 were received from Merck.

2.2 Pulping Method

The pulping processes were performed in 1.3 liter lab-scale autoclave with type of flat-bottom, and wide-mouth flask equipped with a condenser. About 20 g of bamboo fibrous were mixtured with the cooking liquor in the ratio of 15:1. in varied ratio of acetic acid : formic acid as cooking liquor, i.e: 100: 1 (K1); 90:10 (K2); 85:15 (K3); 80:20 (K4); and 75:25 (K5). About 0,5% of HCl was added as the catalyst into the cooking liquor.

The pulping processes were conducted in an hour cooking at temperature of 130°C. When the pulping processes ended, the autoclave was cooled into room temperature. Furthermore, the cooking liquor was colected, separated by using filter paper. The filtered residue was then washed by using distilled water until neutrality, dried and yielded as the amount of pulp.

2.3 Pulp Characterizations

For all experiments the main parameters defining of pulping processes were measured as: pulp yield after oven drying of a pulp into constant weight, and composition (i.e cellulose, hemicellulose, and lignin) by using Chesson methods [37]. In addition, the raw material and product were analyzed by using scanning electron microscopy (SEM) JEOL JSM 6510 LA in order to determine the effect of ratio of acetic acid : formic acid on the pulp morphology and fibre structural details. The samples were pre-treated for SEM analysis by sputter-coating with a thin layer of gold.

3. Results and Discussions

3.1 Cellulose

The results showed that the cellulose content of pulp bamboo ranged from 57 to 66.98% (Figure 1). The highest content of cellulose appeared on the ratio of acetic acid: formic acid 90:10. The same result was reported by Hidayati *et al.* (2017), on the ratio of acetic acid: formic acid 90:10 also produced the highest cellulose content in the pulp of empty palm oil bunches [37]. Increases of formic acid ratio of were not decreases the cellulose content, significantly. The result shows that the Formic acid is able to selectively delignifying of biomass with minor effects on cellulose [37].

Kupiainen *et al.* (2012) reported that the formic acid act simultaneously as an organic solvent and as an acid. In pulping process a significant amount of materials able to dissolved into the cooking liquor at relatively low processing temperatures, even working without mineral acid added as catalyst. During the chemical reaction, depolymerisation of lignin and hemicelluloses occurs by partial, and producing oligomers that are soluble in the liquid mixture [38, 41]. The result showed that, the application of short-chain organic acids (mainly formic acid and acetic) has emerged as an attractive and feasible alternative for the delignification of lignocellulosic materials. Cellulose and liquor were then able to separated by filtration, and the dissolved lignin can be recovered easily by the modification of the liquor pH. Furthermore, the recovery of formic acid enables conducted by using distilation process.

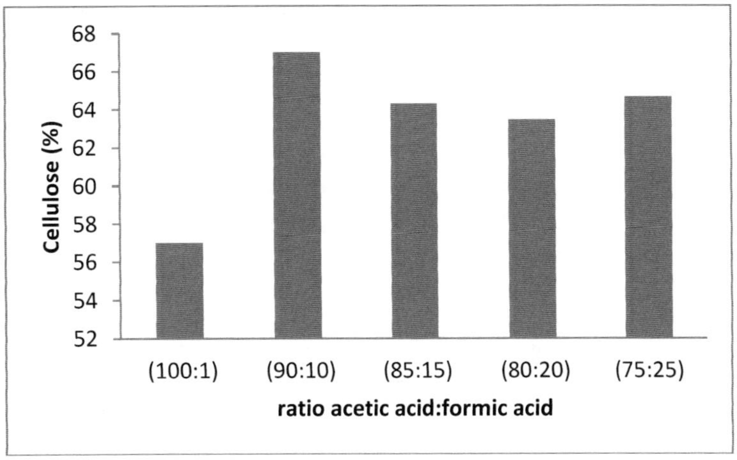

Figure 1. Effect of acetic acid: formic acid ratio on the cellulose contents of bamboo pulp.

3.2 Hemicellulose

The result shows that increase of formic acid ratios can reduce hemicellulose content (Figure 2). The hemicellulose content of bamboo pulp ranged from 7-18.52%. The degradation of hemicellulose occurs due to the acid condition. Sjostom (1981) reported that hemicelluloses are relatively easily hydrolyze into simple sugars by acids since their monomeric components consisting of D-glucose, D-mannose, D-xylose, L-arabinose, and small amounts of L-rhamnose. [41].

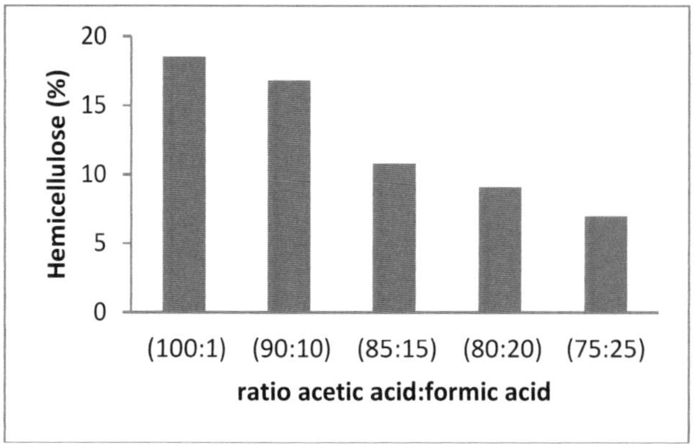

Figure 2. Effect of acetic acid: formic acid ratio on the contents of hemicellulose bamboo pulp.

This result in accordance with Jahan *et a.l* (2006) which was reported that the application formic acid as a cooking liquor in reaction time 2 h produced pulp with hemicellulose content of about 7.8% [43]. High concentrations of formic acid can lead to decrease hemicellulose content [36], [40],[41]; This occurs due to hemicellulose has similar properties with cellulose which that easily degraded in acid solution. Hemicellulose have bound consist of polysaccharides, pectin and lignin and is more soluble than cellulose [43].

The polysaccharide compounds such as cellulose and hemicellulose are glycoside bonds linking the chain of compounds [44]. Glycoside bonds are easy to hydrolyze by acid through a chemical reaction and this condition is accelerated by the heating. Hemicellulose will undergo oxidation and degradation reactions first than cellulose, because the hemicellulosic molecular chains are shorter and branched. Hemicellulose is insoluble in water but soluble in dilute alkaline solutions and more easily hydrolyzed by acids than cellulose [46].

3.3 Lignins

The results showed that the increase of formic acid ratio to a ratio above 10 able to increase the lignin content of bamboo pulp. In this study the resulting lignin content ranged from 3.59 to 6.375%. Lignin is smaller than lignin contained in the bamboo that has not been done pulping process is 22.4% [11]. Delignification process occured due to the acid hydrolysis of α-aryl ether bond [46]. The dissolution of lignin has been ascribed to the cleavage of α-aryl and aryl-glycerol-β-aryl-ether bonds in its molecule [47]. Hydrogen ion concentration plays a very important role in solvent pulping.

The lignin dissolution is expected to be influenced by the acid-catalyzed cleavage of α-aryl and ß-aryl ether linkages in the lignin macromolecule. However, when the formic acid ratio increases the lignin becomes re-condensate and resulting high yielding of lignin. This occurs due to the high concentration of organic acids encourage the reactivation of lignin polymerization that has dissolved in the cooking liquid, and it was making the lignin content of pulp increase [29], [48].

Hidayati *et al.*(2017) stated that the high lignin content of pulp cooking is higher than raw material due to the condensation process so that the lignin attached on the surface of the pulp so that the color becomes darker [35]. The condensation process is formed by combining carbon chains that form longer chains in which the formed compounds are the intermediates of carbonium ions formed on the pulp [46].

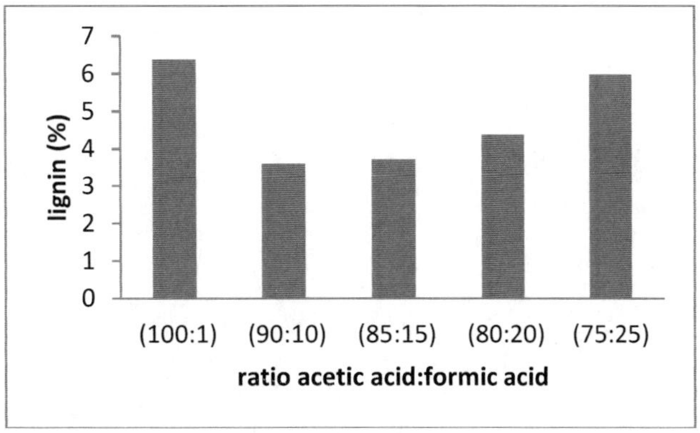

Figure 3. Effect of acetic acid: formic acid ratio on the lignins contents of bamboo pulp.

3.4 Yield

The results showed that increases of formic acid into20 able to decrease the yield of bamboo pulp (Figure 4). In this study the yield of pulp ranged from 36.75 to 44.5%. Yield of pulp decreases due to the degradation of cellulose and hemicelluloses.

Delignification process by formacell has been ascribed to the hydrolysis of α-aryl-ether and lignin-cellulose bonds, and conform to an *pseudo* first-order kinetics reaction. Pulping process by acid solution has properties reactive, so it can break the complex ties of lignin-polysaccharides that resulting the low yield of pulp. According to Fatriasari and Risanto (2011), increase concentrations of cooking liquor tend to decrease the total pulp yield [49].

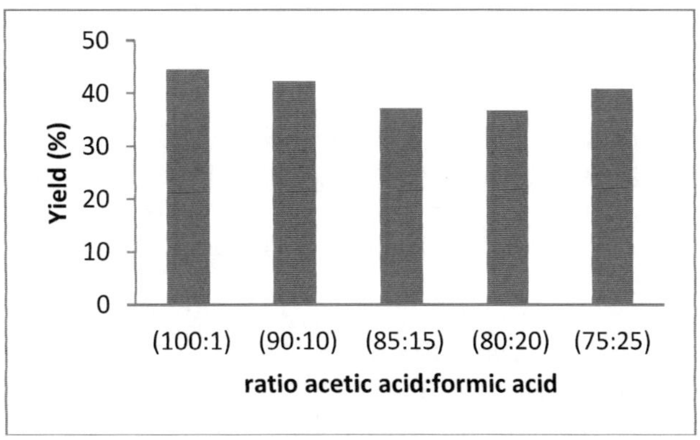

Figure 4. Effect of acetic acid: formic acid ratio on the contents of yields bamboo pulp.

3.5 Assessment the structure of bamboo pulp fibres

Scanning Electron Microscope (SEM) was conducted in order to understand the morphology change of the bamboo pulp fibers. This test was required to estimate the structure and mechanical properties produced by the fibers. The processes of formacell pulping in bamboo produce the morphological changes of bamboo fiber. The analysis was done using 500x magnification, shown on Fig. 5.

The analytical result shos that the fibers undergo randomization from a rigid or crystalline form to randomization. Formic acid as an active agent was able to effectively penetrate into the interior space of the cellulose molecules, thus collapsing the rigid crystalline structure and allowing hydrolysis to occur easily in the amorphous zone as well as in the crystalline zone. The application of formic acid as cooking liquor affected to breaking down the crystalline parts of cellulose [50]. Acid hydrolysis process of

IOP Conf. Series: Materials Science and Engineering **532** (2019) 012024 doi:10.1088/1757-899X/532/1/012024

cellulose at bamboo pulp is heterogeneous and occurs on the structural surface of crystalline lattices of cellulose molecules [52].

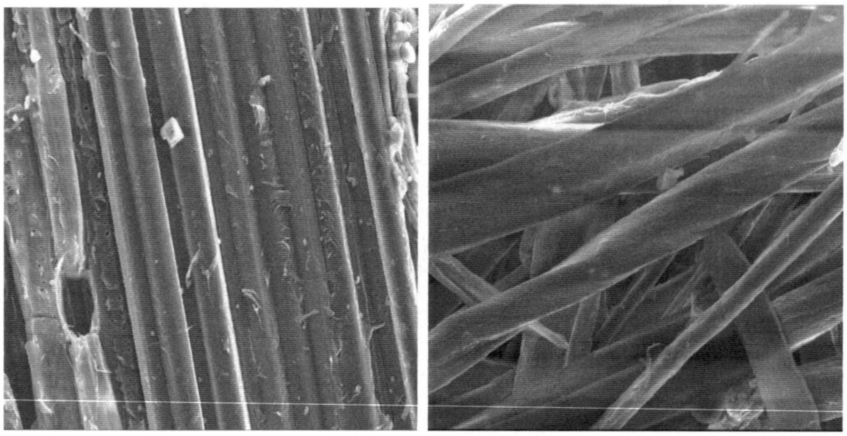

a. Bamboo raw fiber b. Bamboo pulp by acetic acid:formic acid 75:25

Figure 5. Structure of bamboo pulp fibers

4. Conclusions

The results of the study showed that the increasing of the concentration of formic acid able to reduce cellulose, hemicellulose, lignin content and pulp yield of bamboo pulp. The best condition of this research showed on the ratio of acetic acid: formic acid = 90:10. This condition produced the pulp with composition consist are cellulose 66,98%, hemicelluloses 16,83%, and lignin 3,59%, with the yield of pulp is 42,279%.

References

[1] Onilude M A, Ogunwusi A A 2012 The Challenges and Technicalities of Small Scale Pulp and Paper Production in Nigeria In Onyejekwelu J C, Agbeja B O,. Adekunle V A A and Omole A O (eds) Proceedings of 3rd Biennial Conference of Forests and Forest Products Society 205-210.

[2] Sridach W 2010 The environmentally benign pulping process of nonwood fibers Suranaree J. Sci.Technol 17(2) 105-123.

[3] El-Sakhawy M, Fahmy Y, Ibrahim A A, Lonnberg B 1995 Organosolv Pulping: 1. Alcohol of bagasse Cellul. Chem. Technol 29(6) 615-629.

[4] El-Sakhawy M, Lonnberg B, Fahmy Y, Ibrahim A A 1996 Organosolv Pulping: 3. Ethanol pulping of wheat straw Cellul. Chem. Technol. 30(2) 161-174.

[5] Jimenez L, Garci J C, Perez I, Ariza J, Lopez F 2001 Acetone Pulping of Wheat Straw. Influence of Cooking and Beating Conditions on The Resulting Paper Sheets American Chemical Society 40(26) 6201-6206

[6] Ashori A 2006 Non-wood fibers - A potential source of raw material in papermaking Polymer - Plastic Technology and Engineering 45 1133-1136

[7] Viel Q, Esposito A, Saiter J M, Santulli C, Turner J A 2017 Interfacial Characterization by Pull-Out Test of Bamboo Fibers Embedded in Poly(Lactic Acid) *Fibers* *6*(1) No. 7 https: //doi.org/ 10.3390/ fib6010007

[8] Scurlock J M O, Dayton D C, Hames B 2000 Bamboo, an overlooked biomass? Oak Ridge National Laboratory Oak Ridge Tennessee (USA) Environmental Sciences Division Publication 4963

[9] Herliyana E N, Noverita , Lisdar I S 2005 Fungi pada kuning (*Bambusa vulgaris* schardvar. vitata) dan hijau (*Bambusa vulgaris* schardvar vulgaris) serta tingkat degradasi yang diakibatkannnya *JTHH* 18(2) 2-10.

[10] Maoyi F 2006 Bamboo Resources and Utilization in China Research Institute of Subtropical Forestry CAF Fuyang Zhejiang China http://www.bioversityinternational.org/ publications/ Webversion/ 572 /ch24.htm.
[11] Khristova P, Kordsachia O, Daffalla S 2004 Alkaline pulping of *Acacia seyal* Tropical Science 44(4) 207-215 DOI: 10.1002/ts.170.
[12] Batalha L A R., Colodette J L, Gomide J L, Barbosa L C A, Maltha C R A, Borges F J G 2012 Dissolving pulp production from bamboo BioResources 7(1) 640-651
[13] Andtbacka S 2006 A Fibreline Designed for Bamboo Pulping http://www.tappsa.co.za/archive2/Journal_papers/Bamboo_pulping/bamboo_pulping.html.
[14] Adamopoulos S, Martinez M, Ramirez D 2007 Characterization of packaging grade papers from recycled raw materials through the study of fiber morphology and composition Global NEST Journal 1(9) 20-28
[15] Oluwadara A O, Ashimiyu O S, 2007 The relationship between fibre characteristics and pulp-sheet properties of *Leucaenaleucocephala (Lam.)* De Wit Middle-East Journal of Scientific Research 2(2) 63-68
[16] Diaz M J, Alfaro M M, Garcia M E, Eugenio J A, Lopez F 2004 Etanol Pulping from Tagasaste (*Chamaecytisus proliferus L.F. ssp palmensis*) A New Promising Source for Cellulose Pulp Ind. Eng. Chem. Res. 43 : 1875-1881
[17] Ruiz H A 2011 Development and characterization of an environmentally friendly process sequence (Auto hydrolysis and Organosolv) for wheat straw delignification Applied Biochemistry and Biotechnology 164 629-641
[18] Correia V C, Curvelo A A S., Marabezi K, Almeida A E F S, Junior H S 2015 Bamboo cellulosic pulp produced by the ethanol/water process for reinforcement applications Ciência Florestal Santa Maria 25(1) 127-135
[19] Bowyer J L, Haygreen J G, Schmulsky R 2003 Forest Products and Wood Science: An Introduction 4th Ed. Iowa State Press. USA
[20] Sunquist 2000 Organosolv pulping, in: Chemical Pulping, Paper making Science and Technology Book 6B J. Gullichsenand C.J. Fogellholm (eds) FapetOy. Finland 411-427.
[21] Tu Q, Fu S, Zhan H, Chai X, Lucia L A 2008 Kinetic modeling of formic acid pulping of bagasse Journal of Agricultural and Food Chemistry 56(9) 3097-3101.
[22] Li M.F, Sun S N, Xu F, Sun R C 2012 Organosolv Fractionation of Lignocelluloses for Fuels Chemicals and Materials: A Biorefinery Processing Perspective. In: *Biomass Conversion, The Interface of Biotechnology Chemistry and Materials Science* 341-379.
[23] Leponiemi 2008 Non-wood pulping possibilities a challenge for the chemical pulping industry Appita Journal 61(3) 234-243.
[24] Rodríguez and Jiménez 2008 With organic solvents others pulping than alcohols Afinidad 65(535) 188-196.
[25] Lopez F, Alfaro A, Jimenez L, Rodriguez A 2006 Alcohols as organic solvents for the obtainment of cellulose pulp Afinidad 63(523) 174-182.
[26] Lavarack B P, Rainey T J, Falzon K L, Bullock G E 2005 A preliminary assessment of aqueous ethanol pulping of bagasse The Ecopulp Process *International Sugar Journal* 107 611-615.
[27] Shatalov A A, Pereira H 2004 Arundodonax L. Reed:New Perspectives for Pulping and Bleaching. Part 3. Ethanol Reinforced Alkaline Pulping. Tappi Journal 3 (2) 27-31..
[28] Yawalata D, Paszner L, 2004 Anionic effect in high concentration alcohol pulping Organosolv Holzforschung 58(1) 1-6,
[29] Paurjoozi M J M., Rovsshandeh S N, Ardeh S N 2004 Bleachibility of rice straw pulp organosolv Iranian Polymer Journal 13(4) 275-280.
[30] Aziz S, Sarkanen K 1989 rganosolv Pulping, A Review, *Tappi Journal* 72(3) 169– 175.
[31] Muurinen E 2000 Organosolv Pulping (A Review and Distillation Study Related to Peroxyacid Pulping Department of Process Engineering Linnanmaa 1-314.
[32] Mormanee R 2009 Sustainable Utilization of Bamboo for Pulp and Paper Manufacturing in Thailand World Bamboo Congress Forest research and development bureau Royal forest department Thailand

[33] Pan X J, San Y, Ito T 1999 Atmospheric acetic acid pulping of rice straw II: Behavior of ash and silica in rice straw during atmospheric acetic acid pulping and bleaching, Holzforschung 53 (1) 49-55.
[34] Ibrahim M, Sarwar M J, Ali H 2004 Effect of inorganic acid catalyst on the acetosolv pulping of maize stalk Journal Cellulose Chemistry and Technology 38(2) 87-94.
[35] Dominggus Y, Laszio Y 2004 Anionic effect in high concentration alcohol pulping organosolv Holzforschung 58(1) 1-6.
[36] oncalves A R, Denise D, Moriya R, Oliveria L R M 2005 Pulping of sugarcane bagasse and straw and biobleaching of the pulps: conditions parameters and recycling of enzymes. Appita Conference Auckland
[37] Hidayati S, Zuidar A S, Satyajaya W 2017 Effect of acetic acid: formic acid ratio on characteristics of pulp from Oil Palm Empty Fruit Bunches (OPEFB) ARPN Journal of Engineering and Applied Sciences 12(12) 3802-3807
[38] Ferrer A, Vega A, Ligerob P, Rodríguez P 2011 Pulping of Empty Fruit Bunches (EFB) from the Palm Oil Industry by Formic Acid Bioresouces 6(4) 4282-4301
[39] Lam H Q, Bigot Y L, Delmas M, Avigon G 2001 Formic acid pulping of rice straw Ind. Crops Prod. 14(1) 65-71.
[40] Kupiainen L, Ahola J, Tanskanen J 2012 Hydrolysis of organosolv wheat pulp in formic acid at high temperature for glucose production Bioresources Technology 116 29-35.
[41] Wanrosli W D, Zainuddin Z, Law K N, Asro R 2007 Pulp from oil palm fronds by chemical process Industrial Crop and Producs 25 89-94.
[42] Sjöström E 1981 Wood Chemistry: Fundamentals and Applications Academic Press, New York
[43] Jahan M S, Chowdhury D A N, Islam M K, Ahmed F N 2006 Elemental chlorine free and total chlorine free bleaching ofsoda-AQ cotton stalks pulps J Asiatic Soc Bangladesh 32 179–186.
[44] Hidayati S, Zuidar S, Fahreza A 2016 Optimization of Formacell Pulp Production from Oil Palm Bunches (OPB) with Response Surface Method J.Reactor 16(4) 161-171.
[45] Anindyawati T 2009 Prospects and waste lignocellulosic enzymes for bioethanol products LIPI Journal 44(1) 49-56.
[46] Clark J A 1985 *Pulp Technology and Treatment For Paper* Miller Freeman Publications Inc. San Francisco California 1-751.
[47] Fengel D, Wegener G 1984 Wood: Chemistry, Ultrastructure, Reactions de Gruyter Berlin X111 613 S. 1- 729 ISBN 3-11-008481-3
[48] Gierer J 1980 Chemical aspects of kraft pulping Wood Science Technology 14 241-266.
[49] Sarkanen K V 1990 Chemistry of solvent pulping Tappi Journal 73(10) 215–219.
[50] Parajo J C, Alonso J L, Vazquez D, Santos V 1993 Optimization of catalyzed acetosolv fractionation of pine Holzforschung 4 188-196.
[51] Fatriasari W, Risanto L 2011 Kraft Pulp Properties Sengon Wood (*ParaserianthesFalcataria*): Differences Concentrations of Cooking Materials and Bleaching stages Widyariset 14(3) 589-597.
[52] Sun Y, Lin L 2010 Hydrolysis behavior of bamboo fiber in formic acid reaction system J Agric Food Chem 58(4) 2253-9 doi: 10.1021/jf903731s.
[53] Gan Q, Allen S J, Taylor G 2002 Kinetic dynamics in heterogeneous enzymatic hydrolysis of cellulose: an overview, an experimental study and mathematical modeling Process Biochem. 38 1003–1018.

IOP Conf. Series: Materials Science and Engineering **532** (2019) 012025 doi:10.1088/1757-899X/532/1/012025

Electrical Resistivity Imaging (ERI) and Induced Polarization (IP) Survey to Solve Water Drought Problem at Alor Gajah, Melaka, Malaysia

A Suryadi[1], Batara[2], S N Amir[2]

[1] Geological Engineering Program, Faculty of Engineering, Universitas Islam Riau, Pekanbaru, Riau
[2] AMT Geo Services, Selangor, Malaysia

adisuryadi@eng.uir.ac.id

Abstract. Water drought become serious problem during dry season for rice field at Alor Gajah, Melaka, Malaysia. Electrical Resistivity Imaging (ERI) surveys have been conducted in order to locate and delineate groundwater resources by distribution of subsurface resistivity. The survey of ERI was carried out using ABEM Terrameter SAS 400 and ABEM Selector ES10-64 with pole-dipole configuration of electrode. At the same time, Induced Polarization (IP) survey also carried out to support the result of ERI. There are three line profiles with 400 m length and 5 m electrode spacing named PS1, KRBG2 and BB3. From three line profiles, two of them show that the distribution of low resistivity value (20 – 500 Ωm) are located at depth ranging from 20 to 70 m. The low resistivity values are associated with water saturated zone. This result was supported by IP's results that show at that particular depth the value of chargeability ranging 0 to 4 ms. Meanwhile, line profile BB3 indicate low resistivity value (0.5 – 500 Ωm) but the chargeability value are high (10 – 75 ms) interpreted as clay layer.

Keyword: Drought, Electrical Resistivity Imaging, Induced Polarization, Groundwater, Alor Gajah, Melaka

1. Introduction

Water is an essential need for earth life including human, animal and plant. Generally, water can be divided into two types they are surface water and groundwater. Every water on the surface is classified as surface water like water in a river, lake, wetland or ocean. It different with groundwater which present beneath earth's surface in pore space and in the fractures of rock formation. In comparison, groundwater is much better than surface water in quantity and quality. During dry season, surface drought usually occur at Alor Gajah, Melaka, Malaysia which is become serious problems for rice field. Groundwater become the best alternative as water resources when surface water is not sufficient to prevent agriculture failure [1].

Groundwater sources is trapping in aquifers that led to a need of intensive study to locate the aquifers. Electrical Resistivity Imaging (ERI) is widely used in groundwater exploration and environmental survey [2–8]. By using ERI, the distribution of resistivity value of subsurface can be determined to produce two-dimensional modelling. Besides that, IP method to measure chargeability of subsurface

IOP Conf. Series: Materials Science and Engineering **532** (2019) 012025 doi:10.1088/1757-899X/532/1/012025

layer become important to correlate with ERI. Correlation between ERI and IP raising the accuracy of aquifer layer interpretation.

Study area located at three irrigation scheme in Alor Gajah, Melaka, Malaysia that are irrigation scheme of Padang Sebak II Gadek, irrigation scheme of Lanjut Manis and irrigation scheme of Parit China (Figure 1). This study held to solve water drought problem during dry season. The aim of this study is to find aquifer layer as alternative water sources. From that aim, tubewell location and depth will propose to fulfill water needed at the study area.

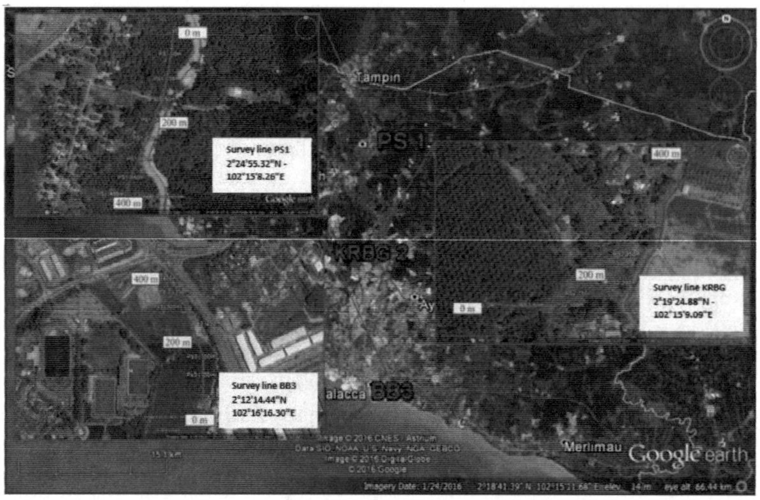

Figure 1. Map of study area that show the location of survey line

2. Geology Regional

Generally, geology regional of study area and surrounding consist of intrusive granite as basement rock, metasediment and marine deposit with aged from Devonian to Ordovisian [9]. First location at irrigation scheme of Padang Sebak II Gadek consist of intrusive granite. Second location at irrigation scheme of Lanjut Manis composed of sandstone and mudstone that overlay graphitic schist and phyllite from howthorndern. At third location is irrigation scheme of Parit China consists of marine and continental deposits such as mudstone, siltstone, sandstone, peat and gravels. Geological maps of study area can be seen at Figure 2.

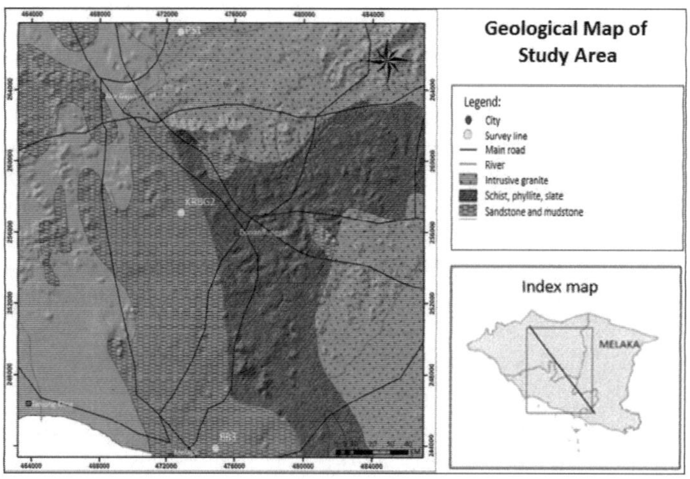

Figure 2. Geological map of study area

IOP Conf. Series: Materials Science and Engineering **532** (2019) 012025 doi:10.1088/1757-899X/532/1/012025

3. Material and Method

ERI and IP Survey was conducted using ABEM SAS 4000 terrameter and Lund electrode selector system ES10-64 (Figure 3). Basically, resistivity measured by using 4 electrodes whereas 2 current electrode and 2 potential electrode. Apparent resistivity (ρa) value can be calculate with equation between current (I) and voltage (V):

$$\rho a = k\, V/I$$

k is geometry factor that depend on configuration. In this study electrode configuration used is pole-dipole (Figure 3). For pole-dipole configuration factor geometry (k) calculate with formula below:

$$k = 2\pi \frac{b(a+b)}{a}$$

a is distance between P1 to P2; b is distance between C1 to P1.

In other hands, IP survey carry out with injecting current through electrode and potential difference (voltage) measured between potential electrodes. The voltage did not drop instantaneously to zero when the current was turned off. Instead, the potential difference dropped sharply at first, then gradually decay to zero after a given interval of time. From that chargeability can be calculate with formula:

$$Chargeability = V_t / V_0$$

Where V_t is voltage after interval time t and V_0 is voltage during current injecting.

Data collection taken by using multi-electrodes with 61 electrodes. Electrode spacing used is 5 meters with 400 m length of survey line in straight length arrangement. The electrode selector system automatically select the relevant four active electrode for each measurement of resistivity data. The data were processed by using inversion software RES2DINV [10,11] to produce two-dimensional model of resistivity value (ERI) and chargeability (IP).

Figure 3. ERI Survey equipment

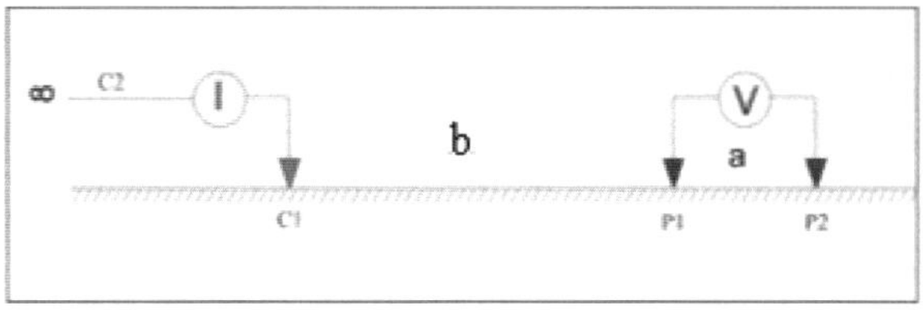

Figure 4. Pole-dipole electrode configuration

4. Result

Three survey lines has processed to produce inversion model of resistivity. The result is representing the true condition of subsurface with shown variation of resistivity value. The different of resistivity value indicate the different of subsurface material.

4.1. Survey line PS1

Survey line PS1 conducted at irrigation scheme of Padang Sebak II with line orientation N30^{0}E - N210^{0}E. Based on the result of inversion model, the apparent resistivity (a) value ranging from 0.5 to 1000 Ωm. A large variation of apparent resistivity value indicating heterogeneous layer of subsurface. There are three layers interpreted from line PS1. The highest resistivity value (500 - 1000 Ωm) represented by red and purple in colour at depth more than 30 m interpreted as granite which is basement rock of study area. This layer is overlay by less resistive layer with resistivity value 30 – 500 Ωm (light green to orange in colour) from 100 m depth up to surface. This layer interpreted as sandstone. Another layer is located within sandstone layer that shown as low resistivity value ranging from 0.5 – 10 Ωm (dark blue to light blue) interpreted as Clay layer. Meanwhile, IP line PS1 also have chargeability value variation with the highest value is 75 msec and the lowest value is 0.00 msec. Low chargeability value found at depth 25 m and below has value at 0.00 – 2.00 msec (dark blue to blue), this value can be associated with aquifer, presented in Fig. 5.

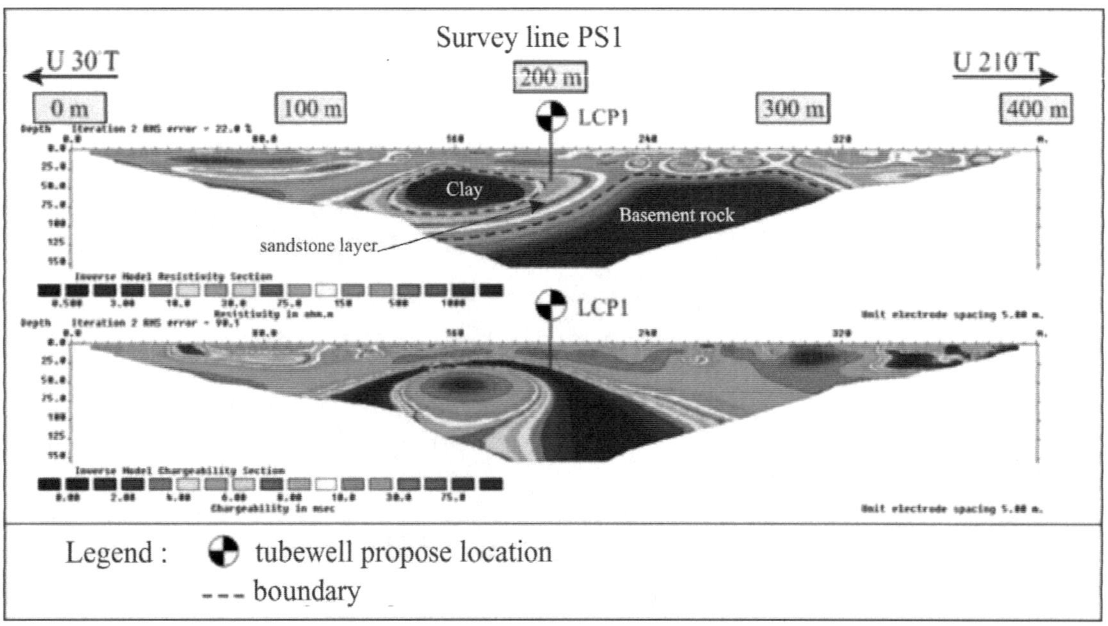

Figure 5. Inversion model of ERI and IP survey line PS1

4.2 Survey line KRBG2

This line survey conducted at irrigation scheme of Lanjut Manis Center of Melaka Area with line orientation N210^{0}E - N30^{0}E where at 0 to 80 m line survey is assigned at palm oil plantation with high topography. Apparent resistivity value that shown from inversion model has value with range 0.5 – 1000 Ωm. Based on large variation of apparent resistivity value, it divided into three layers. First layer is interpreted as granite as basement rock with high resistivity value 500 – 1000 Ωm (orange to purple) that shown at 25 m depth below. Second layer with resistivity value ranging from 30 – 500 Ωm (dark yellow to green) is interpreted as sandstone layer at near surface down to 100 m beneath earth. Third layer with lowest resistivity value 0.5 – 10 Ωm (dark blue to light blue) sandwich between sandstone layers is interpreted as clay layer. In other hand, chargeability value also have wide range from 0.00 – 75.00 msec. aquifer that indicated by low chargeability value (0.00 – 2.00 msec) with blue in colour shown at 25m as presented in Fig. 6 below.

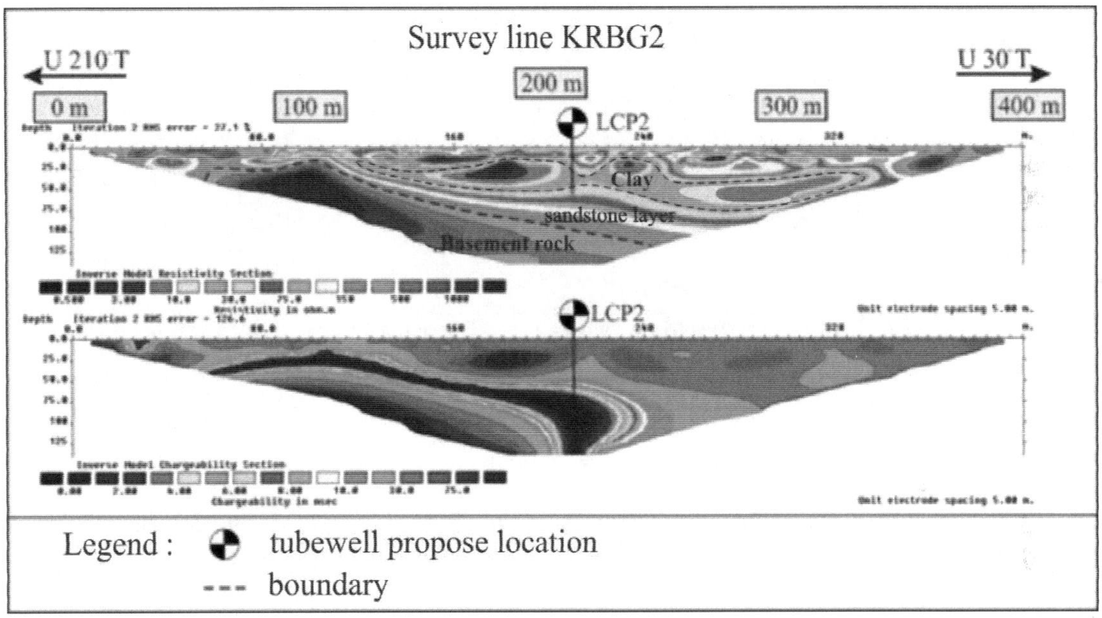

Figure 6. Inversion model of ERI and IP survey line KRBG2

4.3 Survey Line BB3

Survey line BB3 is located at irrigation scheme of Parit China area with line orientation N170^{0}E - N350^{0}E. The data collected at rice field with gentle topography. Result of inversion model shown that apparent resistivity value ranging from 0.5 up to 150 Ωm. it divided into two layers, first layer with resistivity value 30 – 150 Ωm (green to yellow) interpreted as weathered sandstone. This layer found at 50 m below surface. Another layer is interpreted as unconsolidated soil mostly clay layer with resistivity value 0.5 – 30 Ωm (dark blue to light blue) that show at surface to 150 m below. From the chargeability value there is no indicating for aquifer layer because mostly they have high chargeability value (30 – 75 msec), as shown in Fig. 7 and Table 1.

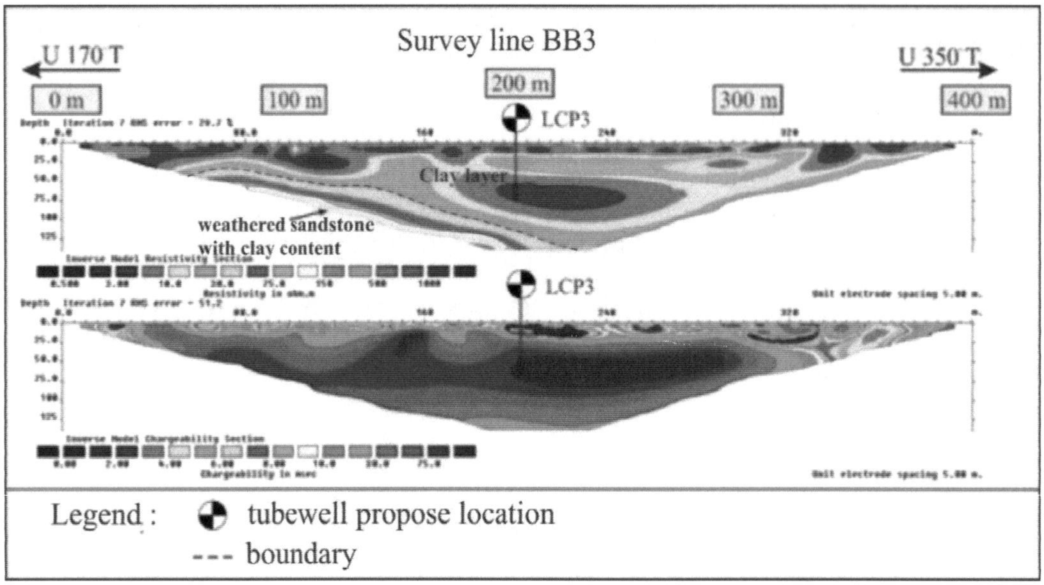

Figure 7. Inversion model of ERI and IP survey line BB3

Table 1. Interpretation of ERI and IP results

Survey line	Depth (m)	Resistivity value (Ωm)	Chargeability value (msec)	Interpretation
PS1	0 – 100	30 – 500	0 – 2	Sand, Sandstone layer
	25 – 75	0.5 – 10	30 – 75	Clay layer
	30 – 150	500 – 1000	0 – 20	Granite (basement rock)
KRBG2	0 – 100	30 – 500	0 – 6	Sand, sandstone layer
	15 – 50	0.5 – 10	30 – 75	Clay layer
	25 – 125	500 – 1000	30 – 75	Granite (basement rock)
BB3	0 – 125	0.5 – 30	6 – 75	Clay layer
	50 – 125	30 – 150	30 – 75	Weathered sandstone (high clays content)

5. Discussion

Based on the result of ERI and IP survey, location of aquifer layer can be predicted and suggestion for tube well location can be determined. At survey line PS1 and KRBG 2, layer with resistivity value 30 – 500 Ωm interpreted as aquifer (sandstone layer). This interpretation supported by low chargeability value (0.00 – 2.00 msec) that always associated with groundwater. For survey line BB3, there is low resistivity value (30 – 150 Ωm) that interpreted as weathered sandstone. But it's not an aquifer layer because that layer have high chargeability value (30 – 75 msec). High chargeability value at this layer interpreted as high clay content within weathered sandstone.

Location and depth of tube well can be suggest by ERI and IP result. For survey line PS1 at irrigation scheme of Padang Sebak II, tube well suggestion located at middle of survey line (200 m from first electrode) with 40 m minimum of depth. Tube well location for survey line KRBG2 at irrigation scheme of Lanjut Manis suggested at 210 m from first electrode with minimum depth is 75 m. While survey line BB3 doesn't have groundwater potential but if needed try to make tube well at 200 m from first electrode with more than 125 m of depth.

IOP Conf. Series: Materials Science and Engineering **532** (2019) 012025 doi:10.1088/1757-899X/532/1/012025

6. Conclusion

Electrical Resistivity Imaging (ERI) and Induced Polarization (IP) method in this study has been successfully used to locate groundwater as alternative water source to solve water drought during dry season at Alor Gajah, Melaka, Malaysia. Aquifer layer at study area characterized by resistivity value ranging from 30 up to 500 Ωm and supported with low chargeability value (0 – 2 msec) that interpreted as sandstone layer. The depth of aquifer layer are around 40 m to 75 as the suggestion to generate tube well at study area.

Acknowledgements

We would like to thank members of JS Heritage Engineering for their cooperation in data acquisition at field. We are thankful to Department of Irrigation and Drainage of Melaka that collaborate with for this study.

References

[1] Putra D B E and Yuskar Y 2017 Pemetaan Airtanah Dangkal Dan Analisis Intrusi Air Laut, Penelitian Terhadap Airtanah Dangkal di Sesa Bantan Tua, Kecamantan Bantan, Kabupaten Bengkalis, Propinsi Riau *Seminar Nasional ke-III Faklutas Teknik Geologi Universitas Padjadjaran*
[2] Hamzah U, Ismail M A and Samsudin A R 2008 Geophysical techniques in the study of hydrocarbon-contaminated soil **54** 133–8
[3] Jumary S Z, Hamzah U and Samsudin A R 2002 Teknik-teknik geoelektrik dalam Pemetaan air masin di Kuala (Mapping of groundwater salinity at Kuala Selangor by geoelectrical techniques)
[4] Hamzah U, Samsudin A R and Malim E P 2007 Groundwater investigation in Kuala Selangor using vertical electrical sounding (VES) surveys *Environ. Geol.* **51** 1349–59
[5] Chandra S, Dewandel B, Dutta S and Ahmed S 2010 Geophysical model of geological discontinuities in a granitic aquifer: Analyzing small scale variability of electrical resistivity for groundwater occurrences *J. Appl. Geophys.* **71** 137–48
[6] Saad R, Nawawi M N M and Mohamad E T 2012 Groundwater detection in alluvium using 2-D electrical resistivity tomography (ERT) *Electron. J. Geotech. Eng.* **17 D** 369–76
[7] Asry Z, Samsudin A R, Yaacob W Z and Yaakub J 2012 Groundwater investigation using electrical resistivity imaging technique at Sg. Udang, Melaka, Malaysia *Bull. Geol. Soc. Malaysia* **58** 55–8
[8] Azhar M A, Suryadi A, Samsudin A R, Yaacob W Z W and Saidin A N 2016 2D Geo-Electrical Resistivity Imaging (ERI) of Hydrocarbon Contaminated Soil *EJGE (Electron. J. Geotech. Eng.* **21** 299–304
[9] Hutchison C S 1989 *Geological Evolution of South-east Asia* (Oxford: Clarendon Press)
[10] Software G 2006 Res2dinv software
[11] Loke M H and Barker R D 1995 Least-square deconvolution of apparent resistivity psuedosection *Geophysics* **60** 1682–90

IOP Conf. Series: Materials Science and Engineering **532** (2019) 012026 doi:10.1088/1757-899X/532/1/012026

Synthesization and characterization of chlorhexidine/calcium phosphate/poly(lactic-*co*-glycolic acid) bone cement

N M Daud[1], Z Tapsir[1], A Idris[1], A K Nasution[2], S Saidin[1,3,*]

[1]School of Biomedical Engineering & Health Sciences, Faculty of Engineering, Universiti Teknologi Malaysia, 81310, UTM Johor Bahru, Johor, Malaysia
[2]Department of Mechanical Engineering, Faculty of Engineering, Muhammadiyah University of Riau, Pekan baru, Indonesia
[3]IJN-UTM Cardiovascular Engineering Centre, Institute of Human Centered Engineering, Universiti Teknologi Malaysia, 81310, UTM Johor Bahru, Johor, Malaysia

syafiqahsaidin@biomedical.utm.my

Abstract. Calcium (Ca) and phosphorus (P) elements are important for bone growth and bone remodeling process while chlorhexidine (CHX) is mainly used as an antiseptic due to its broad antibacterial effects. In this study, CaP nanoparticles were synthesized by a wet chemical route from the precursors of calcium hydroxide (CaOH) and diammonium hydrogen phosphate, $(NH_4)_2HPO_4$. Poly(lactic-*co*-glycolic acid) PLGA polymer was incorporated into the mixture of CaP, acted as a matrix binder to form bone cement composite. Different concentrations of CHX (0.5, 1.0, 1.5 and 2.0 mM) were then added into the CaP/PLGA mixtures (CHX/CaP/PLGA bone cements). The bone cements were then characterized by ATR-FTIR, XRD and SEM analyses. The CHX was successfully incorporated into the bone cements, indicated by the appearance of aromatic amine and alkane C–H stretching vibrations which contributed by the ring and long hydrocarbon chain of CHX. The XRD spectra of bone cements showed poorly crystalline phase of apatite. A porous structure of bone cements was also observed with the dispersion of CaP nanoparticles over the PLGA. The CHX/CaP/PLGA bone cement is intended for bone resorption and bone infection applications that require further exploration on its capability to act as bone cement.

1. Introduction

Calcium phosphate (CaP) compounds are being utilised as artificial bone substitute materials due to its bioactivity, biocompatibility and osteoconductivity properties [1, 2]. The major principle of bone cement is lying on their mechanical strength to reinforce damaged bone structures and their capability to be fabricated as an injectable paste cement [3]. Despite of its excellent biocompatibility, CaP (mostly hydroxyapatite) has low degradation rate which limits bone growth onto biomaterials. Thus, incorporation of natural or synthetic polymers such as poly(lactic-*co*-glycolic acid) (PLGA) into CaP bone cement will increase its degradability [4]. In some cases, bone cement is being used to immobilise metallic implants in the bone. Appropriate degradability of bone cement is crucial to allow direct osseointegration on metallic implant surfaces without interruption on the bone-implant interface.

In orthopaedic surgery, infections are one of the largest problems and often lead to severe pain and loss of bone tissues [5]. While in dental applications, CaP may act as a reservoir for accumulation of periodontal and carious bacteria that often lead to bone resorption and demineralization [6]. There are several strategies to reduce infections including the incorporation of antibacterial agent into bone cements. Calcium phosphate cements, either in a form of hydroxyapatite or brushite had been loaded with antibacterial agents such as gentamycin and chlorhexidine (CHX), either through microcapsule development or simply manually blended cement [7-9]. The cements can provide controlled release of antibacterial agents, through diffusion and/or cement degradation [6].

Chlorhexidine has long been used as skin antiseptic due to its broad antibacterial effects against Gram-positive or Gram-negative bacteria [10]. It has very low toxicity and strong binding to skin, mucus and other structures in the oral cavity due to cationic nature of CHX and mouthwash pH of 5.4 [11]. Previously, CHX has been clarified compatible with osteoblast cell and able to support osteoblastic maturation and mineralization [12]. Therefore, the purpose of this study was to incorporate different concentrations of CHX into CaP/PLGA to form bone cement composite. The characterization analyses were concerning the chemical functionality, crystallinity and morphology of the bone cement using attenuated total reflectance-Fourier transform infrared spectroscopy (ATR-FTIR), X-ray diffractometer (XRD) and scanning electron microscopy (SEM, respectively.

2. Methodology

2.1. *Sample Preparation*

2.1.1. *Synthesization of CaP*
The CaP powders were prepared by titrating 100 mL of 0.6 M calcium hydroxide (CaOH) into 100 mL of 0.2 M diammonium phosphate ($(NH_4)_2HPO_4$) at 40°C under stirring condition. The pH was controlled at 9, by adding either ammonia (NH_4) or nitric acid (HNO_3). The CaP solvent was further stirred at 40°C for 4 hours. The homogenized CaP mixture was sit at room temperature for 24 hours to allow complete reaction and precipitation. The CaP precipitates were then filtered using a filter paper and were dried in an oven at 110°C for 5 hours. Further heat treatment at 1000°C for 4 hours was performed for the calcination purpose. Finally, the CaP was crushed using a mortar and pestle to form fine CaP powders.

2.1.2. *Preparation of CHX/CaP/PLGA bone cement*
The CaP/PLGA bone cement was prepared with a composition of 80 wt% CaP and 20 wt% PLGA. This composition of bone cement was set as a control. Initially, the PLGA was dissolved in dimethyl sulfoxide (DMSO). The synthesized CaP powders were then added into the PLGA solvent and mixed vigorously. The mixture was casted gently into a pellet form and the casted bone cement was stored at room temperature. While the CHX/CaP/PLGA bone cements were fabricated by incorporating 80 wt% CaP and 19 wt% PLGA and 1 wt% CHX. The CHX concentrations were varied to 0.5, 1.0, 1.5 and 2.0 mM.

2.1.3. *Characterization analyses*
The chemical functionalities of the bone cements were verified by ATR-FTIR (Nicolet iS5-IR Spectrometer, Thermo Scientific, USA). The obtained spectra were analyzed at a frequency interval of 500 cm^{-1} to 1000 cm^{-1}. The number of scanning was set at 36 with a resolution of 4. The crystallinity of the bone cements was investigated by XRD (D5000, Siemens, Germany). The analyses were performed at 40 kV and 40 mA to recognize the crystallographic phases and the crystal lattice of the bone cements. The data were recorded in 2θ range from 5° to 70° using CuKα radiation of 1.5406 Å. The morphology of the bone cements was then examined by SEM (Tabletop TM3000, HITACHI, Japan) at 1000× and 5000× magnifications. Each bone cement was sputter coated with an ultrathin platinum conductive film using a sputter coater (Q150R, Quorum Technologies, England) to enhance image resolution and visualization.

IOP Conf. Series: Materials Science and Engineering **532** (2019) 012026 doi:10.1088/1757-899X/532/1/012026

3. Results and discussion

3.1. ATR-FTIR analyses

Figure 1 presents the ATR-FTIR spectra of CHX/CaP/PLGA bone cements. The results showed an appearance of aromatic amine CHX on all bone cements at 1046 cm^{-1}, contributed by the ring of CHX. A strong CHX indicator was observed between 2850 cm^{-1} and 3000 cm^{-1}, which corresponded to alkene C–H stretching vibration, derived from the long hydrocarbon chain of CHX [13]. While, the peaks at 1015 cm^{-1} and 1085 cm^{-1} were derived from the vibrations of phosphate group (PO_4^{3-}), representing the composition of CaP compound [14]. The C–O bond stretching (1270 cm^{-1}) and carbonyl group, C=O (1750 cm^{-1}) were assigned to the PLGA while the broad band at 3000 cm^{-1} – 3600 cm^{-1} region was observed due to stretching vibration of hydroxyl group, OH^-.

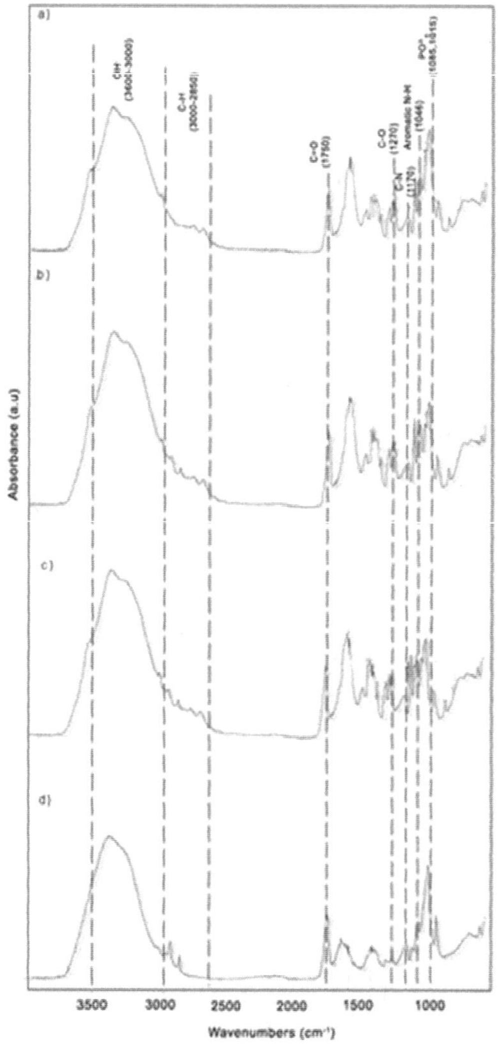

Figure 1. ATR-FTIR spectra of (a) 0.5CHX/CaP/PLGA (b) 1.0CHX/CaP/PLGA, (c) 1.5CHX/CaP/PLGA and (d) 2.0CHX/CaP/PLGA bone cements.

3.2. Surface crystallinity

Figure 2 shows the XRD spectra of CHX/CaP/PLGA bone cements. Most intense peaks were observed at 31.80° (211), 32.10° (112) and 32.85° (300), originated from hydroxyapatite and 25.90° (1010), 28.00° (214) and 31.25° (0210), originated from beta-tricalcium phosphate (β-TCP) based on the JCPDS number of 09-0432 for hydroxyapatite and 09-0169 for β-TCP. The PLGA and CHX were not identified in the XRD patterns due to its non-crystallize form [15]. It is important to point out that the patterns showed an observation of weak hydroxyapatite composition, indicating poorly crystalline phase of apatite [16].

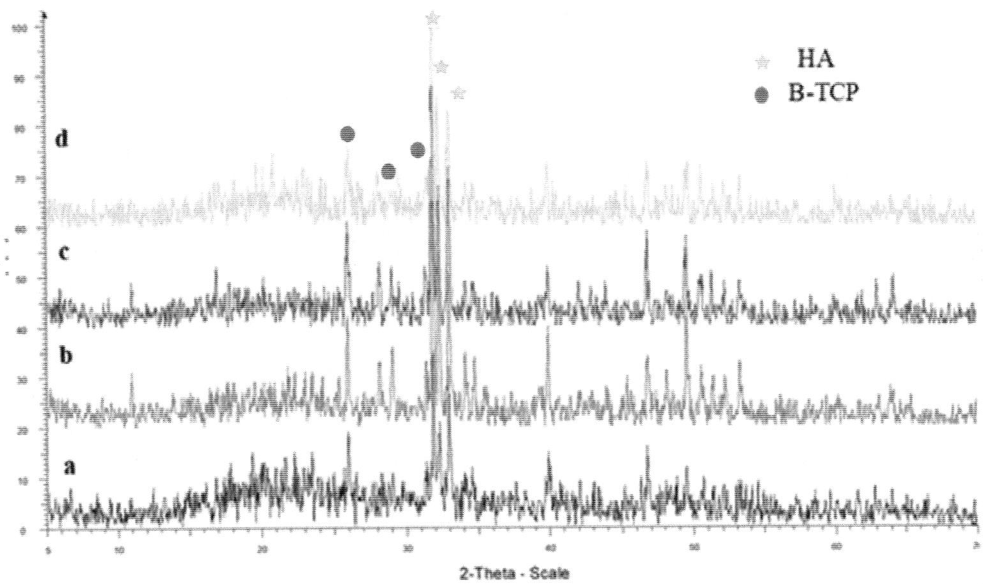

Figure 2. XRD patterns of (a) 0.5CHX/CaP/PLGA (b) 1.0CHX/CaP/PLGA, (c) 1.5CHX/CaP/PLGA and (d) 2.0CHX/CaP/PLGA bone cements.

3.3. Surface Morphology

Figure 3 visualizes the morphology of CaP and CHX/CaP/PLGA bone cements. The bone cements showed a porous structure, essential for the migration and proliferation of cells after a cement restoration [16]. The CaP particles were thoroughly dispersed over the PLGA matrix. The nanosize of CaP is important to improve the mechanical properties of materials and may aid in protein adsorption and cell adhesion as compared to micron size of CaP [17]. The incorporation of PLGA into the CaP mixture has caused the agglomeration of CaP powders within the PLGA compounds. However, the addition of CHX did not markedly change the surface morphology of the bone cement.

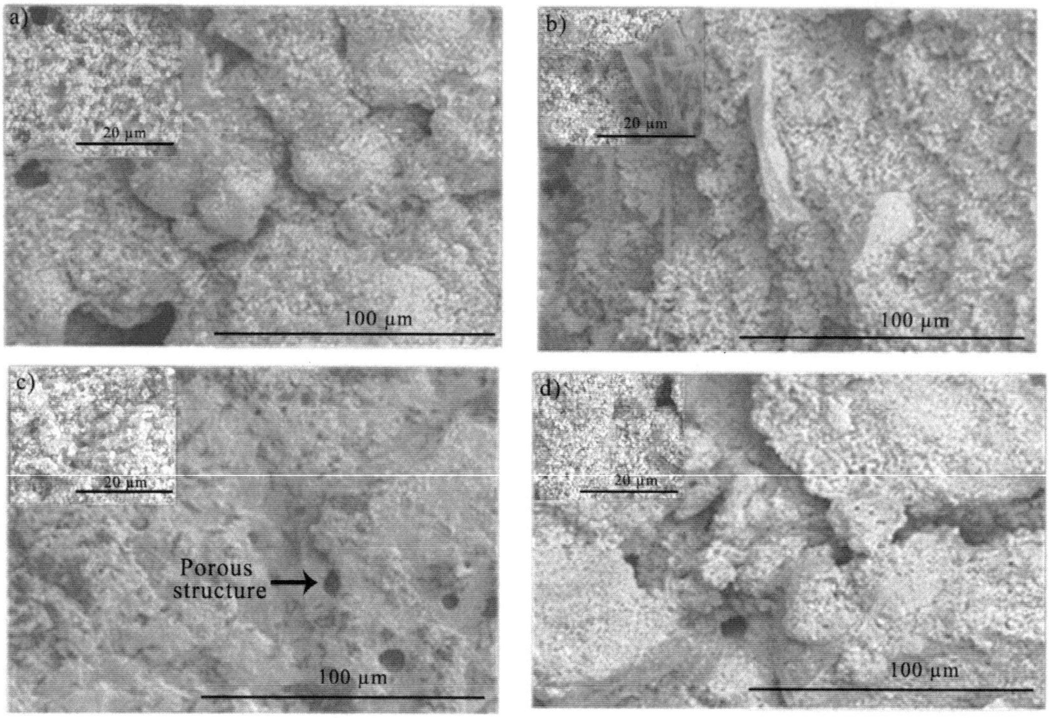

Figure 3. SEM images and EDS analysis of (a) 0.5CHX/CaP/PLGA (b) 1.0CHX/CaP/PLGA, (c) 1.5CHX/CaP/PLGA and (d) 2.0CHX/CaP/PLGA bone cements.

4. Conclusion

In summary, the CHX/CaP/PLGA bone cements were synthesized through a wet chemical route. The incorporation of PLGA and CHX into the CaP are necessary to fabricate antibacterial degradable bone cement, which is preferable for the filling of bone defect areas. Increasing the amount of CHX concentration did not significantly alter the crystallinity and morphology of the bone cements.

Acknowledgements

The authors would like to thank the financial support from the Ministries of Research, Technology and Higher Education Republic of Indonesia through Director of Research and Community Service with the PDUPT contract number of 017/KONTRAK-PENELITIAN/K10/KM/2018. A respective dedication is also given to Malaysia Ministry of Higher Education through the funding of Research University Grant Tier 1 Q.J130000.2545.18H40.

References

[1] Bohner M 2000 *Injury*, **31** D37
[2] Shavandi A, Bekhit A El-D A, Ali M A, Sun Z 2015 *Int. J. Biol. Macromolec.*, **80** 445
[3] Gbureck U, Barralet J E, Spatz K, Grover L M, Thull R 2004 *Biomaterials*, **25 (11)** 2187
[4] Lanao R P F, Leeuwenburgh S C G, Wolke J G C, Jansen J A 2011 *Biomaterials*, **32(34)** 8839
[5] Ruchholtz S, Tager G, Nast-Kolb D 2004 *Der Unfallchirurg*, **107(4)** 307
[6] Xia W, Razi M R M, Ashley P, Neel E A A, Hofmann M P, Young A M 2014 *J. Mater. Chem.*, **2 (12)** 1673
[7] Schnieders J, Gbureck U, Thull R, Kissel T 2006 *Biomaterials*, **27 (23)** 4239

[8] Miola M, Bistolfi A, Valsania M C, Bianco C, Fucale G, Verné E 2013 *Mater. Sci. Eng. C,* **33 (5)** 3025
[9] Young A M, Ng P Y J, Gbureck U, Nazhat S N, Barralet J E, Hofmann M P 2008 *Acta Biomater.,* **4(4)** 1081
[10] Karpiński T, Szkaradkiewicz A 2015 *Eur. Rev. Med. Pharmacol. Sci.,* **19 (7)** 1321
[11] Singh H, Kapoor P, Dhillon J, Kaur M 2014 *Indian J. Dent.,* **5(4)** 199
[12] Mohd Daud N, Nik Malek N A N, Saidin S 2018 *Ann. Anat.*, **220** 29
[13] Daud N M, Bahri I F S, Malek N A N N, Hermawan H, Saidin S 2016 *Colloids Surf. B Biointerfaces*, **145** 130
[14] Hanan M R A, Nasution A K, Hussain R, Saidin S 2018 *Jurnal Teknologi*, **80 (4)** 103
[15] Filho A I, da Silva R V, Bertolo R V 2011 *Dental Press J. Orthod.,* **16 (4)** 55
[16] LeGeros R Z 2002 *Clin. Orthop. Relat. Res.,* **395** 81
[17] Xu Y, Gallert C, Winter J 2008 *Appl. Microbiol. Biotechnol.*, **79 (4)** 687

Optimization of Bioactive Metabolite Production From *Fusarium oxysporum* LBKURCC41

S Marlinda[1], H Y Teruna[1], N W Pratiwi[2], A Ardhi[3], Saryono[1*]
[1]Department of Chemistry, Faculty of Mathematics and Natural Sciences, Universitas Riau
[2]Department of Biology, Faculty of Mathematics and Natural Sciences, Universitas Riau
[3]Department of Food and Agricultural Product Technology, Faculty of Agricultural Technology, Universitas Gadjah Mada

*Corresponding author's email : Saryono@lecturer.unri.ac.id

Abstract. *Fusarium oxysporum* is one of endophyte microbes isolated from dahlia tuber (*Dahlia variabilis*) which has the ability to produce secondary metabolites. In this research, optimization of chemichal and physical fermentation conditions were carried out. Corn, potato, and sweet potato with particle size of 80 mesh were used as corbon sources to produce secondary metabolite for 5, 10, 15, 20, and 25 days respectively with ethyl acetate was used to extract secondary metabolites. The assessment of antimicrobial activity was then performed against *Candida albican, Escherichia coli* and *Staphylacoccus aureus*. A serial of metabolite concentrations 5.7 mg/ml, 3.8 mg/ml, and 1.9 mg/ml were used. The optimum inhibition against microbial pathogen growth was corn for 15 day fementation (C15). The inhibition zone against *Candida albicans, E.coli,* and *S. aureus* were 7.62±0.32, 14.15±0.09, and 15.24±0.24 mm respectively at 5.7 mg/ml metabolite concentration. The result of metabolites screening showed overall extract secondary metabolites comprised terpenoid group. The separation of the active ingredients was performed using both Thin Layer Chromatoghraphy and High Performance Liquid Chromatoghraphy (HPLC) depicted 4 component.

Keywords: *Fusarium oxysporum*, optimization, antimicrobial, secondary metabolite and terpenoid.

1. Introduction

Endophytic fungi is defined as the fungi which spend the whole or part of their lifecycle colonizining inter and/or intracellularly inside the healthy tissues of the host plant, typically causing no apparent symptoms of disease [1]. In the past decade, many valuable bioactive compounds have been successfully discovered from endophytes. During the long period of co-evolution, a friendly relationship was gradually set up between each endophyte and its host plant. Therefore, endopyhtes show abilty to produce same or similar bioactive subtances as those originated from their host plants [2].

Endophytic fungi have also been reported to possess bioactive secondary metabolites and enzymes such as inulinase, chitinase, amylase, and cellulase [3-4]. Endophyte microbial could be classified as bacteria, fungi, and actinomycetes [5]. *Fusarium* sp LBKURCC73 produced cellulase [6], while *Fusarium oxysporum* (LBKURCC6) isolated from pink-flowered dahlia tuber have ability to produce inulinase [7].

Fusarium oxysporum (LBKURCC41) is an endophytic fungi isolated from red-flowered dahlia tuber collected from Bukit tinggi, Sumatera barat [8], it was indentified based on the gene sequenced analyse according to a molecular biological protocol by DNA amplification and sequencing of the ITS region [9]. On the other hand, it has been found ability to produce saponin which produced in Huang medium production with antimicrobial activity [10-11]. The production of secondary metabolite from the *Fusarium oxysporum* can be influenced by medium and time of fermentation. The aim of this research was to discover the best natural carbon resources and time fermentation in order to produce optimum secondary metabolites through corn, potato, and sweet potato.

2. Materials and Methods

Sample Preparation

Corn, potato and sweet potato bought at traditional market in Pekanbaru, Riau, Indonesia were washed and dried at sunlight for 8 h and 40°C for 48h. All sample were blendered separately and sieved for 80 mesh. The powder was stored saparately at room temperature.

Cultivation and preparation of *Fusarium oxysporum* LBKURCC41

Isolate LBKURCC41 *Fusarium oxysporum* was cultivated in potato dextrose agar asecptically and incubated for 4 days at room temperature. Fresh isolate rinsed with NaCl 0.8%, scraped and filtered using sterile glasswoll. Suspencion of spores inoculated into potato dextrose broth at ~7 x 10^{12} spore/ml was determined optical density using spectrophotometer Ultra Viloet-Visible (UV-Vis) Genesis 10 S (*Thermo Scienstific*) then incubated for 4 days at room temperature using *rotary shaker shaking incubator* model LSI 301 6R (Daihan Lab Tech Co. LTD) with150 rpm.

Fermentation

Inoculum starter was placed to Huang production medium with modification in carbon source (corn, potato, and sweet potato) and fementation time 5, 10, 15, 20, and 25 days. After fermentation filtrate was saparated by Buchner funnel through filter paper Whatman No.1. The filtrate was transferred aseptically into a conical flask and stored at 4°C for futher assay. The culture filtrate was extracted 3 times with ethyl acetate. The ethyl acetate phase was saparated from aquos using saparating funnel. Ethyl acetate layer concentrate was evaporated using *vacuum rotary evaporator* Heidolph WB 2000, and stored at room temperature for futher assay.

Antimicrobial assesment

Crude extract was tested for antimicrobial activity using agar well difussion method against *Candida albicans*, Gram-negative bacteria (*Escherichia coli*) and Gram-positive bacteria (*Staphylococcus aureus*). Extract was solved in methanol with a series of concentration of 5.7, 3.8, and 1.9 mg/ml. ketokenazole and amoxan 3.8 mg/ml were used as positive control while methanol as negative control.

Screening of secondary metabolites

Secondary metabolites were screening for their content of terpenoid/steroid, alkaloid, phenolic, flavonoid dan saponin.

Separation Using TLC and HPLC

Extract possessing optimum antimicrobial activity was analyzed further with TLC and HPLC. Aluminum plates pre-coated with silica gel as stationary phase and ethyl acetate 100% as a mobile phase. Chromatograms were observed under UV light, and calculated the retention factor. On the other hand, a number 20 µl extract was analyzed by HPLC Shim-pack VP-ODS (250 x 4,6 mm) using water:methanol sovent with gradient elution system. UV detector was used to determine its maximum wavelenght.

3. Result and Discussion

Fusarium oxysporum growth and fermentation observation

Fusarium oxysporum grew at the second day, produced white mycelia on PDA and purple spore at third day. This inoculum was removed to modified production medium after being incubated for 96 hours. Secondary metabolites produced by *Fusarium oxysporum* were assayed against 3 microbial pathogens. Inhibition zone resulted against *Candida albicans* was showed in Table 1.

IOP Conf. Series: Materials Science and Engineering **532** (2019) 012027 doi:10.1088/1757-899X/532/1/012027

Table 1. Ethyl acetate extract from *Fusarium oxysporum* metabolites and inhibition zone against *Candida albicans*

Carbon source	Fermentation Day	Concentration	Inhibition zone diameter (mm)
Corn	5	Extract 5.7 mg/ml	6.74±0.34[b]
		Extract 3.8 mg/ml	0.00±0.00[c]
		Extract 1.9 mg/ml	0.00±0.00[c]
		Ketoconazole	8.31±0.32[c]
		Methanol	0.00±0.00[c]
Corn	10	Extract 5.7 mg/ml	6.81±0.12[b]
		Extract 3.8 mg/ml	6.14±0.04[c]
		Extract 1.9 mg/ml	0.00±0.00[d]
		Ketoconazole	8.28±0.27[a]
		Methanol	0.00±0.00[d]
Corn	15	Extract 5.7 mg/ml	7.62±0.00[a,b]
		Extract 3.8 mg/ml	7.37±0.15[b]
		Extract 1.9 mg/ml	6.29±0.05[c]
		Ketoconazole	7.87±0.13[a]
		Methanol	0.00±0.00[d]
Corn	20	Extract 5.7 mg/ml	7.22±0.03[b]
		Extract 3.8 mg/ml	6.73±0.00[c]
		Extract 1.9 mg/ml	0.00±0.00[d]
		Ketoconazole	7.37±0.15[a]
		Methanol	0.00±0.00[d]
Corn	25	Extract 5.7 mg/ml	0.00±0.00[b]
		Extract 3.8 mg/ml	0.00±0.00[b]
		Extract 1.9 mg/ml	0.00±0.00[b]
		Ketoconazole	7.39±0.16[a]
		Methanol	0.00±0.00[b]
Potato	5	Extract 5.7 mg/ml	6.54±0.05[b]
		Extract 3.8 mg/ml	0.00±0.00[c]
		Extract 1.9 mg/ml	0.00±0.00[c]
		Ketoconazole	7.54±0.24[a]
		Methanol	0.00±0.00[c]
Potato	10	Extract 5.7 mg/ml	6.91±0.11[b]
		Extract 3.8 mg/ml	6.26±0.01[c]
		Extract 1.9 mg/ml	0.00±0.00[d]
		Ketoconazole	7.09±0.17[a]
		Methanol	0.00±0.00[d]
Potato	15	Extract 5.7 mg/ml	7.39±0.10[b]
		Extract 3.8 mg/ml	6.44±0.03[c]
		Extract 1.9 mg/ml	0.00±0.00[d]
		Ketoconazole	7.76±0.07[a]
		Methanol	0.00±0.00[d]
Potato	20	Extract 5.7 mg/ml	7.18±0.11[b]
		Extract 3.8 mg/ml	6.53±0.40[c]
		Extract 1.9 mg/ml	0.00±0.00[d]
		Ketoconazole	7.88±0.16[a]
		Methanol	0.00±0.00[d]
Potato	25	Extract 5.7 mg/ml	6.56±0.18[b]
		Extract 3.8 mg/ml	0.00±0.00[c]
		Extract 1.9 mg/ml	0.00±0.00[c]
		Ketoconazole	7.43±0.22[a]
		Methanol	0.00±0.00[c]

Sweet potato	5	Extract 5.7 mg/ml	7.42±0.08[b]
		Extract 3.8 mg/ml	6.27±0.06[c]
		Extract 1.9 mg/ml	0.00±0.00[d]
		Ketoconazole	7.63±0.15[a]
		Methanol	0.00±0.00[d]
Sweet potato	10	Extract 5.7 mg/ml	7.77±0.03[b]
		Extract 3.8 mg/ml	6.52±0.04[c]
		Extract 1.9 mg/ml	6.23±0.08[d]
		Ketoconazole	8.09±0.16[a]
		Methanol	0.00±0.00[e]
Sweet potato	15	Extract 5.7 mg/ml	8.01±0.06[a]
		Extract 3.8 mg/ml	7.00±0.13[c]
		Extract 1.9 mg/ml	0.00±0.00[d]
		Ketoconazole	7.76±0.03[a]
		Methanol	0.00±0.00[d]
Sweet potato	20	Extract 5.7 mg/ml	7.20±0.06[b]
		Extract 3.8 mg/ml	6.68±0.03[c]
		Extract 1.9 mg/ml	0.00±0.00[d]
		Ketoconazole	8,41±0,22[a]
		Methanol	0.00±0.00[d]
Sweet potato	25	Extract 5.7 mg/ml	0.00±0.00[b]
		Extract 3.8 mg/ml	0.00±0.00[b]
		Extract 1.9 mg/ml	0.00±0.00[b]
		Ketoconazole	8.41±0.21[a]
		Methanol	0.00±0.00[b]

Means were calculated by one-way ANOVA and separated by Duncan least Significicant Difference at 5% level. Value with same alphabets does not differ significantly. Ketoconaloze as positive control, methanol as negative control.

Based on antifungal activity from secondary metabolites obtained from *Fusarium oxysporum* against *Candida albicans* showed sweet potato for 15 day fermentation possessed highest antifungal activity with inhibition zone diameter 8.01 mm at 5.7 mg/ml.

Inhibition zone diameters of ethyl acetate extract from *Fusarium oxysporum* against *Escherichia coli* is showed in Table 2.

Table 2. Ethyl acetate extract from *Fusarium oxysporum* metabolite and inhibition zone diameters against *Escherichia coli*

Carbon source	Fermentation (Day)	Concentration assay	Averages
Corn	5	Extract 5.7 mg/ml	13.70±0.22[b]
		Extract 3.8 mg/ml	7.79±0.22[c]
		Extract 1.9 mg/ml	7.02±0.23[d]
		Amoxan	14.68±0.25[a]
		Methanol	0.00±0.00[e]
Corn	10	Extract 5.7 mg/ml	13.72±0.36[a]
		Extract 3.8 mg/ml	8.22±0.26[b]
		Extract 1.9 mg/ml	7.24±0.18[c]
		Amoxan	14.04±0.76[a]
		Methanol	0.00±0.00[d]
Corn	15	Extract 5.7 mg/ml	14.15±0.09[a]
		Extract 3.8 mg/ml	10.08±0.62[b]
		Extract 1.9 mg/ml	7.31±0.18[c]
		Amoxan	14.24±0.36[a]
		Methanol	0.00±0.00[d]
Corn	20	Extract 5.7 mg/ml	11.55±0.21[b]

		Extract 3.8 mg/ml	9.97±0.51[c]
		Extract 1.9 mg/ml	8.37±0.51[d]
		Amoxan	13.83±0.15[a]
		Methanol	0.00±0.00[e]
Corn	25	Extract 5.7 mg/ml	11.16±0.39[b]
		Extract 3.8 mg/ml	8.66±0.21[c]
		Extract 1.9 mg/ml	7.37±0.31[d]
		Amoxan	12.80±0.14[a]
		Methanol	0.00±0.00[e]
Potato	5	Extract 5.7 mg/ml	0.00±0.00[b]
		Extract 3.8 mg/ml	0.00±0.00[b]
		Extract 1.9 mg/ml	0.00±0.00[b]
		Amoxan	7.65±0.20[a]
		Methanol	0.00±0.00[b]
Potato	10	Extract 5.7 mg/ml	0.00±0.00[b]
		Extract 3.8 mg/ml	0.00±0.00[b]
		Extract 1.9 mg/ml	0.00±0.00[b]
		Amoxan	7.63±0.27[a]
		Methanol	0.00±0.00[b]
Potato	15	Extract 5.7 mg/ml	0.00±0.00[b]
		Extract 3.8 mg/ml	0.00±0.00[b]
		Extract 1.9 mg/ml	0.00±0.00[b]
		Amoxan	7.87±0.12[a]
		Methanol	0.00±0.00[b]
Potato	20	Extract 5.7 mg/ml	0.00±0.00[b]
		Extract 3.8 mg/ml	0.00±0.00[b]
		Extract 1.9 mg/ml	0.00±0.00[b]
		Amoxan	7.56±0.26[a]
		Methanol	0.00±0.00[b]
Potato	25	Extract 5.7 mg/ml	0.00±0.00[b]
		Extract 3.8 mg/ml	0.00±0.00[b]
		Extract 1.9 mg/ml	0.00±0.00[b]
		Amoxan	7.67±0.16[a]
		Methanol	0.00±0.00[b]
Sweet potato	5	Extract 5.7 mg/ml	7.56±0.33[b]
		Extract 3.8 mg/ml	6.60±0.22[c]
		Extract 1.9 mg/ml	0.00±0.00[d]
		Amoxan	11.49±0.53[a]
		Methanol	0.00±0.00[d]
Sweet potato	10	Extract 5.7 mg/ml	9.96±0.21[b]
		Extract 3.8 mg/ml	7.71±0.29[c]
		Extract 1.9 mg/ml	6.45±0.21[d]
		Amoxan	12.77±0.13[a]
		Methanol	0.00±0.00[e]
Sweet potato	15	Extract 5.7 mg/ml	10.54±0.32[b]
		Extract 3.8 mg/ml	8.45±0.40[c]
		Extract 1.9 mg/ml	6.71±0,06[d]
		Amoxan	11.77±0.21[a]
		Methanol	0.00±0.00[e]
Sweet potato	20	Extract 5.7 mg/ml	8.65±0,33[b]
		Extract 3.8 mg/ml	7.06±0.69[c]
		Extract 1.9 mg/ml	6.56±0.31[c]
		Amoxan	11.37±0.46[a]
		Methanol	0.00±0.00[d]
Sweet potato	25	Extract 5.7 mg/ml	7.42±0.36[b]
		Extract 3.8 mg/ml	6.44±0.21[c]

		Extract 1.9 mg/ml	6.35±0.21[c]
		Amoxan	10.37±0.48[a]
		Methanol	0.00±0.00[d]

Means were calculated by one-way ANOVA and separated by Duncan least Significicant Difference at 5% level. Value with same alphabets does not differ significantly. Amoxsan as positive control, and methanol as negative control.

Based on antibacterial activity from overall extract metabolites isolated from *Fusarium oxysporum* against *E.coli* showed that corn as carbon source for 15 day fermentation indicated highest antibacterial activity with inhibition zone diameter 10.54 mm at 5.7 mg/ml. Inhibition zone diameters against *Staphylococcus aureus* is showed in Table 3.

Table 3. Ethyl acetate extract from *Fusarium oxysporum* metabolite and inhibition zone diameters against *S. aureus*

Carbon source	Fermentation (Day)	Concentration assay	Averages
Corn	5	Extract 5.7 mg/ml	7.09±0.09[b]
		Extract 3.8 mg/ml	6.68±0.22[c]
		Extract 1.9 mg/ml	0.00±0.00[d]
		Amoxan	17.51±0.42[a]
		Methanol	0.00±0.00[d]
Corn	10	Extract 5.7 mg/ml	8.68±0.29[b]
		Extract 3.8 mg/ml	7.56±0.32[c]
		Extract 1.9 mg/ml	6.72±0.25[d]
		Amoxan	17.44±0.32[a]
		Methanol	0.00±0.00[e]
Corn	15	Extract 5.7 mg/ml	15.24±0.42[b]
		Extract 3.8 mg/ml	11.78±0.29[c]
		Extract 1.9 mg/ml	10.86±0.13[d]
		Amoxan	16.09±0.39[a]
		Methanol	0.00±0.00[e]
Corn	20	Extract 5.7 mg/ml	12.54±0.29[b]
		Extract 3.8 mg/ml	10.26±0.29[c]
		Extract 1.9 mg/ml	9.03±0.24[d]
		Amoxan	15.53±0.38[a]
		Methanol	0.00±0.00[e]
Corn	25	Extract 5.7 mg/ml	9.98±0.53[b]
		Extract 3.8 mg/ml	8.01±0.21[c]
		Extract 1.9 mg/ml	6.77±0.25[d]
		Amoxan	16.63±0.29[a]
		Methanol	0.00±0.00[e]
Potato	5	Extract 5.7 mg/ml	11.34±0.49[b]
		Extract 3.8 mg/ml	9.57±0.09[c]
		Extract 1.9 mg/ml	7.93±0.08[d]
		Amoxan	15.83±0.14[a]
		Methanol	0.00±0.00[e]
Potato	10	Extract 5.7 mg/ml	11.63±0.34[a]
		Extract 3.8 mg/ml	9.64±0.39[c]
		Extract 1.9 mg/ml	8.18±0.20[d]
		Amoxan	15.76±0.36[a]
		Methanol	0.00±0.00[e]
Potato	15	Extract 5.7 mg/ml	12.20±0.57[b]
		Extract 3.8 mg/ml	8.74±0.27[c]
		Extract 1.9 mg/ml	7.36±0.19[d]
		Amoxan	15.64±0.34[a]

IOP Conf. Series: Materials Science and Engineering **532** (2019) 012027 doi:10.1088/1757-899X/532/1/012027

		Methanol	0.00±0.00[e]
Potato	20	Extract 5.7 mg/ml	8.57±0.31[b]
		Extract 3.8 mg/ml	7.45±0.27[c]
		Extract 1.9 mg/ml	7.08±0.15[c]
		Amoxan	14.27±0.28[a]
		Methanol	0.00±0.00[e]
Potato	25	Extract 5.7 mg/ml	7.40±0.29[b]
		Extract 3.8 mg/ml	6.98±0.19[b,c]
		Extract 1.9 mg/ml	6.62±0.33[c]
		Amoxan	13.64±0.29[a]
		Methanol	0.00±0.00[d]
Sweet potato	5	Extract 5.7 mg/ml	7.54±0.44[b]
		Extract 3.8 mg/ml	6.97±0.52[b,c]
		Extract 1.9 mg/ml	6.59±0.51[c]
		Amoxan	13.30±0.13[a]
		Methanol	0.00±0.00[d]
Sweet potato	10	Extract 5.7 mg/ml	8.69±0.31[b]
		Extract 3.8 mg/ml	7.66±0.26[c]
		Extract 1.9 mg/ml	7.57±0.07[c]
		Amoxan	14.44±0.10[a]
		Methanol	0.00±0.00[d]
Sweet potato	15	Extract 5.7 mg/ml	8.88±0.35[b]
		Extract 3.8 mg/ml	7.95±0.25[c]
		Extract 1.9 mg/ml	7.68±0.41[c]
		Amoxan	14.60±0.57[a]
		Methanol	0.00±0.00[d]
Sweet potato	20	Extract 5.7 mg/ml	9.05±0.35[b]
		Extract 3.8 mg/ml	8.37±0.54[b,c]
		Extract 1.9 mg/ml	7.75±0.09[c]
		Amoxan	14.78±0.56[a]
		Methanol	0.00±0.00[d]
Sweet potato	25	Extract 5.7 mg/ml	7.06±0.51[b]
		Extract 3.8 mg/ml	6.81±0.08[b]
		Extract 1.9 mg/ml	0.00±0.00[c]
		Amoxan	13.18±0.52[a]
		Methanol	0.00±0.00[c]

Means were calculated by one-way ANOVA and separated by Duncan least Significicant Difference at 5% level. Value with same alphabets does not differ significantly. Amoxsan as positive control, and methanol as negative control.

Based on antimicrobial activity from overall extract metabolites were isolated from *Fusarium oxysporum* toward *S. aureus*, corn as carbon source for 15 day fermentation posses highest antibacterial activity with inhibition zone diameter 15.24 mm at 5.7 mg/ml..

The result of metabolite screening was carried out toward overall ethyl acetate extract isolated from *Fusarium oxysporum* (Table 4).

Table 4. Screening metabolit test of ethyl acetate extract isolated from *Fusarium oxysporum*

Carbon source	Fermentation (day)	Metabolites test				
		Alkaloid	Phenolic	Flavonoid	Saponin	Terpenoid
Corn	5	_	_	_	_	+
	10	_	_	_	_	+
	15	_	_	_	_	+
	20	_	_	_	_	+
	25	_	_	_	_	+
Potato	5	_	_	_	_	+

	10	–	–	–	–	+
	15	–	–	–	–	+
	20	–	–	–	–	+
	25	–	–	–	–	+
Sweet potato	5	–	–	–	–	+
	10	–	–	–		+
	15	–	–	–	–	+
	20	–	–	–	–	+
	25	–	–	–	–	+

Metabolites isolated from *Fusarium oxysporum* achieved maximum production at 15 day fermentation and decreased at 20 and 25 day fermentation, which was assumed to have started death phase [12]. Production of secondary metabolites by microorganisms was often influenced by primary metabolism. The composition and medium concentration were closely linked with the metabolic producing capacities by microorganism and greatly influenced the biosynthesis of bioactive molecules. There were many factors playing important to produce secondary metabolite such as carbon source and fermentation time [13]. Overally, metabolite extract with corn as carbon source and 15 day fermentation (C15) showed better activity than other extracts. Inhibition zone diameter lower than triterpenoid saponin was isolated from endophytic fungi *Fusarium* sp from *Panax notogingseng* plant [14].

Based on screening metabolite *Fusarium oxysporum* have the ability to produce secondary metabolite, overally comprise terpenoid group through mevalonate pathway. Mostly, crude extract from *Fusarium oxysporum* showed antimicrobial activty againts Gram positive bacteria of genera *Staphylococcus aureus*. Previously Ibrahim et al (2016) reported endhopytic fungus *Fusarium* sp isolated from the roots *Mentha longifolia* have the ablity to produce two new tetracyclic terpenoid, and antibacterial acitvity [15]. *Fusarium chlamydosporium* isolated from leaves *Anvillea garcinii* produced aminobenzamide derivative with potent cytotoxic and antimicrobial agent (Ibrahim *et al.*, 2018) [16].

C15 metabolite has been analyzed both thin layer chromatography and high performance liquid chromatography and depicted 4 component with Rf 0,3 ; 0,7; 0,84 ; dan 0,93 respectively using ethyl acetate 100% as mobile phase for TLC and water and methanol for HPLC with t_R 16.86, 17.62, 18.23 and 21. 54 min. Every peak comfirm one compound, it was showed at Fig 1.

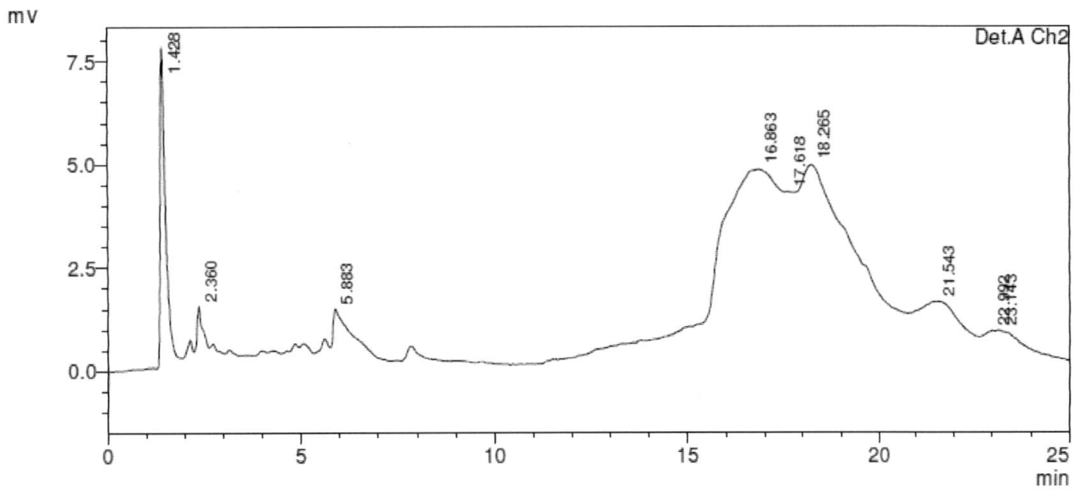

Figure 1. Chromatogram HPLC C15.

4. Conclusion

Fusarium oxysporum had ability to produce secondary metabolites. Based onoptimization of carbon source and time of fermentation, corn is better carbon source than other medium for 15 day fermentation. Screening metabolite showed that C15 comprised terpenoid. Chromatogram analyzed using TLC and HPLC achieved 4 component.

Acknowledgements
This study was funded by postgraduate grant research from the Ministry of Research, Technology and Higher Education of the Republic of Indonesia year 2017 on behalf of Prof. Dr. Saryono, M. Si.

Reference

[1] Simarmata, Rumella, Sylvia Lekatompessy, and Harmastini Sukiman. 2007. *Berkala Penelitian Hayati.* 13(April):85–90.

[2] Zhao, J., T. Shan, Y., Mou, &. Zhou, L. 2011. *Mini Review in Medicinal Chemistry* 11:159–68.

[3] Vagelas, I., & Gowen, S. R. 2012. *Pakistan Journal Phytophatol*. 24 (1): 32-38.

[4] Yasinok., Eryasin, A., Sahin, F, I., & Haberal, M. 2008. *Tarim Bilimeri Dergisi.* 14(4):374–80.

[5]] Radji, Maksum. 2005. *Majalah Ilmu Kefarmasian.* 2(3):113–26.

[6] Saryono., Finna, P., Nurmala, S., Wahyu, P. W., & Aulia, A. 2018. *Research Journal of Chemistry and Environtment.*

[7] Devi, S., Lutfhin, I.,Ardhi, A., Pratiwi, N. W., & Saryono. 2018. *Research Journal of Chemistry and Environtment.*

[8] Lorenita, M., Haryani, M., Puspita, F & Trihartomo, D. 2013. *Journal of Agriculrural Technology.* 9(3):585–90.

[9] Saryono, Hendris, S., Fitryah, D., Jose, C., Nugroho, T. T. & Ardhi. 2015. *Journal of Chemical and Pharmaceutical Research.* 7(9):201–8.

[10] Fitriyah, D., Christine, J., & Saryono. 2013. *Journal Indonesia. Acta* 3(2).

[11] Shinta, D. Y.,Yusmarini., Sonata, H.,Teruna, H. Y., & Saryono. *Journal of Chemical and Pharmaceutical Research.* 7(9S): 239-245.

[12] Pavani, M., Sankar, G., Prabhakar, T., Bhavani, A., & Sravani, P. 2014. *International Journal of Pharmacheutical Science Review and Research..* 33: 192-196.

[13] Khatab, A. I., Babiker, E. H., & Saeed, H. A. 2014. *International Current Pharmacheutical Journal.* 5(3): 27-32.

[14] Jin, Z., Gao, L., Zhang, L., Liu, T., Yu, F., Zhang, Z., Guo, Q., & Wang, B. 2016. *Natural Prodcut Research.* 31(22): 2700-2703.

[15] Ibrahim, S. R. M., Mohamed, G. A., & Ross, S. A. 2016. *Phytochemsistry Letters.* 15: 125-130.

[16] Ibrahim, S. R. M., Mohamed, G, A., Alhaidari, R. A., Zayed, M. F., Elkholy, A. A., Elkhayat, E. S., & Ross, S. A. 2018. *Bioorganic and Medicinal Chemistry.* 26: 786-790.

Isolation and Identification of Bacteria and Actinomycetes Isolated from Wilting Banana Plants (*Musa* Sp.)

Saryono[1]*, P. Finna[1], P. Usman[2], P. N. Wahyu[3], A. Aulia[4]
[1]Department of Chemistry, Faculty of Mathematics and Natural Sciences Universitas Riau
[2]Department of Agriculture, Faculty of Agriculture Universitas Riau
[3]Department of Biology, Faculty of Mathematics and Natural Sciences Universitas Riau
[4]Department of Food and Agricultural Product Technology, Faculty of Agricultural Technology, Universitas Gadjah Mada

*Corresponding author's email: saryono@lecturer.unri.ac.id

Abstract. In Indonesia, wilting disease in banana due to *Fusarium* has spread in several areas. In addition, fungi, bacteria, and actinomycetes are also known to cause the disease. In this research, isolation and identification of bacteria and actinomysetes causing the wilting of rot and rhizosphere were done. Identification used several stages of biochemical tests. There were 40 bacterial isolates with 9 generas, which were *Pseudomonas* sp., *Xanthomonas* sp., *Caulobacter* sp., *Erwinia* sp., *Eschericia* sp., *Ralstonia* sp., *Neisseria* sp., *Staphylococcus* sp., and *Proteus* sp. Whilst 18 isolates of actinomycetes were obtained with 7 generas, including *Nocardia* sp., *Actinobiospora* sp., *Nocardiopsis* sp., *Streptomyces* sp., *Streptoverticillium* sp., *Streptosporangium* sp., and *Microbispora* sp.

Keywords: actinomycetes, bacteria, banana, isolation, wilt

1. Introduction

Banana (*Musa paradisiaca*) in Indonesia is widely used in everyday life as food. All of the banana plant can be utilized so that the availability of healthy banana plants is needed in the community. However, this is constrained because many banana plants are affected with the disease. The famous disease is the wilt caused by pathogen microbes. In Indonesia, *Fusarium* wilt disease has spread in several areas such as Jawa Barat, Jawa Timur, Sumatera Utara, and Riau (Nasir and Jumjunidang, 2003). Poerwanto (2013) proved from 2009-2012 the amount of banana production was decreased 24% due to the disease. This is affected to export the banana from Indonesia. The wilted on the banana is caused by microbes that takes nutrients from it. Previously, Saryono *et al.* (2018) have isolated fungi caused wilting banana plants and succesed to get 4 genera of fungi, including *Trichoderma*, *Fusarium*, *Aspergillus*, and *Penicillium*.

Microbes was attacked banana plants from out to inside of cells on the stump. This disease can also spread to other healthy or injured plants through rhizosphere soil or tools. Besides of the fungus, bacterial caused wilt on banana plants and rhizosphere soil identified as *Proteus* sp., *Erwinia* sp., *Klebsiella* sp., *Staphylococcus* sp., *Corynebacterium* sp., *Bacillus* sp., and *Cellulomonas* sp. [9, 15]. The pathogen bacterial abilities in producing some hydrolytic enzymes such as cellulase, amylase, protease, and inulinase on cells of plants, which would be of great potential when applied to the biotechnology industry.

Actinomycetes are also found from rhizosphere soil of culture plant that can potentially produce antibiotics by inhibiting the growth of gram-positive bacterial [2]. Sudarma (2010) proved actinomycetes from the rhizosphere soil of wilting banana abilities to produced antibiotics. Actinomycetes produced bioactive compounds highly such as antivirus and anticancer, while in agriculture as herbicides, insecticides, and antiparasitic compounds [19]. Some bioactive compounds namely streptomycin, neomycin, chloramphenicol, and tetracycline could be produced from *Streptomyces* [20]. Therefore, isolation and identification of bacterial and actinomycetes from rhizosphere soil and stump of wilting banana in Panam, Pekanbaru. The results of isolated were then test using the biochemical test.

2. Material and methods

Samples was collected in jalan Kartama, Panam, Pekanbaru by purposive sampling method. Stump of banana plants samples showing the presence of disease were cut into small pieces. The surface of stump washed using flowed water, sterilized by soaking them in 70% ethanol for 3 minutes. And then, the stump were soaked in 0.2% $HgCl_2$ solution for few minutes. Samples were subsequently soaked back in 70% ethanol for 1 minute and rinsed with demineralized water twice. The stump were cut into smallest in blender into powder. Samples was added into 9 ml of 0.8% NaCl in test tube to dilution till 10^{-2}. Samples soil from rizhosphere wilting banana was taken from the plants grew. The samples taken were weighed for 1 g and put into a test tube containing 9 ml of 0.8% NaCl (10^{-1}), homogenized using a vortex, then repeated until 10^{-2} dilution. All dilution were micropippetted and inoculated into a petri dish containing NA and SCA with a spread method. Samples were incubated at 37°C for 2 days and 10 days, respectively. The growing colonies were separated based on different colonies, colors, and the shape of colonies. The single colony of actinomycetes and bacterial were subcultured on other NA and SCA media. And then, the isolates continued to be put under macroscopic and microscopic identification.

2.1. Morphological characterization of bacterial and actinomycetes

Characterization of growth colonies was such as: pinpoint/punctiform (point), small, moderate, large; shape: circular, irregular, spindle, filamentous, and rhizoid; elevation: flat, raised, convex, and umbonate. The growth of isolates were observed such as philliform, echinulate, beaded, arborescent, plumose, papilliate, and villose in straight NA and SCA media. The patterns of colony isolates were observed such as philliform, echinulate, beaded, effuse, spreading, plumose, and rhizoid in slant NA and SCA medium. Observations were also made on NB and SCB media to observe the growth of isolates based on O_2 requirements, with anaerobic facultative patterns (uniform turbidity and flocculent growth), aerobic (pellicle), microaerophilic (ring), and anaerob (sediment). Otherside, bacterial and actinomycetes isolates were also identified by gram-stain test for microscopic identification.

For acid-resistant coloration test, actinomycetes and bacterial isolates were added on glass preparations, spilled with carbolfuchsin and heated for 5 minutes. The preparate was washed and soaked in 96% alcohol for 15 minutes, then rinsed with flowing water. After that, the preparate was soaked with methylene blue for 2 minutes and rinsed back with flowing water. The observation was taken under a microscope. Positive test showed by red cells and negative test showed by blue or colorless.

2.2. Biochemical Test

2.2.1. Catalase test

The isolates was put into preparate test. The 3% H_2O_2 was dropped on the isolate. Positive test showed if gas formed on the preparate.

2.2.2. Starch hydrolysis test

The media was prepared by mixture 10 g soluble starch, beef extract 3 g, and 1 L demineralized water. The media was sterilized and put into a petri dish. The isolates were streaked into the media with a diffusion method [7] and incubated at 37°C for oberve the clear zone after being added with iodine solution.

2.2.3. Carbohydrate fermentation test

Phenol-red lactose/glucose/sucrose broth medium was consisted of lactose/glucose/sucrose each 10 g, trypton 10 g, NaCl 5 g, and phenol red 0.0180 g. The mixture was itaken put into test tube and durham tube was adjust carefully for observe gas formed during incubation at 37°C. The color change into yellow showed a positive fermentation reaction by producing acid. The test was purposed for observe the ability of bacteria in fermentation carbohydrate.

2.2.4. Casein hydrolysis test

Skim milk was weighed 28 g and put into demineralized water 1 L. The isolates were inoculated into the media with a diffusion method [7] during the incubation at 37°C. If a clear zone was formed then the test was positive.

2.2.5. Gelatin hydrolysis test

Gelatin media were made with the following composition of 15 g gelatin, 15 g agar, 4 g pepton, 1 g yeast extract in 1 L of demineralized water. The patterns of growth were observed, such as filiform, echinulate, beaded, arborescent, plumose, papilliate, and villose. The gelatin hydrolysis test was observed by incubating the media contained isolate into the freezer for 1 h. The test was positive if the media melted.

2.2.6. Methylene blue reduction test

The isolates was inoculated into NB media were added with 5 mL of 10% methylene blue solution. The homogenized mixture was stored for 60 min at 37°C and then observed the changes blue of color media. The change from blue into clear indicated that the isolates were abilities of reducing methylene blue [7].

2.2.7. Bromocresol Purple Milk Solids Glucose Agar (BCPM) Hydrolysis Test

The isolates were grown on media agar composed of mix solution A (1000 mL demineralized water, 80 g skim milk powder, 2% bromcresol purple (1.6% dissolved in 70% alcohol) mixed in Erlenmeyer and sterilized in autoclave), Solution B (40 g glucose and 200 ml demineralized water sterilized in autoclave), and solution C (30 g agar and 800 mL demineralized water, sterilized in autoclave). The isolates was inoculated with a diffusion method [7] during the incubation at 37°C. If the medium turned purple, it indicated the isolate hydrolyzed the skim milk, but if the medium color changed into yellow, it indicated the isolate hydrolyzed glucose.

3. Results and Discussion

In total isolates obtained were 40, which 20 isolates were isolated from the soil and 20 other isolates from the stump. Otherside, total isolates of actinomycetes obtained were 18, each 9 isolates from the soil and 9 other isolates from the stump. Bacterial isolates contained gram-positive and negative cells, however the cells of actinomycetes isolates were gram-positive. To determine the genera of bacterial and actinomycetes isolates, some of biochemical tests were performed, and matched on Bergeys manual book. There are results of morphology test and some of biochemical tests from one genus of bacterial and one genus of actinomycetes (Figure 1 and Figure 2).

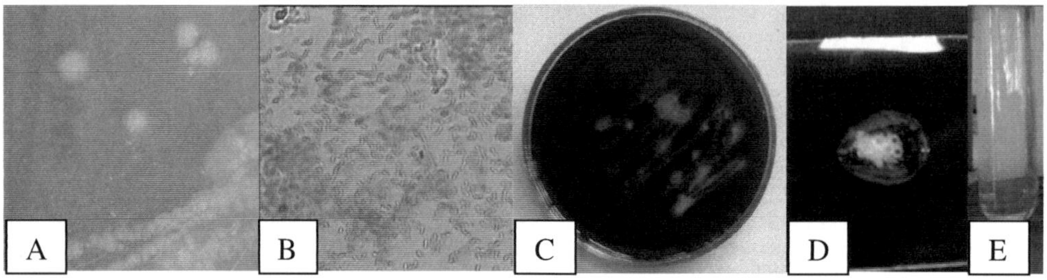

Figure 1. Genus *Pseudomonas* sp. (A) *Pseudomonas* colonies, (B) cell form (40x), (C) starch hydrolysis test, (D) catalase test, and (E) methylene blue reduction test .

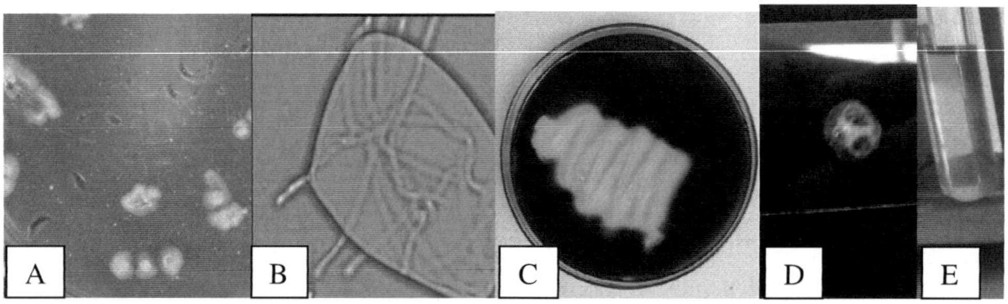

Figure 2. Genus *Streptomyces* sp. (A) *Streptomyces* colonies, (B) cell form (40x), (C) starch hydrolysis test, (D) catalase test, and (E) methylene blue reduction test.

Based on the results of morphological and biochemical tests, it could be determined the genera of isolates. All actinomycetes isolates were positive in gram-stain tests. Observations of bacterial and actinomycetes isolates can be seen in the following of tables.

Table 1. Morphological characterization of actinomycetes isolates

No.	Size	Morfology	Models	Elevation	Color of colony	Gram	Cell form	Genus (sp.)
1	Small	Rhizoid	Undulate	Raised	Brown	Positive	Filamentous	*Nocardia*
2	Small	Circular	Entire	Convex	Brown	Positive	Filamentous	*Actinobiospora*
3	Moderate	Rhizoid	Serrate	Umbonate	Brown	Positive	Filamentous	*Nocardiopsis*
4	Moderate	Circular	Entire	Convex	Beige	Positive	Filamentous	*Streptomyces*
5	Small	Circular	Undulate	umbonate	Beige	Positive	Filamentous	*Streptoverticillium*
6	Small	Rhizoid	Entire	Convex	Brown	Positive	Filamentous	*Streptosporangium*
7	Moderate	Circular	Undulate	Raised	Beige	Positive	Filamentous	*Microbispora*

Actinomysetes were indicated as rod-shaped, gram-positive, anaerobic or facultative aerobic. The actinomycetes structure was consisted of soft filaments often called hypha or mycelia, as found in the fungus, and had conidia [29]. Actinomycetes a group of gram-positive bacteria and is widely distributed in the soil. Actinomycetes could be divided into two groups, namely *Streptomyces* and *rare-actinomycetes*. *Rare-actinomycetes* used to refer to a genus other than *Streptomyses*. This group growth was slower and relatively difficult with *Streptomyses* group. As in the results of this study, actinomycetes isolates have difficult form when grown in SCA media.

Other *Streptomyses*, there are genera namely *Nocardia*, *Micromonospora*, *Microbispora*, and *Streptosporangium* [17]. *Streptomyses* sp. in antagonistic test was unable against *Fusarium oxyporum* with highest value 82,59% [23]. Sunaryanto *et al.* (2010) isolated actinomycetes with genera *Streptomyces* and *Nocardia* from sediments. Actinomycetes in sediments were spreaded highly than in population in soil, while bacterial populations were more dominant in sediments. Sulistiyani and Nunuk (2011) revealed that *Staphylococcus* and *Streptococcus* were released antibiotic-resistant pathogens, and also produced antibiotics to inhibiting the growth of *S. aureus* and *C. albicans*. Based on previous research, it can be stated that the actinomycetes from wilting banana plants were not entirely a pathogen.

Actinomycetes has many varians of morphology, from the form of the circular/coccus (*Micrococcus*) and the rod-coccus (*Arthrobacter*), hyperfragment (*Nocardia*, *Rothia*), to the species with different branching mycelium (*Micromonospora* and *Streptomyces*). Other examples, *Actinobacteria*, *Actinoplanetes*, *Nocardioforms*, and *Streptomyces* were difficult phylogenics. Actinomysetes be able to saprophytes, but other of genus also parasite or symbiotic mutualism in plants and animals. Firstly, actinomysetes was classified in the fungus group, because the appearance of morphology and its development were similar to the fungus. However, with the advanced science, actinomisetes morphology was known to be closer to bacteria. From the size of the cell, its spores and miscellaneous, actinomisetes were categorized as bacteria that had the same characteristics as bacteria. Chitin and cellulose as constituents of fungi cell walls were not present in actinomycetes. Actinomycetes cell wall composed sugar polymers, amin-sugars, and some amino acids such as in cell of gram-positive bacteria. Actinomycetes was usually as a group of gram-positive bacteria that contained high levels of Guanin (G) and Citosin (C) in their DNA (> 55%) with the ability formed hypha branches during their optimized stages (Budiyanto and Farhan, 2012) . The color of the actinomycetes colonies were varied, because of the difference in compound of pigment in the cell, corresponding to each type of actinomycetes [26].

Table 2. Morphological characterization of bacterial isolates

No.	Size	Morfology	Models	Elevation	Color of colony	Gram	Cell form	Genus (sp.)
1	Small	Circular	Entire	Convex	Yellow	Negative	Bacil	*Pseudomonas*
2	Small	Circular	Entire	Convex	Yellow	Negative	Bacil	*Xanthomonas*
3	Moderate	Circular	Entire	Flat	Beige	Negative	Bacil	*Caulobacter*
4	Moderate	Circular	Entire	Convex	White-milk	Negative	Bacil	*Erwinia*
5	Small	Circular	Lobate	Flat	White	Negative	Bacil	*Eschericia*
6	Small	Circular	Entire	Convex	Yellow	Negative	Bacil	*Ralstonia*
7	Moderate	Circular	Entire	Convex	Yellow	Negative	Bacil	*Neisseria*
8	Moderate	Irregular	Lobate	Flat	White	Negative	Coccus	*Staphylococcus*
9	Small	Circular	Lobate	Flat	White	Negative	Bacil	*Proteus*

In this research, 9 genera of bacteria were obtained namely *Pseudomonas* sp., *Xanthomonas* sp., *Caulobacter* sp., *Erwinia* sp., *Eschericia* sp., *Ralstonia* sp., *Neisseria* sp., *Staphylococcus* sp., and *Proteus* sp. In addition, 7 genera actinomycetes were obtained namely *Nocardia* sp., *Actinobiospora* sp., *Nocardiopsis* sp., *Streptomyces* sp., *Streptoverticillium* sp., *Streptosporangium* sp., *Microbispora* sp. Msogoya *et al.* (2012) and Habiba *et al.* (2002) also isolated *Proteus* sp., *Erwinia* sp., *Klebsiella* sp., *Staphylococcus* sp., *Corynebacterium* sp., *Bacillus* sp., and *Cellulomonas* sp. from banana plants and its rhizosphere soil. Winarni (2013) successed to isolate *Pseudomonas* sp. from seeds of rice and soybean, *Pseudomonas* and *Xanthomonas* were pathogens in plants. *Pseudomonas* are cream-yellowish and circular.

Proteus is a group of Enterobacteriaceae which normal flora in humans and animals, also be found in the environment (Arisandi, 2016). *Ralstonia* sp. caused blood disease in banana plants [8]. On the other hand, Zhou *et al.* (2012) proved *Pseudomonas* as antagonistic bacteria against *Ralstonia.* Abidin *et al.*, (2015) proved a *Pseudomonas* sp. as antagonist against *Sclerotium rolfsii* Sacc. *Pseudomonas* sp. can released antibiotic, siderophores, and other secondary metabolites that inhibited fungal activity. In addition, according to Herdyastuti *et al.* (2010), *Pseudomonas* also produced chitinase highly.

Banana plants-wilting bacteria are capable of producing a variety of extracellular enzymes with different functions, including to degraded plant cell wall components and became to pathogens. Another virulent factor was acidic extrapolysaccharides from bacterial pathogen with high molecular weight caused blocked of the xylem and inhibited its function [8]. Susanna (2006) isolated bacteria from the rhizosphere of banana plants, which could be as antagonistic to *Fusarium oxysporum* f.sp. cubense. Based on this, the probability of isolates obtained from this study was not entirely of a pathogenic. According to Djatnika and Wakiah (1995), *Pseudomonas fluorescens* reduced the number of wilting disease by 85.7% in tissue culture of banana plants. Hedges and Messens (1990) stated that *Pseudomonas fluorescens* produced pyoverdine, pyrrolnitrin, and pyoluterin in inhibiting pathogens.

4. Conclusions

We have isolated actinomycetes and bacterial from wilting banana plants with 7 genera of actinomycetes and 9 genera of bacterial. Genera of actinomycetes are following: *Nocardia* sp., *Actinobiospora* sp., *Nocardiopsis* sp., *Streptomyces* sp., *Streptoverticillium* sp., *Streptosporangium* sp., and *Microbispora* sp. In the other hand, genera of bacterial are following: *Pseudomonas* sp., *Xanthomonas* sp., *Caulobacter* sp., *Erwinia* sp., *Eschericia* sp., *Ralstonia* sp., *Neisseria* sp., *Staphylococcus* sp., and *Proteus* sp.

Acknowledgement
This research was funded by Professor Grant of Universitas Riau on behalf of Prof. Dr. Saryono, M.Si.

References

[1] Abidin, Z., Luqman, Q., dan Abdul, L., A. 2015. Pengaruh bakteri *Bacillus* sp. dan *Pseudomonas* sp. terhadap pertumbuhan jamur pathogen *Sclerotium rolfsii* Sacc. penyebab penyakit rebah semai pada tanaman kedelai. *Jurnal HPT. 3(1): ISSN 2338-4336.*

[2] Ambarwati dan Eni, P. 2010. Keanekaragaman *Streptomyces* yang berasosiasi dengan rizosfer jagung (*Zea mays*). *Seminar Biologi. 513-517.*

[3] Arisandi, F. 2016. Isolasi dan identifikasi bakteri pada sampel tanah di Dukuh Ngantru, Desa Sekaran, Kecamatan Kasman, Kabupaten Bojonegoro. *Repository.* Institut Pertanian Bogor.

[4] Budiyanto, M., A., K. dan Farhan, M. Peranan Bakteri Actinomycetes dalam Industri Antibiotik *Role of Bacteria Actinomycetes in Industrial Antibiotics. Jurnal Online Biosains. 1: 71-85.*

[5] Butar-butar. 2014. Isolasi dan uji potensi mikrob pelarut posfat. Universitas Sumatera Utara. *Skripsi.*(online) http://repository.usu.ac.id/bitstream/123456789/40668/6/Chapter%20II. pdf, tanggal akses 11 April 2016.

[6] Djatnika dan Wakiah. 1995. Pengendalian Biologi Penyakit Layu Fusarium pada Pisang dengan Beberapa Isolat *Pseudomonas fluorescens*: 422 – 425. Kong. Nas. XII dan Seminar Ilmiah PFI, Mataram. 27 – 29 September.

[7] Devi, S., Illiyin, L., Ardhi, A., Pratiwi, N., W., and Saryono. *Optimization of process parameters for inulinase production from endophytic fungi Fusarium solani LBKURCC67, Humicola fuscoatra LBKURCC68 and Fusarium oxysporum LBKURCC69. Res. J., Chem. Environ.*

[8] Edy, N., Siti, S., Christianti, S., dan Jaka, W. 2011. Characterization and rapid detection of blood disease bacterium on banana and plantain. *Jurnal Perlindungan Tanaman Indonesia. 17(1): 26-30.*

[9] Habiba, U., Reja, S., Saha, M. L., and Khan MR. 2002. Endogenous bacterial contamination during in vitro culture of table banana: identification and prevention. *Plant Tissue Culture. 12 (2): 117 –124.*

[10] Hafsan, Eka S., dan Mashuri, M. 2015. *Penuntun Praktikum Mikrobiologi.* Fakultas Sains dan Tekonologi. UIN-Alauddin, Makassar.

[11] Hedges, R.W. & E. Messens. 1990. *Genetic Aspects of Rhizosphere Interactions., In* J.M. Lynch (ed.). The Rhizosphere. John Wiley & Sons Ltd. England. pp. 129 – 176.

[12] Herdyastuti, N., *et. al.* 2010. *Chitinolytic bactery activity isolated from the mud fields. Berk. Panel. Hayati. 15: 107-111.*

[13] Indrayoga, P. M., I Made Sudarma, dan Ni Made Puspawati. 2013. Identifikasi jenis dan populasi jamur tanah pada habitat tanaman kubis (*Brassica oleracea* L.) sehat dan sakit akar gada pada sentra produksi kubis di Kecamatan Baturiti Tabanan. *E. Jurnal Agroteknologi Tropika. 2(3).*

[14] Magnet, M. H., Dayanidhi, S., dan Zakaria, A. 2013. *Isolation and identification of bacteria and fungi from soil samples of different industry side in Dhaka city, Bangladesh. International Journal of Innovative Research and Development. 2(8): 338-339.*

[15] Msogoya, T., H. Kanyagha, J. Mutigitu, M. Kulebelwa, dan Delphina Mamiro. 2012. *Identification and management of microbial contaminants of banana in vitro culture. Journal of Applied Biosciences. 55: 3987-3994.*

[16] Nasir, N., Jumjunidang, Eliesti, F., dan Meilda, Y. 2003. Penyakit layu panama pada pisang: Observasi ras 4 *Fusarium oxysporum* f. sp. *cubense* di Jawa Barat. *J. Hort. 13(4): 269-275.*

[17] Nurkanto, A. dan Agusta, A. 2015. Identifikasi molecular dan karakterisasi morfo-fisiologi actinomysetes penghasil senyawa antimikroba. *Jurnal Biologi Indonesia. 11(2): 195-203.*

[18] Poerwanto, R. 2013. Pengembangan manajemen budidaya tanaman untuk peningkatan mutu. (online) *manajemenproduksitanaman.*

[19] Rabah, F. L., Elshafei A., Saker M., Cheikh B., Hocine H., 2007. *Screening, isolation, and characterization of novel antimicrobial producing actinomycetes, strain RAf10. Biotechnology, 6(4): 489–496.*

[20] Raharini, A., O., Retno, K., dan Khamdan, K. 2012. Penggunaan *Streptomyces* sp. sebagai biokontrol penyakit layu pada tanaman cabai merah (*Capsicum annuum* L.) yang disebabkan oleh *Fusarium oxysporum* f.sp. *capsici. AGROTROP. 2(2) 161-168.*

[21] Retnowati, Y., Nurhayati, B., dan Nona, W., P. 2011. Pertumbuhan bakteri *Staphylococcus aureus* pada media yang diekspos dengan infus daun sambiloto (*Andrographis paniculata*). *Saintek. 6(2).*

[22] Saryono, Piska, F., Sari, N., Pratiwi, N., W., and Ardhi, A. 2018. *Morphological identification and hydrolytic enzymeproducing abilities of fungi associated with wilting banana plants (Musa sp.). Res. J. Chem. Environ.*

[23] Sudarma, I., M. 2010. Seleksi dan pemanfaatan actinomycetes sebagai mikroba antagonis yang ramah lingkungan terhadap *Fusarium oxysporum* f.sp. cubense secara *in vitro. ECOTROPHIC. 5(2); 104-107.*

[24] Sulistiyani, T., R. dan Nunuk, W. 2011. *Isolation, selection, and molecular identification of antibiotic producing actinomycetes. Widyariset. 4(3): 541-548.*

[25] Sunaryanto, R., *et. al.* 2010. *Isolation and Characterization of antimicrobial substance from marine Streptomyces* sp. *Microbiology Indonesia. 4(2): 84-89.*

[26] Susilowati, D., N., Ratih, D., H., dan Erny, Y. 2007. Isolasi dan karakterisasi aktinomisetes penghasil antibakteri Enteropatogen *Escherichia coli* K1.1, *Pseudomonas pseudomallei* 02 05, dan *Listeria monocytogenes* 5407. *Jurnal AgroBiogen. 3(1): 15-23.*

[27] Susanna. 2006. *The Use of Antagonistic Bacterial as Biocontrol Agents to Wilt Disease (Fusarium oxysporum f.sp.cubense) on Banana. J. Floratek. 2: 114-121.*

[28] Winarni, I. 2013. Isolasi dan karakterisasi bakteri patogen pada benih padi dan kedelai. *Jurnal Matematika, Sains, dan Teknologi. 14(2): 135-141.*
[29] Wulandari, S. dan Nanik, S. 2016. *Influence of media to isolate growth of actinomycetes code al35 and optimization of antibactery metabolite production based on fermentation time and pH. Media Farmasi. 13(2): 186-198.*
[30] Zhou, T., *et. al.* 2012. *Isolation and characterization of Pseudomonas brassicacearum J12 as an antagonist against Ralstonia solanacearum and identification of its antimicrobial components. Microbiological Research. 167: 388-394.*